Lecture Notes of the Institute for Computer Sciences, Social Informatics and Telecommunications Engineering 381

More information about this series at http://www.springer.com/series/8197

Xi Wu · Kaishun Wu · Cong Wang (Eds.)

Quality, Reliability, Security and Robustness in Heterogeneous Systems

16th EAI International Conference, QShine 2020
Virtual Event, November 29–30, 2020
Proceedings

Editors
Xi Wu
Chengdu University of Information Technology
Chengdu, China

Cong Wang
City University of Hong Kong
Hongkong, China

Kaishun Wu
Shenzhen University
Shenzhen, China

ISSN 1867-8211 ISSN 1867-822X (electronic)
Lecture Notes of the Institute for Computer Sciences, Social Informatics and Telecommunications Engineering
ISBN 978-3-030-77568-1 ISBN 978-3-030-77569-8 (eBook)
https://doi.org/10.1007/978-3-030-77569-8

This Springer imprint is published by the registered company Springer Nature Switzerland AG
The registered company address is: Gewerbestrasse 11, 6330 Cham, Switzerland

Preface

We are delighted to introduce the proceedings of the sixteenth edition of the European Alliance for Innovation (EAI) International Conference on Heterogeneous Networking for Quality, Reliability, Security and Robustness (Qshine 2020). This conference brought researchers, developers, and practitioners around the world to disseminate, exchange, and discuss all recent advances related to heterogeneous networking, particularly with regard to quality, experience, reliability, security, and robustness.

The technical program of Qshine 2020 consisted of 20 full papers, which were selected from the 49 submitted papers. Aside from the high-quality technical paper presentations, the technical program also featured a keynote speech, which was given by Prof. Xiaoming Fu from the University of Gottingen, Germany. Because of the COVID-19 pandemic, the conference was, unfortunately, held virtually in the cyberspace, with all the presentations, including the keynote speech, given online. During the conference there were more than 200 academics in attendance.

Coordination with the steering chairs, Prof. Bo Li and Prof. Imrich Chlamtac, was essential for the success of the conference. We sincerely appreciate their constant support and guidance. It was also a great pleasure to work with such an excellent Organizing Commitee team, including Dr. Ao Feng, Min Chen, Dr. Changming Zhao, Dr. Wenzao Li, Prof. Jianbing Ma, and Jing Peng; we are grateful for their hard work in organizing and supporting the conference. We would like to thank the Technical Program Committee (TPC), led by our TPC Co-chairs, Prof. Sheng Xiao, Prof. Baowei Wang, and Prof. Zhipeng Yang, who completed the peer-review process for the technical papers and put together a high-quality technical program. We are also grateful to the Conference Manager, Natasha Onofrei, for her support and all the authors who submitted their papers to the Qshine 2020 conference.

We strongly believe that the Qshine conference provides a good forum for all researchers, developers, and practitioners to discuss all science and technology aspects that are relevant to heterogeneous networking. We also expect that the future editions of the Qshine conference will be as successful and stimulating as Qshine 2020, as indicated by the contributions presented in this volume.

Xi Wu
Kaishun Wu
Cong Wang

Organization

Steering Committee

Bo Li	Hong Kong University of Science and Technology, Hong Kong
Imrich Chlamtac	University of Trento, Italy

Organizing Committee

General Chair

Xi Wu	Chengdu University of Information Technology, China

Technical Program Committee Chair and Co-chairs

Kaishun Wu	Shenzhen University, China
Cong Wang	City University of Hong Kong, China
Sheng Xiao	Hunan University, China
Baowei Wang	Nanjing University of Information Science and Technology, China
Zhipeng Yang	Chengdu University of Information Technology, China

Sponsorship and Exhibit Chair

Wenzao Li	Chengdu University of Information Technology, China

Local Chair

Jing Peng	Chengdu University of Information Technology, China

Workshops Chair

Changming Zhao	Chengdu University of Information Technology, China

Publicity and Social Media Chair

Min Chen	Chengdu University of Information Technology, China

Publications Chair

Jianbing Ma	Chengdu University of Information Technology, China

Web Chair

Ao Feng	Chengdu University of Information Technology, China

Technical Program Committee

Fangming Liu	Huazhong University of Science and Technology, China
Jinsong Han	Zhejiang University, China
Yang Qin	Harbin Institute of Technology, China
Xiaojiang Chen	Northwest University, China
Xu Chen	Sun Yat-sen University, China
Lin Wang	Vrije Universiteit Amsterdam, Netherlands
Reza Malekian	Malmö University, Sweden
Sherali Zeadally	University of Kentucky, USA
Haiyang Wang	University of Minnesota Duluth, USA
Yifeng Zheng	CSIRO Data61, Australia
Xingliang Yuan	Monash University, Australia
Hai Liu	The Hang Seng University of Hong Kong, Hong Kong
Zhenjiang Li	City University of Hong Kong, Hong Kong
Yang Liu	City University of Hong Kong, Hong Kong
Yu Guo	City University of Hong Kong, Hong Kong
Lei Xu	City University of Hong Kong, Hong Kong
Jinghua Jiang	City University of Hong Kong, Hong Kong
Wei Wang	Hong Kong University of Science and Technology, Hong Kong
Wenping Liu	Hubei University of Economics, China
Jiang Xiao	Huazhong University of Science and Technology, China
Yajin Zhou	Zhejiang University, China
Fei Chen	Shenzhen University, China
Kaishun Wu	Shenzhen University, China
Yuedong Xu	Fudan University, China
Zhenkui Shi	Guangxi Normal University, China
Helei Cui	Northwestern Polytechnical University, China
Ao Feng	Chengdu University of Information Technology, China
Ke Jia	Chengdu University of Information Technology, China

Contents

Network Reliability and Security

Emerging Applications

Network Reliability and Security

Research and Application of Visual SLAM Based on Embedded GPU

Tianji Ma, Nanyang Bai, Wentao Shi, Lutao Wang(✉), and Tao Wu

Chengdu University of Information Technology, Chengdu, Sichuan, China
{wanglt,wut}@cuit.edu.cn

Abstract. In automatic navigation robots, robotic autonomous positioning is one of the most difficult challenges. Simultaneous Localization and Mapping (SLAM) technology can incrementally construct a map of the robot's moving path in an unknown environment while estimating the position of the robot in the map, providing an effective solution for robots to fully navigate autonomously. The camera can obtain corresponding two-dimensional digital images from the real three-dimensional world. These images contain very rich color, texture information and highly recognizable features, which provide indispensable information for robots to understand and recognize the environment based on the ability to autonomously explore the unknown environment. Therefore, more and more researchers use cameras to solve SLAM problems, also known as visual SLAM.

Visual SLAM needs to process a large number of image data collected by the camera, which has high performance requirements for computing hardware, and thus its application on embedded mobile platforms is greatly limited. In this regard, this paper uses embedded hardware equipped with embedded GPU, combines CUDA-based GPU parallel computing and visual SLAM algorithm, finally, designs a parallelization scheme based on embedded GPU.

Keywords: Visual-SLAM · Embedded · Parallel computing · CUDA · GPU

1 Introduction

1.1 Background

In order to achieve fully autonomous work in an unknown environment, mobile robots must solve two basic problems of positioning themselves and perception of the environment. Simultaneous Localization and Mapping (SLAM) was first proposed by Smith [1] and applied in the field of robotics. It combines the robot's self-positioning and map construction into one. The goal is to make the robot locate itself through the movement of the robot without the prior information of the environment, and then establish a real-time map of the environment based on the sensor data, at the same time the robot's motion trajectory is accurately estimated.

At present, SLAM has relatively mature applications. For example, sweeping robots, drones, Augmented Reality (AR), Virtual Reality (VR), etc. Autonomous driving and

X. Wu et al. (Eds.): QShine 2020, LNICST 381, pp. 3–21, 2021.
https://doi.org/10.1007/978-3-030-77569-8_1

accurate 3D reconstruction are also in rapid development. According to different sensors used, SLAM can be divided into visual SLAM and laser SLAM. Laser SLAM uses LiDAR (Light Laser Detection and Ranging) as a sensor, and the collected data is called Point Cloud data, which contains accurate angle information and distance information. The distance measurement using LiDAR is more accurate, and the error model is relatively simple. At the same time, LiDAR has the advantages of being insensitive to light. Compared with visual SLAM, laser SLAM's related theoretical research is relatively mature, but the sensors are expensive. Visual SLAM can use a variety of cameras: monocular camera, stereo camera, and depth camera as sensors. These cameras are cheaper than LiDAR and are widely used in various fields of society. At the same time, rich color information, texture information, and more recognizable image features can be obtained from the images captured by the camera. Therefore, visual SLAM has gradually become the main research direction for solving SLAM problems, but its disadvantage is that real-time processing of a large amount of image data requires high computing resources, which brings real-time operation on embedded platforms and mobile platforms a great challenge. Compared with the computing resources of high power consumption PC platforms, embedded platforms and mobile platforms generally have low power consumption, and the computing resources are also greatly restricted. Therefore, it is an important direction of the research to use limited computing resources to efficiently execute algorithms of visual SLAM on embedded platforms.

1.2 Main Research Content

Thanks to the rapid development of parallel technology, the performance of processors suitable for parallel computing is also rapidly improving, which makes it possible to double the operating efficiency of the algorithm. In recent years, GPU computing performance has achieved rapid growth. Its computing performance, especially parallel computing performance, is far stronger than that of CPU. Researchers have gradually discovered the potential of GPU parallel computing. In order to provide a more friendly interface for researchers and developers to use GPU to solve problems, in 2006, NVIDIA Corporation released CUDA (Compute Unified Device Architecture), a general-purpose parallel computing platform and programming model, as an "engine" to drive GPU to solve complex computing problems, which is more efficient than CPU. After more than ten years of development, CUDA has been widely used in the field of image processing. XianLou [2] uses CUDA to accelerate the processing of image segmentation algorithms based on normalization, and Chengyao [3] uses CUDA to optimize image feature extraction and realizes the real-time stitching of panoramic video, which overcomes the shortcomings of high power consumption, non-real-time and low stability that used to rely on post-processing.

The current development of visual SLAM has been relatively mature, and there are various types of solutions, including sparse method, semi-dense method, and dense method, as well as feature point method based on image features and direct method based on image grayscale. The execution efficiency, positioning accuracy, and robustness of these algorithms perform well in specific experimental environments. However, most of these algorithms are performed on desktop-level high-power platforms, and there is very little work to solve visual SLAM problems for embedded platforms. Embedded platforms have many advantages such as low power consumption, miniaturization, low cost, and high reliability, but their performance is far inferior to high-power PC platforms. Due to visual SLAM has high requirements for computing resources and correspondingly high requirements for hardware computing performance, embedded platforms and their performance are easily ignored by visual SLAM researchers.

With the development of embedded hardware, high-performance embedded hardware with integrated GPU has emerged. Since there are a lot of image processing and pose estimation operations in visual SLAM, these operations consume a lot of computing resources. Therefore, GPU parallel computing can be used to accelerate processing. The real-time processing performance of visual SLAM in embedded systems can be effectively improved through the combination of high-performance embedded processing hardware and algorithm optimization, which is beneficial to the mobility, miniaturization, and low energy consumption applications of visual SLAM technology.

In summary, our work mainly studies how to use GPU parallel computing to accelerate processing on embedded hardware to overcome the computational complexity of visual SLAM. Although there have been some works that use GPU to accelerate parallel calculation of certain algorithms in SLAM, such as Wu C [4] and Rodriguez Losada [5] have implemented beam adjustment and ICP (Iterative Closest Point) algorithms on GPU, but these works are all performed on desktop GPUs.

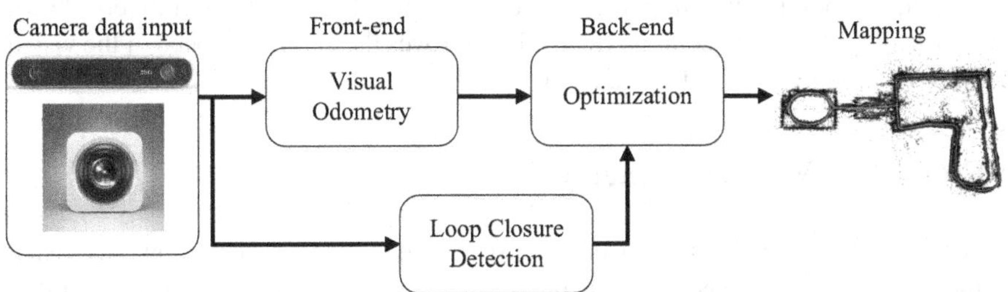

Fig. 1. A block diagram of a typical visual SLAM system.

Figure 1 illustrates a typical visual SLAM system structure diagram, including five parts: visual sensor data, front-end (also called visual odometer), back-end, mapping and loop closure detection. The vision sensors input the images, and then system performs feature extraction and matching on these input images at the front end, and then roughly estimates the position of the feature points and the robot, and then transfers the estimated result to the back end, and executes graph optimization to get a more accurate result. In this way, it is possible to locate and then build a map, and at the same time transfers the optimized result to the closed-loop inspection to eliminate the accumulated error of the robot moving for a long time, and then uses the result for tracking. Among these five parts, the front-end and the back-end are important parts in charge of processing data, and these two parts consume a lot of computing resources.

Our work focuses on the front-end, illustrating the main key technologies of feature extraction, feature matching and the principle of relevant algorithms. We combine the selected scheme and the computing performance of embedded hardware, and then select the most appropriate GPU parallel computing method to optimize and improve the visual SLAM processing performance, operating efficiency and ensure good positioning accuracy.

The theoretical basis and related work studied in this paper are as follows:

Firstly, the overall framework of visual SLAM is introduced. The responsibilities and functions of each module in the framework are described.
Then the main method of the visual SLAM front-end, the feature point method, is introduced. In this paper, we use ORB (Oriented FAST and Rotated BRIEF) features as the front-end implementation method which has the fastest calculation speed on the basis of meeting the accuracy of feature detection, to ensure the fast processing of the embedded platform.
Finally, the parallel mechanism is analyzed on the CUDA-based GPU hardware architecture and programming model. On this basis, the detailed parallel analysis of the key technologies of the selected scheme is carried out. A reasonable parallelization scheme was designed, and GPU parallelized visual SLAM system was built on embedded development board NVIDIA Jetson TX2.

In order to evaluate the performance of the system, relevant experiments were carried out on the data set. The results show that the whole system is in good working condition. In addition, by counting the time overhead of executing data set, it is intuitively shown that the use of GPU parallelization effectively improves the operating speed of the visual SLAM system on the embedded platform.

2 Front-End Visual Odometry

The front-end is at the lower level in the visual SLAM system, also known as visual odometry (VO) [6]. For visual odometry, its focus is on the frame-to-frame motion between adjacent images. When the sensor data module transmits the image frame sequence (i.e. video stream) to the visual odometry, its function is to extract the key information of adjacent image frames to roughly estimate the camera movement in advance to provide better results for the back-end. At present, there are two main methods of visual odometry, feature point method and direct method. In this paper, we use the feature point method.

2.1 Feature Point Method

The front-end based on the feature point method is a classic method of visual odometry. It uses the redundancy of the image to detect and extract feature points from the preprocessed input image, and then performs feature matching to estimate the camera motion trajectory. Therefore, it avoids processing the complete image containing a large amount of redundant information, and greatly reduces the amount of calculation while preserving the important information of the image. And it runs stably and is not sensitive to lighting and dynamic objects, so it is widely used in visual SLAM. For the visual odometry of the feature point method, one of the keys is to use feature detection algorithms to extract the best features from a frame of images. At present, the development of image feature detection algorithms is relatively mature. Commonly used feature detection algorithms: SIFT [7], SURF [8], ORB, AKAZE [9]. For details, please check the relevant literature.

2.2 ORB Feature Detection Algorithm

ORB algorithm was proposed by Ethan Rublee [10] and others in 2011. It combines an improved FAST (Features From Accelerated Segment Test) corner detection algorithm and a direction normalized BRIEF (Binary Robust Independent Elementary Features) feature descriptor algorithm. The ORB feature detector will detect FAST corner points in each layer of the image Gaussian pyramid, and use Harris corner scores to evaluate the detection points to select the highest quality feature points. Since the original BRIEF feature descriptor is very sensitive to rotation, the ORB algorithm is improved. ORB features have scale invariance, rotation invariance, and certain affine invariance.

3 GPU Parallel Accelerated Visual SLAM

3.1 GPU Hardware Features

In recent years, with the rapid development of science and technology, the problems faced by many research fields have become larger and the corresponding requirements for computing performance have become higher and higher. However, as the manufacturing technology gradually approached its limit, the growth rate of CPU computing performance has gradually slowed down. Even if CPU manufacturers represented by Intel and AMD have introduced multi-core architecture CPU to make up for the limit of single-core performance improvement, their performance still cannot meet the needs of the market. For the GPU, driven by the market urgent need for real-time and high-definition 3D image rendering, GPU has gradually developed into a highly parallel, multi-threaded, multi-core processor architecture like today, with huge computing power and extremely high memory bandwidth. Its computing performance is far stronger than the CPU (see Fig. 2).

The reason behind this huge computing performance gap is that the difference in hardware structure between GPU and CPU. Let's start this topic from the "core" perspective. First of all, the CPU is composed of several cores optimized for sequential serial processing. While the GPU is composed of thousands of smaller, more efficient cores, which are specifically designed to handle multiple tasks at the same time, and can efficiently handle parallel tasks. In other words, although each core of the CPU is extremely powerful in processing tasks, it has fewer cores and does not perform well in parallel computing. In contrast, although the computing power of each core of the GPU is not powerful, it has a large number of cores, which can handle multiple computing tasks at the same time, and is well competent for parallel computing.

The different hardware features of GPU and CPU determine their application scenarios. The CPU is the core of the computer's operation and control, and the GPU is mainly used for graphics and image processing. The form of the image presented in the computer is a matrix. Our processing of the image is actually to operate various matrices

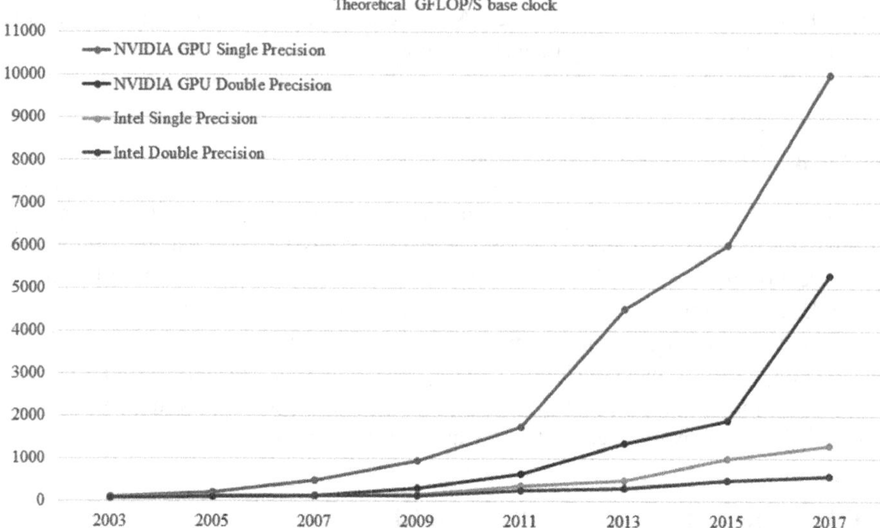

Fig. 2. Shows the intuitive comparison of single-precision and double-precision floating-point computing capabilities between CPU and GPU. It can be foreseen that the computing difference between CPU and GPU will become larger and larger in the short term.

for calculations, and many matrix operations can actually be parallelized, which makes image processing fast.

Now we compare the features of CPU and GPU from the perspective of data processing. The CPU needs strong versatility to handle a variety of different data types, such as integers, floating point numbers, etc., and it must be good at handling a large number of branch jumps and interrupt handling caused by logical judgments. So the CPU is actually a powerful processing unit. It can handle many things properly. Of course, we need to give it a lot of hardware resources for it to use, which makes the CPU impossible to have too many cores. The GPU is facing a highly unified, independent, large-scale data and a pure computing environment that does not need to be interrupted. Although the processing power of its core is far less powerful than that of the CPU, the GPU has a lot of cores, which makes up for the lack of single-core computing power and supports parallel computing.

It can be seen from the structure (see Fig. 3) that a large part of the CPU is used for caching and control, and there are relatively few arithmetic logic units, while the GPU is the exact opposite, and the computing units occupy the vast majority.

In the early days, it was very inconvenient for researchers to use GPU to perform calculations in the field of non-graphics rendering, because GPU is dedicated to graphics rendering and has streamlined rendering pipelines. Therefore, general-purpose computing programs can only be encapsulated into rendering programs and embedded in these pipelines before they can be executed by the GPU. With the increasing demand for general-purpose computing, in order to provide researchers and developers with a more friendly interface to use GPU to solve problems, NVIDIA Corporation released CUDA in 2006, a general-purpose parallel computing platform and programming model, as

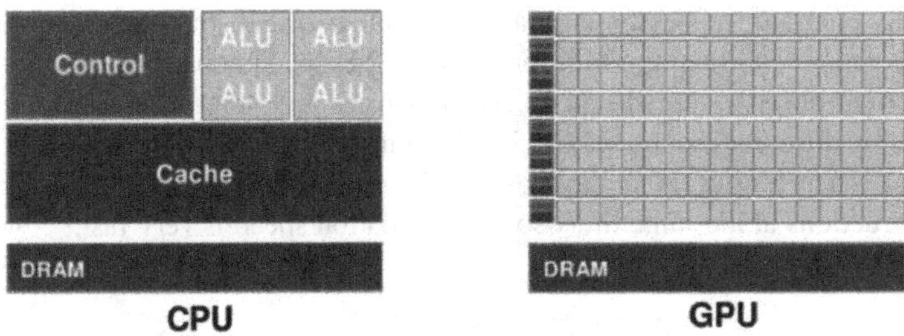

Fig. 3. Diagram of CPU and GPU structure.

an "engine" to drive the GPU to efficiently solve complex computing problems. Now this kind of general-purpose parallel computing is widely used in various industries and fields, including deep learning that has developed rapidly in recent years.

In the current computer architecture, in order to complete CUDA parallel computing, the GPU alone cannot complete the computing task. The CPU must be used to cooperate to complete a high-performance parallel computing task. Generally speaking, the parallel code performs on the GPU and the serial code performs on the CPU. This is heterogeneous computing. Specifically, heterogeneous computing means that processors of different architectures cooperate with each other to complete computing tasks. The CPU is responsible for the overall program flow, and the GPU is responsible for the specific calculations. When each thread of the GPU completes the calculations, the results are copied to the CPU to complete a computing task (see Fig. 4).

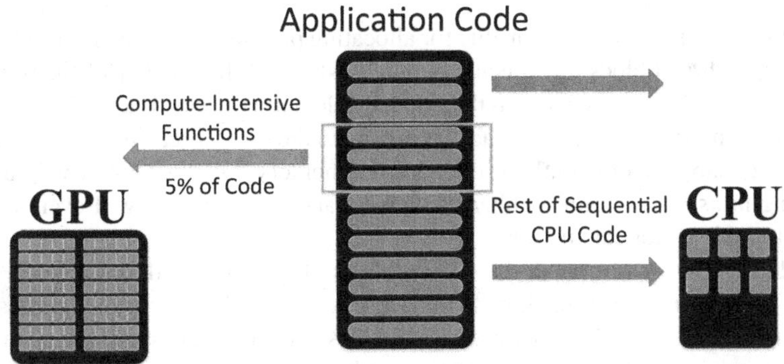

Fig. 4. The intensive calculation code (about 5% of the code amount) is completed by the GPU, and the remaining serial code is executed by the CPU.

3.2 CUDA Hardware Model

A simplified diagram of the GPU hardware architecture that supports CUDA is shown in Fig. 5. The most basic processing unit is the Streaming Processor (SP), also known as CUDA-CORE, which is responsible for the execution of each specific instruction. The

GPU parallel computing is essentially a large number of SP simultaneous processing tasks. The core unit is Streaming Multiprocessor (SM), also known as GPU core, which consists of multiple SPs, thread schedulers, memories and other units. The number of SMs owned by different models of GPU and the number of SPs contained in each SM are different. Therefore, a GPU may have thousands of SPs. In theory, these SPs can execute instructions at the same time, so the calculation speed is very fast.

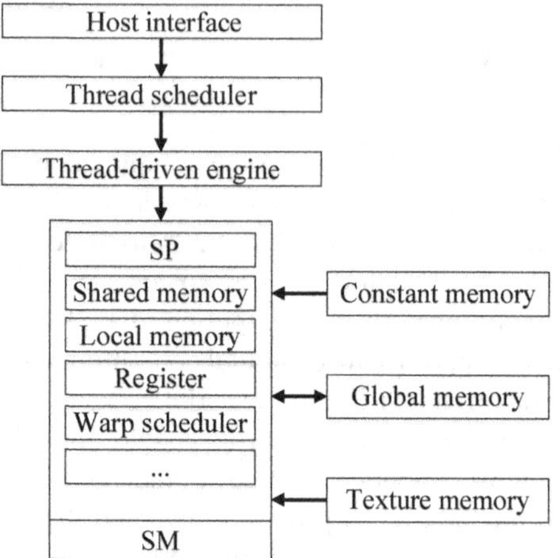

Fig. 5. GPU hardware architecture diagram.

The thread scheduler is responsible for allocating parallel tasks to SM. Each SM can start multiple thread blocks to execute parallel tasks, and the number of them is allocated by the developer according to their actual calculation needs. Each thread block is composed of multiple threads. Threads belonging to the same thread block can communicate and collaborate efficiently through shared memory. Finally, CUDA will allocate computing tasks to a certain number of threads, and these threads will eventually be mapped to each SP for calculation.

In addition, in order to meet the diverse needs of graphics rendering, the GPU has multiple types of memory for threads to access its stored data. These memories have their own characteristics. The characteristics of these memories are summarized in Table 1. Making good use of these memories can reduce unnecessary data access time and improve calculation speed.

3.3 CUDA Programming Model

The CUDA software environment is constantly updated with the development of the GPU hardware architecture. The latest version is CUDA 11.0. The functions provided by the entire environment are getting more and more powerful, and the interface becomes very friendly. CUDA supports C/C++, Python, JAVA and other high-level languages, and the corresponding programs are executed by the CPU and GPU. The CUDA parallel

Table 1. The characteristics of GPU memories.

Memory	Cache	Access permission	Life cycle
Register	×	Thread private	Thread
Local memory	×	Thread private	Thread
Shared memory	×	Shared within thread block	Block
Global memory	×	Device read/write	Grid
Constant memory	√	Device read only	Grid
Texture memory	√	Device read only	Grid

program executed on the GPU is also called a kernel function, which is specifically used to complete GPU parallel computing tasks, and its definition is also different from ordinary functions.

The following will introduce a typical CUDA program execution procedure (see Fig. 6). Each grid contains several thread blocks, which are composed of several threads. All thread blocks in the grid can be parallelized, and all threads in the thread block can also be parallelized, so the degree of parallelization is high.

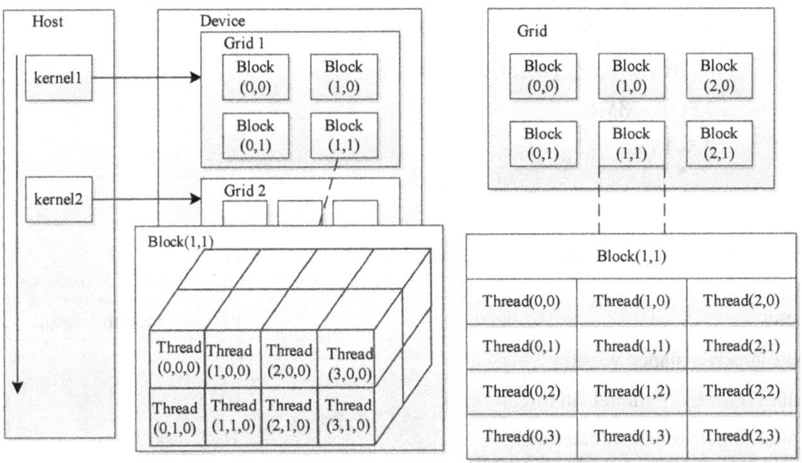

Fig. 6. Host refers to the host side, representing the CPU, and Device refers to the device side, representing the GPU. Grid is the outermost thread organization structure in CUDA.

The program is first executed from the Host side, the serial program performs initialization work, and the kernel function is started after the data and storage space required for the execution of the kernel function are allocated. And then the Device side will generate a large number of threads based on various variable parameters set in the kernel function, and these threads will be organized into thread blocks. Subsequently, the thread blocks will be allocated to the SM for parallel execution. Each thread block is divided into several groups of threads when executed on the SM, and each group of threads will eventually be mapped to a group of SPs in the SM for parallel calculation. After the kernel function is executed, the serial program will copy the calculation result from Device to Host. Then prepare for the next execution of the kernel function, or complete and end the execution of this CUDA program.

3.4 Experimental Hardware

This time, the high-performance embedded platform we used is Jetson TX2. It is a powerful multi-core mobile SOC released by NVIDIA in 2017, mainly for smart terminal devices such as smart robots, drones, unmanned driving, smart cameras, and portable medical equipment. Its CPU has a total of 6 cores, including 4 Cortex-A57 and 2 customized Denver cores. In addition, TX2 is a heterogeneous system, which integrates a Pascal architecture GPU with 256 CUDA cores. Its main performance indicators are shown in Table 2. The computing performance and related parameters of its GPU are shown in Table 3.

Table 2. TX2 main performance indicators.

Parameter	Performance indicator
CPU	ARM Cortex-A57(quad-core) 2 GHz + NVIDIA Denver2(dual-core) 2 GHz
GPU	256 CUDA-COREs Pascal 1300 MHz
Memory	8 GB 128-bit LPDDR4 1866 MHz 58.3 GB/s
Storage	32 GB eMMC
TDP	7.5–15w

Table 3. GPU related parameters of TX2.

Parameter	Performance indicator
Computing performance version	6.2
Maximum number of threads in thread block	1024
Maximum dimension of thread block	3
Maximum dimension of thread block in x and y direction	1024
Maximum dimension of thread block in z direction	64

3.5 Front-End Parallelization

For the front-end based on the feature point method, the main processing procedures are image feature extraction and matching. They consume more than half of the computing resources and are calculated for images, so this part is particularly suitable for parallelization. In the following subsections, we will analyze how the procedures of image feature extraction and matching can be parallelized. Then we parallelize the relevant parts by CUDA, and finally test the execution efficiency of GPU parallelization through experiments.

Front-End Parallelizable Analysis. After the front-end obtains a frame of images transmitted from the visual sensor, it constructs an image Gaussian pyramid based on the original image first. Afterwards, the key points and feature vectors are extracted from each image layer of the image pyramid to ensure that the ORB features are scale-invariant. Finally, all key points and feature vectors extracted from each image layer will be mapped to the original image, but this makes the image features of each original image too dense and repetitive. Therefore, it is necessary to delete the repeated feature points and perform non-maximum suppression on the rest of feature points to ensure that the distribution of the feature points is relatively uniform and to improve the effect of image matching. The main calculation procedure of the feature extraction (see Fig. 7) is as follows:

1. Construct image Gaussian pyramid.
2. Perform FAST key points detection in each image layer of the pyramid.
3. Perform coordinate normalization in image layers of different sizes.
4. Delete duplicate FAST key points. Compare each key point with the corresponding key point in the upper and lower adjacent image layers at the same scale, and keep the key point with the largest response value.
5. Non-maximum suppression. Each key point is compared respectively with 26 adjacent points in the image layer where it is located and in the upper and lower adjacent image layers at the same scale. Only when the response value of this key point is greater than the other key points, will it be kept, otherwise it will be deleted.
6. Sort key points according to FAST and Harris response values [11]. Select the top N best feature points, and the N value is preset according to requirements.
7. Assign the direction to each key point, and calculate the BRIEF descriptor to complete the extraction of ORB features.

Analyze the parallelization of the above steps, we can get the conclusion:

1. In the FAST key points detection procedure, there is no data communication between the image layers of the Gaussian pyramid, so FAST key point detection can be performed in parallel in each image layer.
2. FAST key points detection only has data association with each pixel and its neighboring pixels in the image, and the detection procedure is exactly the same, so it can be executed in parallel on a large scale.
3. Duplicate point deletion and non-maximum suppression are both related to the feature point and the image where it is located, and also related to the neighborhood feature points in the upper and lower adjacent images at the same scale. Therefore, the Gaussian pyramid of the three image layers can be input at the same time and calculated in parallel in the same way.

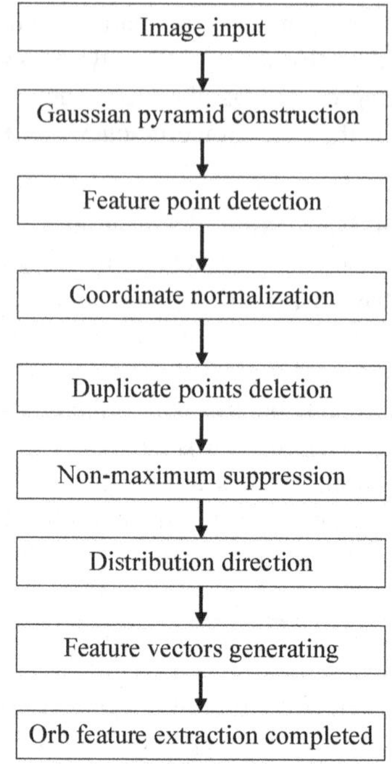

Fig. 7. ORB feature extraction flow chart.

4. For each key point, the calculation of its direction is based on the data association between the key point and the independent local image information in the image where it is located, so it can be calculated in parallel.

After the image feature extraction is completed, feature matching is needed to find the same feature point pair in the two images. There are many methods for feature matching. Due to the ORB feature vector (also known as feature descriptor) is a binary string, Hamming distance can be used to describe the similarity of a pair of feature vectors. After that, the same feature point pair can be found by the method of brute-force matching. Hamming distance can be expressed as:

$$D(V_1, V_2) = \sum_{i=0}^{255} x_i \oplus y_i \tag{1}$$

V_1, V_2 are two ORB feature vectors, $V_1 = x_0x_1 \cdots x_{255}$, $V_2 = y_0y_1 \cdots y_{255}$. The process of calculating the Hamming distance is to perform an exclusive OR operation on each bit of the two feature vectors. The smaller the value of $D(V_1, V_2)$, the higher the similarity of the two feature vectors; and the larger the value of x, the lower the similarity of the two feature vectors. In addition, a threshold K needs to be set. If $D(V_1, V_2)$ is greater than the threshold, the corresponding feature vector and feature point should be deleted. Finally, a rough matching point pair can be obtained by means of the method of brute-force matching.

With the matched point pairs, a change model can be established through the RANSAC (Random Sample Consensus) algorithm to describe the change relationship between the points in the two images:

$$[x_1, y_1, 1]^T = M[x, y, 1]^T \tag{2}$$

The transformation matrix M is written as:

$$M = \begin{bmatrix} a_{11} & a_{12} & a_{13} \\ a_{21} & a_{22} & a_{23} \\ 0 & 0 & 1 \end{bmatrix} \tag{3}$$

Suppose P is the coordinate data set of the point pair obtained by rough matching, then select 3 pairs of matching points from P to calculate the transformation matrix M through Eq. 2. Then use the remaining point pairs in P to verify the accuracy of the transformation matrix M, and count the number of matching points that conform to the model. Repeat the above procedures, the final selected transformation model should have the most matching points.

Analyze the parallelization of the above steps, we can get the conclusion:

1. The calculation of the similarity of two ORB feature vectors has no data communication, so it can be processed in parallel.
2. When calculating the Hamming distance, the exclusive OR operation of each bit of the two feature vectors is only related to the binary value of the bit itself, so this procedure can be processed in parallel.
3. The RANSAC algorithm selects 3 pairs of points from the set P of matching points, which are independent for matching, and there is no data communication, so this procedure can be processed in parallel.
4. When calculating the transformation matrix M, it is only related to the matching points used, so this procedure can be processed in parallel.
5. Verify the accuracy of each transformation matrix M is only related to all rough matching point pairs, so this procedure can be processed in parallel.

The FAST Key Point Detection Parallelization Design. Map each image layer in the image pyramid to a thread, and each thread block allocates 32×8 threads, the number of thread blocks is:

$$N = \frac{W + blockDim.x - 7}{blockDim.x} \times \frac{H + blockDim.y - 7}{blockDim.y} \tag{4}$$

In Eq. 4, N represents the total number of thread blocks required to perform this task, *blockDim.x* represents the number of threads in the x direction, *blockDim.y* represents the number of threads in the y direction, W and H are the width and height of the image respectively. In order to prevent access conflicts in parallel threads, each thread in parallel will perform FAST key point detection on a pixel in the image. If it conforms the FAST key point, the coordinate of the pixel is saved in the global memory, and the total number of detected FAST key points is recorded and saved in the global memory.

In addition, a corresponding size of global storage space is allocated for each layer of image and initialized to zero to store the response value of key points, if a certain pixel is selected as the FAST key point, its response value is saved in the corresponding position of the allocated space.

Coordinate Normalization. In order to store the updated response value of the key point, the global memory is allocated the same size as the storage space of the original image, and each thread block is allocated 256 threads. The total number of thread blocks is the same as Eq. 4. Each thread in parallel calculates the normalized coordinate of a key point and updates the response value of the feature point after the normalized coordinate in the allocated space.

Duplicate Points Deletion and Non-maximum Suppression. The coordinate value and the response value of a feature point to be detected, the response value of the feature points in the neighborhood around the feature point to be detected, and the response value of the neighborhood feature points in the upper and lower scale image of the image layer where the feature point to be detected is located are transferred as parameters to the CUDA kernel function that performs duplicate points deletion and non-maximum suppression. And each thread block allocates 256 threads, the number of thread blocks is:

$$N = \frac{\overline{N}_{fast} + blockDim.x - 1}{blockDim.x} \tag{5}$$

In Eq. 5, N represents the total number of thread blocks required to perform this task, $\overline{N}_{fast}$ is the total number of FAST key points of the image layer, each thread in parallel calculates one key point. Read the response value of all key points in the neighborhood of the key point in the three image layers. If the response value of the key point is the largest, keep this key point and delete the two key points at the corresponding positions in the upper and lower adjacent image layers at the same scale, otherwise delete itself. At the same time, the number of key points remaining after duplicate points deletion and non-maximum suppression operations should be recorded.

Filter Key Points. Use CUDA kernel function to sort the key points and corresponding response values remaining after the previous operation. Here directly use the parallel sorting algorithm in the Thrust library, and then according to requirement, select the top N feature points with the highest response value for subsequent calculations.

Distribution Direction. Calculate the direction of each feature point in parallel, and allocate 32 × 8 threads per thread block, the number of thread blocks is:

$$N = \frac{N_{fast} + blockDim.y - 1}{blockDim.y} \tag{6}$$

In Eq. 6, N represents the total number of thread blocks required to perform this task, N_{fast} is the total number of FAST key points. According to reference [12], each thread in parallel calculates the first-order moment m_{10}, m_{01}, and the origin moment m_{00} of the feature point. Then save the results in shared memory to improve the efficiency of repeated access to data. In addition, thread 0 in each thread block is used to calculate the direction of the feature point according to reference [12], and the result is saved in the global memory after the calculation is completed.

Generating Feature Vectors. Transfer relevant parameters to the CUDA kernel function, and then start the kernel function to extract the ORB feature vector. And each thread block allocates 32 × 8 threads, the number of thread blocks is:

$$N = \frac{L + blockDim.x - 1}{blockDim.x} \times \frac{N_{fast} + blockDim.y - 1}{blockDim.y} \tag{7}$$

In Eq. 7, N_{fast} is the total number of FAST key points, and L is the length of the ORB feature vector. The feature vector is generated according to the direction of the key point and the sampling mode of the neighborhood of the image block in each thread block.

Feature Matching. Calculate the Hamming distance between two feature vectors in parallel by the method of brute-force matching. Each thread calculates the Hamming distance between a feature vector in the current image and all feature vectors in another image. Save the index value and Hamming distance value of the two feature vectors with the smallest distance in the global memory. Each thread block is allocated 256 threads, and the total number of thread blocks is:

$$N = \frac{N_{vec} + blockDim.x - 1}{blockDim.x} \tag{8}$$

In Eq. 8, N_{vec} is the total number of feature vectors in the current image. When the calculation is completed, the results saved in the global memory are transferred back to the memory on the Host side. Then filter out the mismatched results through a threshold to obtain rough matching results. When the rough matching is completed, all the saved data is transferred from the Host side memory to the Device side constant memory to provide data for the RANSAC algorithm.

RANSAC. First set the number of iterations K, and then randomly generate K groups of random numbers, each group contains three different random numbers. Since these random numbers will not be changed in subsequent calculations, these random numbers are transferred from the Host side memory to the Device side constant memory. Thereby, the data access speed can be improved through the cache in the constant memory.

Transfer relevant parameters to the CUDA kernel function, and then start the kernel function to calculate the transformation matrix M. Thread 0 in each thread block reads a set of random numbers from the constant memory to determine the three sample numbers. Then the feature point coordinates corresponding to the sample number can be read from the constant memory. Each thread in parallel uses the coordinates of the three feature points randomly selected to calculate the transformation matrix M according to Eq. 2 and save the result in the global memory.

Then transfer relevant parameters for the second CUDA kernel function, and start the kernel function to verify the accuracy of the transformation matrix M. Each thread in parallel reads the coordinates of the feature points from the constant memory to verify the accuracy of a transformation matrix M, and it is necessary to count the number of feature points that conform to the transformation matrix M. The best transformation matrix M is selected according to the number of feature points previously counted.

3.6 Front-End Parallelization Efficiency Test

In order to verify that the GPU parallelization method used in this paper can effectively accelerate the processing speed of the front-end, the parallelization method is tested on the EuRoC MAV Dataset [13]. In the data sets, the data sets those name starts with MH are the videos shot indoors. Easy, medium, difficult represents the complexity of the scene in the video. As the complexity increases, high-speed moving scenes, scenes with strong lighting changes, and scenes with fast turning will appear in the video. These scenes will affect the accuracy of the estimated camera movement trajectory. During the experiment, the time cost of executing each set of data sets was recorded under the conditions of only using CPU and using GPU acceleration respectively. Time unit is minutes (min) (Table 4).

Table 4. The time overhead of executing data set

Data set	CPU only	CPU + GPU
MH_01_easy	7:51	4:56
MH_02_easy	6:31	3:53
MH_03_medium	5:41	3:31
MH_04_difficult	4:07	2:30
MH_05_difficult	4:46	2:47

From the results, the time cost of running the data set after using GPU parallelization on TX2 is greatly reduced, reducing the time by about half. Therefore, the use of GPU parallelization can significantly improve the processing performance of the visual SLAM front-end.

3.7 Movement Trajectory Estimation Test

Under the conditions of using only CPU and using GPU acceleration respectively, the camera motion trajectory is estimated through the visual SLAM algorithm. Due to the limited length of the paper, we only introduce the results of testing MH_04_difficult (see Fig. 8) and MH_05_difficult (see Fig. 9). In the two figures, the dotted line represents the actual value of the camera's motion trajectory, the blue line represents the estimated value of the trajectory when GPU acceleration is enabled, and the green line represents the estimated value without GPU acceleration. We can roughly see that whether GPU acceleration is enabled has no effect on the accuracy of the estimated camera movement trajectory. This conclusion can also be obtained from Table 5 and Table 6.

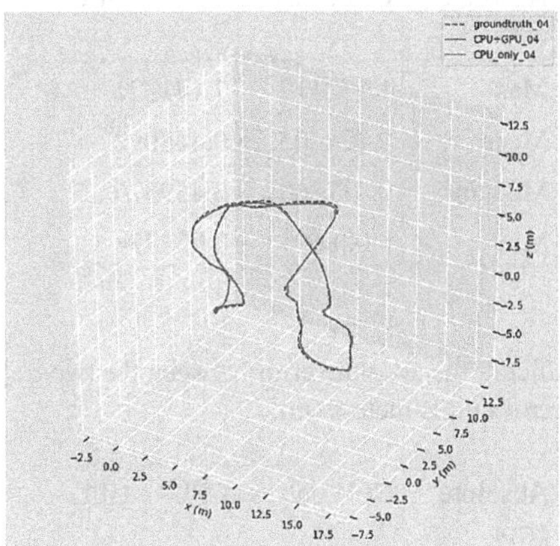

Fig. 8. Whether GPU acceleration is enabled has no effect on the accuracy of the estimated camera movement trajectory.

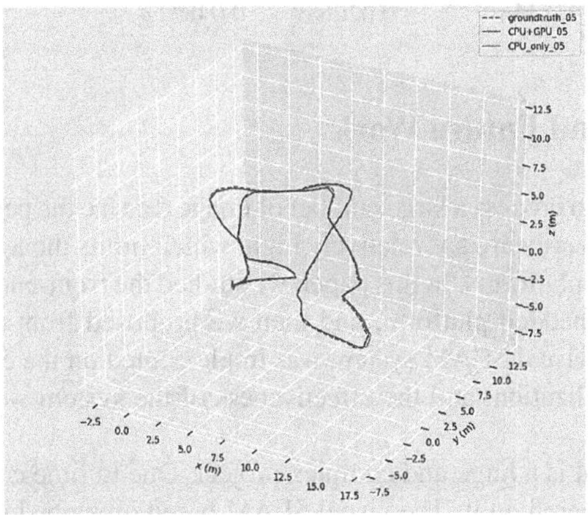

Fig. 9. Whether GPU acceleration is enabled has no effect on the accuracy of the estimated camera movement trajectory. However, at some corners of the trajectory, the result obtained by turning on GPU acceleration is closer to the actual value.

Table 5. In the MH_04_difficult, the absolute error between the two trajectory estimates and the groundtruth. The absolute error unit is meters (m).

Absolute error	CPU only	CPU + GPU
Max	1.132117	1.141322
Mean	0.424403	0.448092
Median	0.417360	0.439376
Min	0.028691	0.053778

Table 6. In the MH_05_difficult, the absolute error between the two trajectory estimates and the groundtruth. The absolute error unit is meters (m).

Absolute error	CPU only	CPU + GPU
Max	1.246482	1.224756
Mean	0.408896	0.421434
Median	0.387428	0.396226
Min	0.020689	0.048313

4 Conclusion and Future Work

Visual SLAM needs to process a large number of image data, so the performance requirements of computing hardware are relatively high, which limits the application of visual SLAM on embedded platforms. In this paper, we studied the front-end problem of visual SLAM based on embedded platform, and then we proposed front-end parallelization scheme. Finally, the visual SLAM system was implemented on the embedded platform through GPU parallelization, and the effectiveness of the system was verified through the data sets.

The visual SLAM is a huge and complex project. Due to time constraints, we have not done enough research on it. The visual SLAM based on embedded GPU studied in this paper can be further explored from the following two aspects:

1. With the advancement of technology, the performance of embedded and other miniaturized mobile platforms will become more powerful, such as higher performance GPU, or high performance FPGA. These hardware devices can make visual SLAM algorithms more efficient.
2. When the camera moves too fast, the image texture information collected by the camera is not rich enough, and the scene illumination changes drastically, the visual SLAM will have large estimation errors or data loss. For this, inertial measurement unit can be used for the multi-sensor fusion to compensate disadvantages of visual sensor.

Acknowledgement. This work was supported by the National Key Research and Development Program of China (No. 2018YFC1507005).

References

1. Smith, R.C., Cheeseman, P.: On the representation and estimation of spatial uncertainly. Int. J. Robot. Res. **5**, 56–68 (1986)
2. XianLou, H., ShuangYuan, Y.: Image segmentation based on Normalized Cut and CUDA parallel implementation. In: 5th IET International Conference on Wireless, Mobile and Multimedia Networks (ICWMMN 2013), Beijing, pp. 209–214 (2013)
3. Chengyao, D., Jinlin, Y.: Real-time splicing of panoramic video with GPU acceleration and L-ORB feature extraction. Comput. Res. Dev. **54**(6), 1316–1325 (2017)
4. Changchang, W., Agarwal, S.: Multicore bundle adjustment. In: IEEE Computer Vision and Pattern Recognition, pp. 3057–3064. Colorado Springs (2011)
5. Rodriguez, L.D., Segundo, P.S.: GPU-mapping: robotic map building with graphical multiprocessors. IEEE Robot. Autom. Mag. **20**(2), 40–51 (2013)
6. Fraundorfer, F., Scaramuzza, D.: Visual odometry: Part II: Matching, robustness, optimization, and applications. IEEE Robot. Autom. Mag. **19**(2), 78–90 (2012)
7. Lowe, D.G.: Distinctive image features from scale-invariant keypoints. Int. J. Comput. Vis. **60**(2), 91–110 (2004)
8. Bay, H., Tuytelaars, T., Van Gool, L.: SURF: speeded up robust features. In: Leonardis, A., Bischof, H., Pinz, A. (eds.) ECCV 2006. LNCS, vol. 3951, pp. 404–417. Springer, Heidelberg (2006). https://doi.org/10.1007/11744023_32
9. Alcantarilla, P.F., Nuevo, J.: Fast explicit diffusion for accelerated features in nonlinear scale spaces. In: Proceedings of the British Machine Vision Conference (2013)
10. Rublee, E., Rabaud, V.: ORB: an efficient alternative to SIFT or SURF. In: IEEE International Conference on Computer Vision, Barcelona, pp. 2564–2571 (2011)
11. Corner Detection. http://en.wikipedia.org/wiki/Corner_detection/. Accessed 10 Aug 2020
12. Xiang, G., Tao, Z.: Fourteen Lectures on Visual SLAM: From Theory to Practice. 2nd. Publishing House of Electronics Industry, Beijing (2019)
13. ASL Datasets. https://projects.asl.ethz.ch/datasets/doku.php?id=kmavvisualinertialdatasets. Accessed 10 Aug 2020

Hardware Trojan Detection Method Based on Multi-featured GEP

Huan Zhang[1], Jiliu Zhou[1](✉), Xi Wu[2], and Yi Zhang[1]

[1] College of Computer Science, Sichuan University, Chengdu 610065, China
zhoujl@cuit.edu.cn
[2] School of Computer Science, Chengdu University of Information Technology, Chengdu 610225, China

Abstract. In the hardware Trojan detection method, destructive reverse engineering can most precisely restore the original circuit of the chip to be detected, but this method is a huge amount of work, high cost, long life cycle. In this paper, we proposed a multi-featured GEP technology, non-destructive reverse engineering of the chip using various data obtained from bypass detection, in order to restore the actual circuit of the hardware, or at least find out the unknown circuit design.

Keywords: Hardware Trojans · Trojan horse detection · Multi-featured · GEP

1 Introduction

In today's hardware Trojan detection technology [1–7], destructive reverse engineering [1, 3, 8–10] can most precisely restore the original circuit of the chip to be detected, but this technology is a huge amount of work, high cost, long life cycle, and the chip has been scrapped after detection, can only be used for sampling or replication of a class of chips, bypass detection [11–16] technology analyzes the bypass circuit signal, such as timing, power, electromagnetics, heat, etc., determine whether or not they contain Trojans, no damage to the chip, relatively small amount of data is needed, lower cost, is the most important and effective method.

Some of the work is done by evolutionary algorithms to study evolutionary hardware [17–23]. This paper proposes a multi-featured GEP evolutionary algorithm, use a single circuit component or group of circuit structures as a node, use an operator to describe a node's multi-features, use GEP to evolved circuits close to the original circuit, then determine the structure of the original circuit.

2 Brief Introduction of GEP

GEP (Gene Expression Programming) [24] combines the advantages of GA and GP, follow the basic steps of Evolutionary Computation (EC). The basic composition of its genetic material is two kinds of symbols, that is terminators and functions. A gene

X. Wu et al. (Eds.): QShine 2020, LNICST 381, pp. 22–37, 2021.
https://doi.org/10.1007/978-3-030-77569-8_2

consists of a linear, fixed-length string of symbols, code for expression trees with different sizes and shapes, a chromosome can consist of a single gene or multiple genes, each chromosome is decoded to map as a candidate solution for problem response.

F is the set of functions and T is the set of terminals, the genes of GEP are composed of a head and a tail. The head contains symbols that represent both F and T, whereas the tail contains T. For each problem the length of the head h is chosen, whereas the length of the tail e is a function of h and the number of arguments of the function with more arguments n (also called maximum arity) and is evaluated by the equation:

$$e = h \times (n - 1) + 1 \tag{1}$$

For example, consider a gene for which the set of functions F = {+, −, *, /, Q} and the set of terminals T = {a, b}. In this case n = 2, if we chose an h = 10, then e = 11, thus the length of the gene is 21, one such gene is shown below (the tail is shown in bold):

$$+Q - /b * abQbababab baaab$$

It codes for the following expression tree (ET):

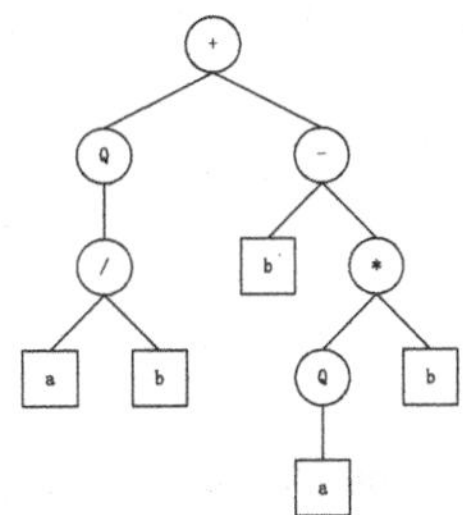

The algebraic expression:

$$\sqrt{a/b} + b - \sqrt{a} * b$$

Tail excess symbols are discarded directly without use. This allows GEP to use fixed-length encoding to express different sizes and shapes of expression trees.

3 Use GEP to Represent Circuit

GEP has done well in mining association rules, clustering, classification rules, time series predictions, sunspot predictions [25–29].

As can be seen from the data structure of GEP, GEP can solve tree structure problems very well, Other words, circuits that can be represented as a tree-shaped structure with n leaf nodes, can be described directly with GEP. The logic circuit of the 6-in/1 output in Fig. 1(a) can be easily represented as a tree structure in Fig. 1(b), It replaces the logical gate function with the logical symbol, in GEP, Its corresponding effective gene is:

And, And, And, A, Not, And, And, B, C, Not, E, F, D

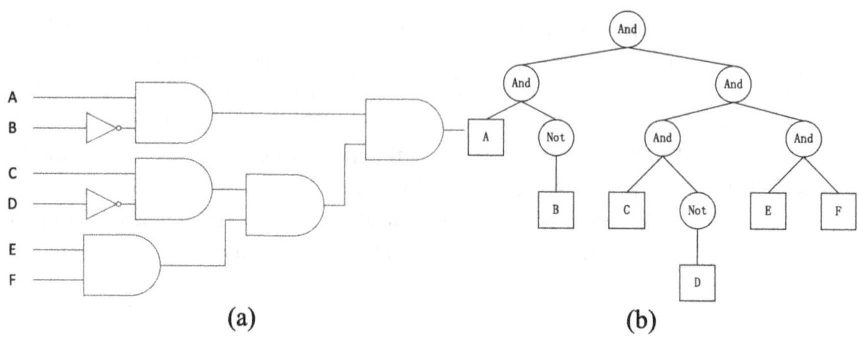

Fig. 1. 6-in/1-out circuit

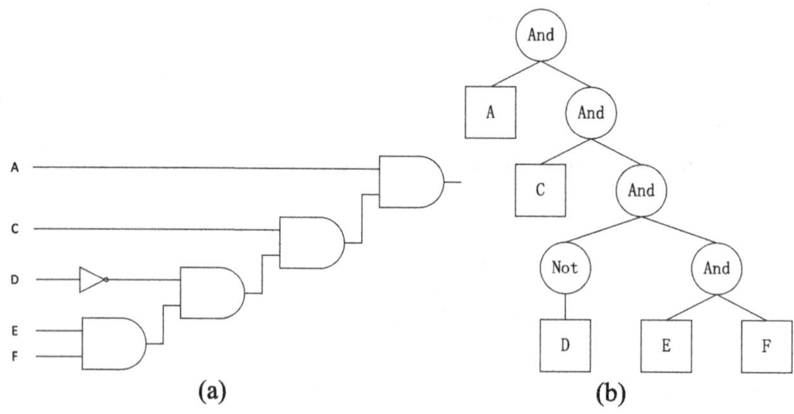

Fig. 2. Isomorphic circuit of Fig. 1

But, if only logical values are used to represent the circuit, there are too many isomorphic situations, for example, the circuit in Fig. 1 can be replaced with the circuit in Fig. 2.

The corresponding valid gene is:

$$\text{And, A, And, C, And, Not, And, D, E, F}$$

The two circuits are fully equivalent in logic values, both is:

$$Y = ACD'EF$$

And you can see, in the second circuit, input B is not used at all. That is, input B in the first circuit does not actually affect the output.

This example is just for the logic value of the circuit, In fact, for the circuit bypass information detection, there is a similar situation. The test results for any single bypass information, we can get a number of different circuit structures. This article refers to this situation as isomorphic.

Thus, only a logical value is used to learn a certain kind of bypass information to describe the circuit, so much Isomorphism that the circuit structure could not be confirmed. In this paper, a multi-featured GEP algorithm is proposed, use the same structure in GEP to represent multiple circuit features, effectively reduces the number of isomorphic circuits, and use this algorithm to detect hardware trojans.

4 Multi-featured GEP

In this paper, the logic value of the circuit or any kind of bypass information such as voltage, current, etc. is called a characteristic value. Isomorphism for a characteristic value, essentially, the characteristics of the circuit components on this characteristic value are superimposed on each other, allows several different circuit structures to exhibit similar or even identical characteristics after the characteristic values are superimposed on each other, this makes it impossible to represent the corresponding circuit by the result of a particular characteristic value.

While we are doing the test, multiple feature values are detected at the same time, the detection results of each feature value can result in multiple isomorphic circuits, but since the overlay characteristics of different feature values cannot be identical, these homogeneous circuits cannot be exactly the same, and the same part of which is the possible real circuit.

The Multi-featured GEP algorithm can be illustrated by a Not-Gate designed by a triode in Fig. 3.

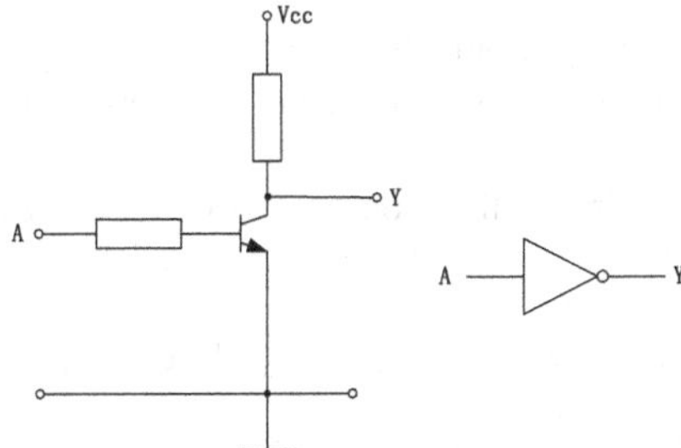

Fig. 3. A Not Gate Designed by Triode

- Feature Value 1:

  In digital logic terms, the description of this circuit is: Y = !A.

- Feature Value 2:

  In terms of voltage, the description of this circuit is: $V_y = V_A < V_{SH}\ ?\ V_{cc}\colon V_{CES}$.

  Among them: VSH represents high-level threshold, that is, the lower potential limit of "1" in the circuit. VCES represents a saturation voltage drop of the tripole.

- Feature Value 3:

In terms of current, the description of this circuit is: Cy = N * CA, where N is the magnification of the current.

In addition, there are a variety of other bypass information detection items, like delay, spectrum, etc.

The above 3 characteristic values, using any one alone can not determine the lower circuit structure, however, if the three cases are combined, the circuit structure can be determined.

In GEP, an operator represents only one calculation, a GEP individual can only represent a description of a feature value, can get an unlimited number of isosome circuits, it is difficult to determine the actual circuit structure.

This paper proposes a method to combine the test results of multiple characteristic values on a basic circuit into a single function representation, combine into a composite function, i.e. have a function represent multiple calculations, use GEP's functional evolutionary ability to evolve representations close to the original circuit.

In detail, we include these multiple detection values in a function, the input of the function is multiple feature values, the output is a vector, like this non-gate circuit. in GEP's expression tree, it is still represented by "Not.", but the implication becomes the calculation of the below vector:

$$\mathrm{Not}(\mathrm{A1},\ \mathrm{A2},\ \ldots,\ \mathrm{An}) = [\mathrm{F1}(\mathrm{A1}),\ \mathrm{F2}(\mathrm{A2}),\ \ldots,\ \mathrm{Fn}(\mathrm{An})]$$

Where Ak (k is 1,..., n) is the input value of some kind of feature detection, Fk (Ak) (k s 1, ..., n) is the result of this feature value detection corresponding to the input value Ak.

For example, corresponding to this non-gate circuit, the symbol Not indicates the following meaning:

$$\mathrm{Not}(\mathrm{A},\ \mathrm{V_A},\ \mathrm{C_A}) = [!\mathrm{A},\ \mathrm{V_A} < \mathrm{V_{SH}}?\,\mathrm{Vcc} : \mathrm{V_{CES}},\ \mathrm{N} * \mathrm{C_A}]$$

Where A represents the logical value (1 or 0) represented by the voltage entered at point A, The VA represents the input voltage of point A and the CA represents the input current of point A.

Thus, when using GEP evolution, a single symbol Not can also represent multiple feature value tests that are not associated with each other.

5 Experiments

5.1 Experimental Settings

This paper has designed 4 experiments (Table 1).

1. Parameters

- The circuit parameters are:
  - Output m = 1
  - The number of features is k = 1, 2, 2, 3
  - 3 features are used: feature 1 is logical, feature 2 is voltage value, feature 3 is current value.
- As a comparison experiment, the parameters used are exactly the same:

Table 1. Experiment parameters

Parameter	Value
Fitness	>=1
Selection mode	Tournament, size = 3
Population size	10000
Head length	20
Tail length	21
Chromosome length	1
Mutation rate	0.05
Insert rate	0.1
Root insert rate	0.01
One-point cross rate	0.1
Two-point recombination rate	0.1
Input number	4
Output number	1
Function set	Not, And, Or

2. Fitness function

The characteristic data are logical data, voltage data and current data, which have their respective Fitness function:

a. Logical Fitness function:

$$F_1 = 1 - \sum\nolimits_{i=1}^{N} |y_i - y|/N \tag{2}$$

N is the test data count, yi is the logical value calculated based on the test data after decoding the individual, y is the output logic values for test data. Because of the logic value, so the worst case is that each output after the individual decodes is the opposite of the test value, that is $|y_i - y| = 1$, so the value of the F1 is in [0,1].

b. Voltage Fitness function

$$F_2 = 1 - (\sum\nolimits_{i=1}^{N} |y_i - y|/N)/(V_{CC} - V_{DD}) \tag{3}$$

N is the test data count, yi is the voltage value calculated based on the test data after decoding the individual, y is the output voltage values for test data. The worst case is, each test output value is either the highest level or the lowest level, and each output that is decoded by an individual is the opposite of the test value that is $|y_i - y| = V_{cc} -$ VDD, so the value of the F2 is in [0,1].

c. Current Fitness function

$$F3 = 1 - SSE/SST \tag{4}$$

$$SSE = \sum\nolimits_{i=1}^{m} (y_i - \hat{y}_i)^2$$

$$SST = \sum\nolimits_{i=1}^{m} (y_i - \bar{y}_i)^2$$

N is the test data count, $\hat{y}_i$ is estimate of yi calculated from Ti using formula, $\overline{y}_i$ is the average of the y, SSE is Residual Sum of Squares, SST is Sum of Squares of Deviations, F3 is the square of the multiple correlation coefficient in statistics, its value is also in [0,1].

d. Individual's Fitness

According to the previous algorithm description, the individual's fitness should be a combination of the three, then the individual's fitness is:

$$F = C_1 * F_1 + C_2 * F_2 + C_3 * F_3 \tag{5}$$

C1, C2, C3 is weights of 3 feature values in final fitness.

5.2 Experimental Results

Figure 4 is a circuit that has no Trojan.

Its boolean expression is

$$Y = A + BC + BD \tag{6}$$

The circuit dose the computation

$$Y = \begin{cases} 0, \ (ABCD) < 5 \\ 1, \ else \end{cases} \tag{7}$$

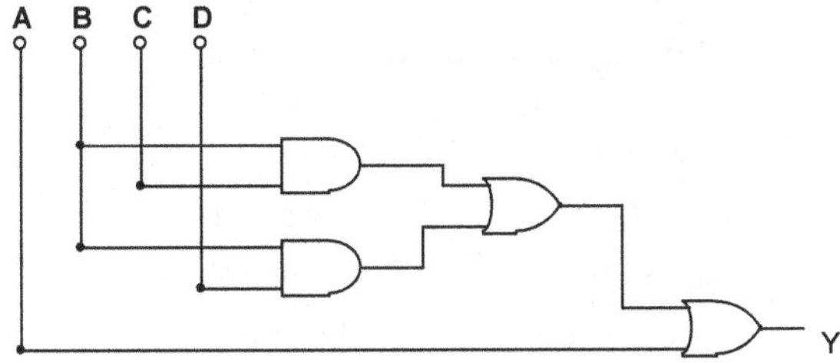

Fig. 4. Circuit has No Trojan

After several logic gates added to the circuit of the Fig. 1, it becomes a circuit has trojan (Fig. 5). In the new circuit that Y will also get a value of 0 when (ABCD) = 7:

$$Y = \begin{cases} 0, \ (ABCD) < 5 \ or \ (ABCD) = 7 \\ 1, \ else \end{cases} \tag{8}$$

Its Boolean expression becomes:

$$Y = A + BCD' + BC'D \tag{9}$$

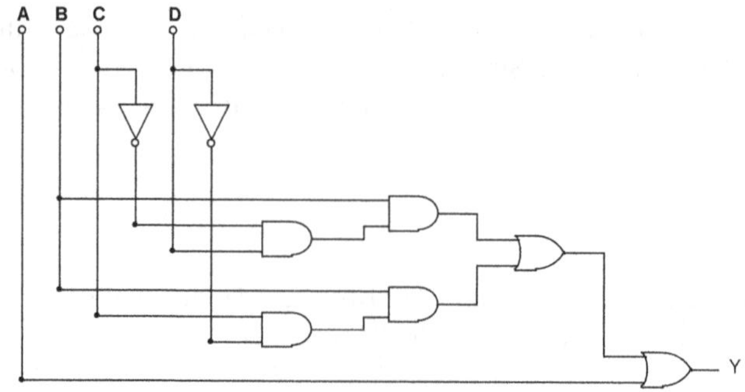

Fig. 5. Circuit with Trojan

The activate gene of the circuit is: Or, A, Or, And, And, B, And, B, And, Not, D, C, Not, C, D

If the pins are so many that we can only test part of the values, the input value to activate the trojan (ABCD) = $(0111)_2$ may be missed at this time. In the following experiment, the input value $(0111)_2$ will not be provided, and the output of this input value will be determined by the evolved circuit.

Using different combinations, we designed 4 sets of experiments. Considering that the logic values are required to be correct in the circuit first, the voltage and current values must be based on the correct logic values in order to make sense. The individual's fitness is a combination of several data, in which the proportion of logical value is larger.

- **Experiment 1** (Table 2)

Table 2. Setting of the Experiment 1

Parameter	Value
Logic Gate	And, or, not
Values Provided	Logic Values
Fitness function	$F = F_1$
Exercise Count	100
Trojan Discovered Count	0

None of the 100 times exercise is able to discover the Trojan circuit if only the logical values were provided. Figure 6 shows some typical results.

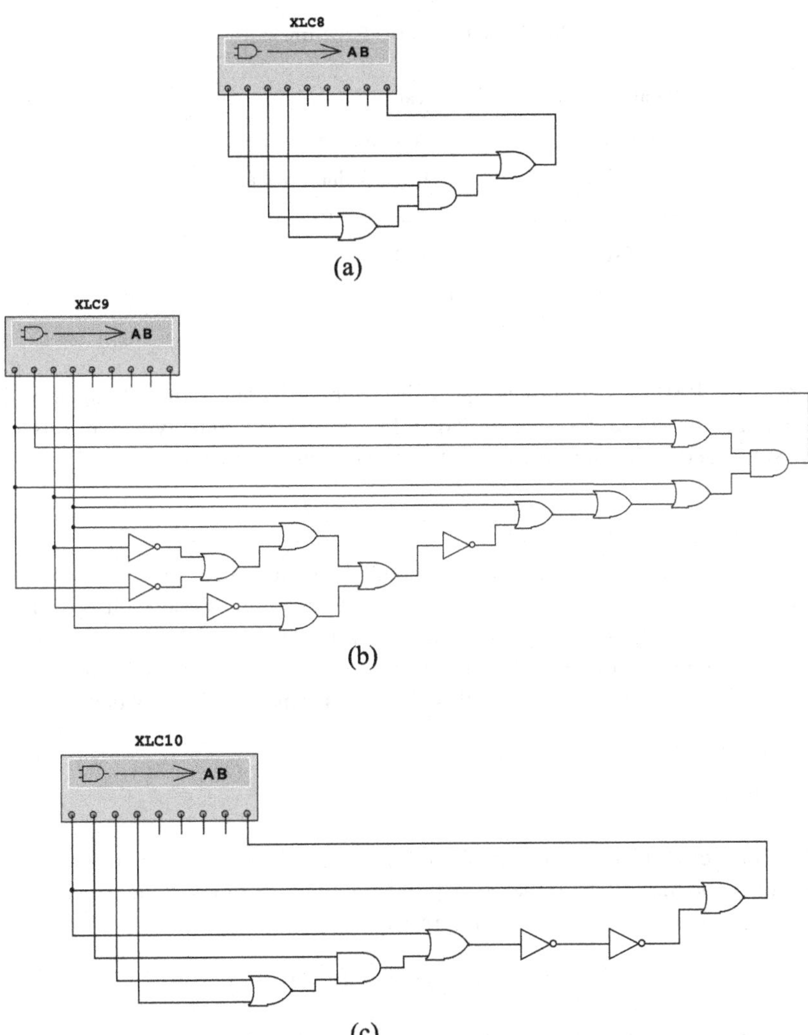

Fig. 6. Some Circuits evolved by Multi-featured GEP in Experiment 1

The simplified Boolean Expression of all the circuits above is:

$$Y = A + BC + BD$$

- **Experiment 2** (Table 3)

Table 3. Setting of the Experiment 2

Parameter	Value
Logic Gate	And, or, not
Values Provided	Logic Values, Voltage Values
Fitness function	$F = 0.8 * F1 + 0.2 * F2$
Exercise Count	100
Trojan Discovered Count	0

None of the 100 times exercise is able to discover the Trojan circuit if both the logical values and the voltage values were provided. Figure 7 shows some typical results.

The simplified Boolean Expression of all the circuits above is:

$$Y = A + BC + BD$$

The Trojan still cannot be discovered although both the logical value and the voltage value were provided. The reason is in digital circuits, the logic value itself is expressed in terms of voltage values, for example, a voltage value less than 3 V is considered to be 0, a voltage value greater than 3 V is considered to be 1. So the choice of logical value and voltage value as feature values on this issue is as same as only provided the logical value.

- **Experiment 3** (Table 4)

The Trojan circuit was discovered 72 times among the 100 times exercise when the logical values and the voltage values were provided. Figure 8 shows some typical results.

Both circuits' simplified Boolean Expression is

$$Y = A + BCD' + BC'D$$

The equivalent circuit has been discovered although the original circuit has hidden.

- **Experiment 4** (Table 5)

The Trojan circuit was discovered 67 times among the 100 times exercise when three feature values were all provided. Figure 9 shows a different result.

The simplified Boolean Expression is:

$$Y = A + BCD' + BC'D$$

A circuit equivalent to the original circuit has been found.

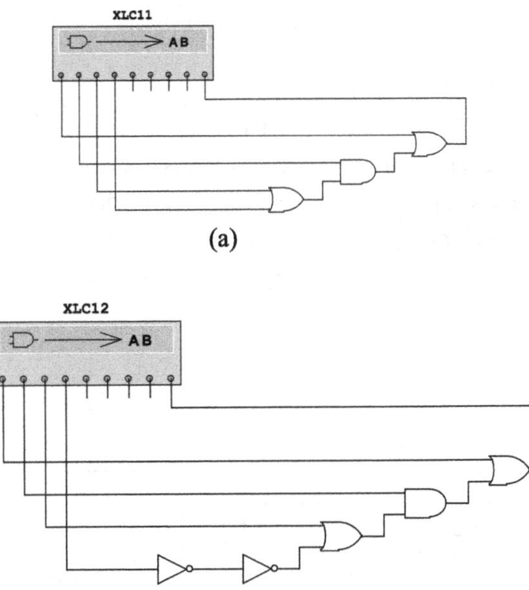

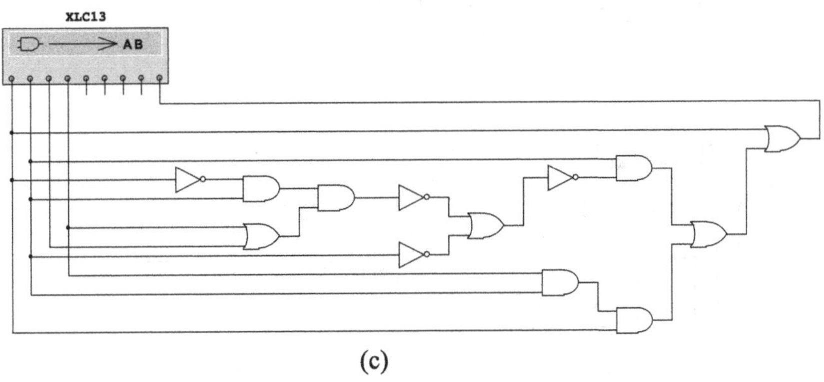

(c)

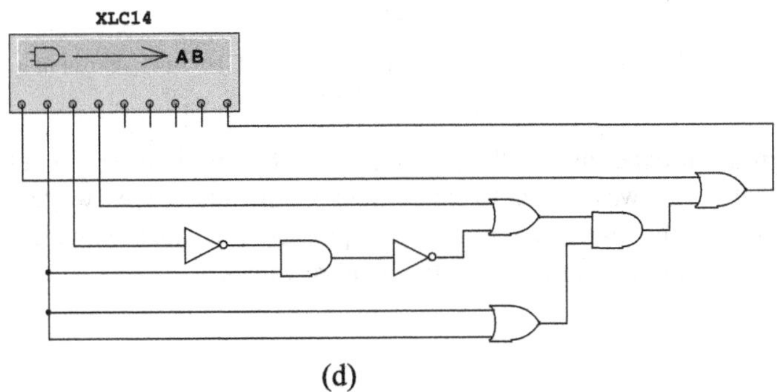

(d)

Fig. 7. Some Circuits evolved by Multi-featured GEP in Experiment 2

Table 4. Setting of the Experiment 3

Parameter	Value
Logic Gate	And, or, not
Values Provided	Logic Values, Current Values
Fitness function	F = 0.8 * F1 + 0.2 * F3
Exercise Count	100
Trojan Discovered Count	72

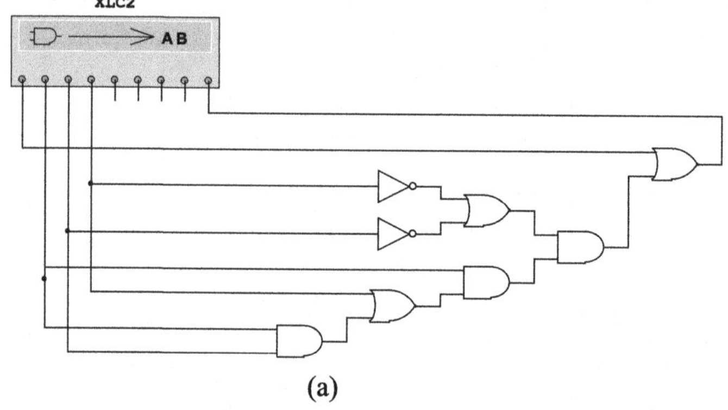

(a)

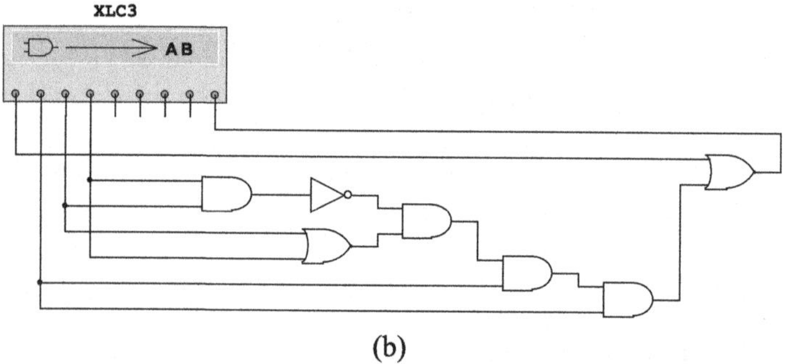

(b)

Fig. 8. Some Circuits evolved by Multi-featured GEP in Experiment 3

In this group of experiments, three feature values were used, but the efficiency of discovering the Trojan was similar to experiment 3 which used only two feature values. The reason is since the logical value itself is expressed in terms of voltage values, then the three feature values are equal to the two feature values.

Table 5. Setting of the Experiment 4

Parameter	Value
Logic Gate	And, or, not
Values Provided	Logic Values, Voltage Values Current Values
Fitness function	F = 0.6 * F1 + 0.2 * F2 + 0.2 * F3
Exercise Count	100
Trojan Discovered Count	67

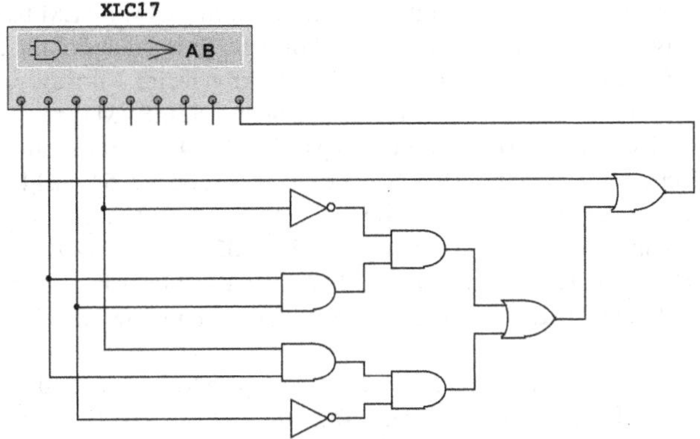

Fig. 9. Typical Circuit evolved by Multi-featured GEP in Experiment 4

6 Conclusions

This paper has proposed multi-featured GEP algorithm, when multiple feature values were fused into the same operator, there is a certain probability that GEP can automatically discover the evolutionary power of mathematical formulas to discover Trojan circuits. The fewer features used, the more efficient GEP evolves, but farther away the conclusion is from the real circuit. On the otherwise, the more features used, the less efficient GEP evolves, but the conclusion is closer to the real circuit. However, if there is a direct conversion relationship between the multiple feature values used, the characteristic values can be considered one, and this situation does not increase the evolutionary accuracy of GEP.

Acknowledgment. This work was supported by the National Key Research and Development Program of China (Grant No. 2018YFB1502803).

References

1. Xiao, K., Forte, D., Jin, Y., et al.: Hardware Trojans: lessons learned after one decade of research

2. Antonopoulos, A., Kapatsori, C., Makris, Y.: Trusted analog/mixed-signal/RF ICs: a survey and a perspective. IEEE Des. Test **34**(6), 63–76 (2017)
3. Bhunia, S., Hsiao, M.S., Banga, M., et al.: Hardware Trojan attacks: threat analysis and countermeasures. Proc. IEEE **102**(8), 1229–1247 (2014)
4. ACM Trans. Des. Autom. Electron. Syst. (TODAES) **22**(1), 6:1–6:23 (2016)
5. Tehranipoor, T., Koushanfar, F.: A survey of hardware Trojan taxonomy and detection. IEEE Des. Test **27**(1), 10–25 (2010)
6. Jian-Feng, Z., Gang, S.: A survey on the studies of hardware Trojan. J. CyberSecur. **2**(1), 74–90 (2017)
7. Bhasin, S., Regazzoni, F.: A survey on hardware Trojan detection techniques. In: Proceedings of the 2015 IEEE International Symposium on Circuits and Systems (ISCAS). Lisbon, Portugal, pp. 2021–2024 (2015)
8. Courbon, F., Loubet-Moundi, P., Fournier, J.J.A., et al.: SEMBA: a SEM based acquisition technique for fast invasive hardware Trojan detection. In: Proceedings of the 2015 European Conference on Circuit Theory and Design (ECCTD), Trondheim, Norway, pp. 1–4 (2015)
9. Bao, C.X., Forte, D., Srivastava, A.: On application of one-class SVM to reverse engineering-based hardware Trojan detection. In: Proceedings of the 2014 15th International Symposium on Quality Electronic Design (ISQED), Santa Clara, USA, pp. 47–54 (2014)
10. Bao, C.X., Forte, D., Srivastava, A.: On reverse engineering-based hardware Trojan detection. IEEE Trans. Comput. Aided Des. Integr. Circuits Syst. **35**(1), 49–57 (2015)
11. Agrawal, D., Baktir, S., Karakoyunlu, D., et al.: Trojan detection using IC fingerprinting. In: Proceedings of the 2007 IEEE Symposium on Security and Piracy (SP 2007), Berkeley, USA, pp. 296–310 (2007)
12. Xiao, K., Zhang, X.H., Tehranipoor, M.: A clock sweeping technique for detecting hardware Trojans impacting circuits delay. IEEE Des. Test **30**(2), 26–34 (2013)
13. Aarestad, J., Acharyya, D., Rad, R., et al.: Detecting Trojans through leakage current analysis using multiple supply pad IDDQ. IEEE Trans. Inf. Forensics Secur. **5**(4), 893–904 (2010)
14. Nowroz, A.N., Hu, K.Q., Koushanfar, F., et al.: Novel techniques for high-sensitivity hardware Trojan detection using thermal and power maps. IEEE Trans. Comput. Aided Des. Integr. Circuits Syst. **33**(12), 1792–1805 (2014)
15. Zhou, B.Y., Adato, R., Zangeneh, M., et al.: Detecting hardware Trojans using backside optical imaging of embedded watermarks. In: Proceedings of the 2015 52nd ACM/EDAC/IEEE Design Automation Conference (DAC 2015), San Francisco, USA, pp. 1–6 (2015)
16. He, J.J., Zhao, Y.Q., Guo, X.L., et al.: Hardware Trojan detection through chip-free electromagnetic side-channel statistical analysis. IEEE Trans. Very Large Scale Integr. (VLSI) Syst. **25**(10), 2939–2948 (2017)
17. Higuchi, T., Murakawa, M., Iwata, M., et al.: Evolvable hardware at function level. In: Proceedings of the IEEE International Conference on Evolutionary Computation, pp. 187–192 (1997)
18. Higuchi, T., Iwata, M., Kajitani, I., Iba, H., Hirao, Y., Furuya, T., Manderick, B.: Evolvable hardware and its application to pattern recognition and fault-tolerant systems. In: Sanchez, E., Tomassini, M. (eds.) TEH 1995. LNCS, vol. 1062, pp. 118–135. Springer, Heidelberg (1996). https://doi.org/10.1007/3-540-61093-6_6
19. Vassilev, V., Job, D., Miller, J.: Towards the automatic design of more efficient digital circuits. In: Proceedings of the 2nd NASA/DOD Workshop on Evolvable Hardware, pp. 151–160 (2000)
20. Timothy, G.W., Peter, J.B.: Towards development in evolvable hardware. In: Proceedings of the 3rd NASA/DOD Workshop on Evolvable Hardware Pasadena, pp. 241–250 (2002)
21. Erba, M., Rossi, R., Liberali, V., et al.: Digital filter design through simulated evolution. In: Proceedings of the European Conference on Circuit Theory and Design, pp. 389–393 (2001)

22. Hemmi, H., Mizoguchi, J., Shimohara, K.: Development and evolution of hard ware behaviors. In: Proceedings of the Artificial Life IV, pp. 250–265 (1994)
23. Hounsell, B., Arslan, T.: A novel evolvable hardware framework for the evolution of high performance digital circuits. In: Proceedings of the Genetic and Evolutionary Computation Conference, pp. 525–529 (2000)
24. Ferreira, C.: Gene expression programming: a new adaptive algorithm for solving problems. Complex Syst. **13**(2), 87–129 (2001)
25. Ferreira, C.: Gene Expression Programming: Mathematical Modeling by an Artificial Intelligence. Springer, New York (2002)
26. Zuo, J., Tang, C., Zhang, T.: Mining predicate association rule by gene expression programming. In: Meng, X., Su, J., Wang, Y. (eds.) WAIM 2002. LNCS, vol. 2419, pp. 92–103. Springer, Heidelberg (2002). https://doi.org/10.1007/3-540-45703-8_9
27. Zuo, J., Tang, C.-J., Li, C., Yuan, C., Chen, A.-L.: Time series prediction based on gene expression programming. In: Li, Q., Wang, G., Feng, L. (eds.) WAIM 2004. LNCS, vol. 3129, pp. 55–64. Springer, Heidelberg (2004). https://doi.org/10.1007/978-3-540-27772-9_7
28. Zhou, C., Xiao, W.M., Tirpak, T.M., et al.: Evolution accurate and compact classification rules with gene expression programming. IEEE Trans. Evol. Comput. **7**(6), 519–531 (2003)
29. Lopes, H.S., Weinert, W.R.: EGIPSYS: an enhanced gene expression programming approach for symbolic regression problems. Int. J. Appl. Math. Comput. Sci. **14**(3), 375–384 (2004)

Sleep Apnea Monitoring System Based on Channel State Information

Xiaolong Yang(✉), Xin Yu, Liangbo Xie, Mu Zhou, and Qing Jiang

School of Communication and Information Engineering, Chongqing University of Posts and Telecommunications, Chongqing, China
yangxiaolong@cqupt.edu.cn

Abstract. Sleep apnea is an important factor that affects human health. Traditional approaches based on wearable devices or pressure sensor devices are too expensive to be suitable for daily use, which also don't consider the impact on the breathing frequency when the human body turns over or gets up. In this paper, we propose a system based on WiFi to monitor sleep apnea state. Firstly, we use linear fitting to eliminate the phase errors of the receiving antennas, and wavelet transform to remove the noise of signal amplitude. Secondly, we combine the short-time Fourier transform and sliding window method to segment the signal. Finally, the features such as the variance of the phase difference between antennas are extracted, and the neural network model is built to identify apnea state, so as to eliminate interference caused by changes in sleep postures. Experiment results show that the detection accuracy rate for sleep apnea is over 95.6%. Our system can be a daily apnea monitoring approach and provide health reference for users.

Keywords: WiFi · Channel state information · Sleep apnea

1 Introduction

Sleep quality is closely related to human health. Some commercial devices collect data from headphones or wristbands worn by users to analyze the quality of sleep [1–3].

In recent years, because of its non-contact and high privacy characteristics, radar has been widely used in breathing detection. WiKiSpiro [4] monitors breathing in real time by combining a depth camera and a radar system. WiSpiro [5] is a system that uses frequency modulated continuous wave (FMCW) to re-construct the thorax and abdomen movement, and maps it to the breathing process through the training process. The Vital-Radio system [6] for monitoring breath and heart rate proposed by Adib et al. uses a bandwidth of 5.46 GHz to 7.25 GHz. However, radar system equipment is not only expensive to manufacture but also requires additional customized hardware equipment and has a high operating frequency.

Compared with radar systems, radio frequency identification (RFID) has the characteristics of simple structure and equipment, and high recognition rate. Therefore, it is also used for vital signs monitoring. Tagbreathe [7] attaches a lightweight RFID tag to

X. Wu et al. (Eds.): QShine 2020, LNICST 381, pp. 38–47, 2021.
https://doi.org/10.1007/978-3-030-77569-8_3

the user's clothing, analyzes the low-level data obtained by the RFID reader, and uses phase information to estimate the respiratory rate. With the popularity of WiFi devices, WiFi-based respiratory detection technology is gradually becoming a research hotspot. Ubibreathe [8] uses the received signal strength indication (RSSI) on the WiFi device for breath estimation. A bandpass filter with a cut-off frequency of 0.1 Hz to 0.5 Hz is used to filter the received RSSI signal. The Fourier transform is used to estimate the breathing frequency. However, it only provides accurate results when the human brings the WiFi device close to the chest.

Compared with the superimposed energy information provided by RSSI, channel state information (CSI) not only contains carrier amplitude information, but also provides phase information of each carrier. The finer granularity provides higher possibility for more accurate breath detection research. Liu [9] uses the CSI amplitude and phase difference to capture the tiny movements caused by breathing and heartbeat. Phasebeat [10] uses wavelet transform to decompose and reconstruct the respiratory signal and heart rate signal in the CSI signal. TR-BREATH [11] combines Root-Music and other algorithms to analyze the time-reversed resonance intensity to estimate the respiratory frequency.

In view of the above-mentioned shortcomings, the detection system for sleep proposed in this paper has strong privacy and low price, and is suitable for daily detection systems. Firstly, we construct the signal model of the received signal and modify the received CSI phase information. Secondly, we eliminate the noise, interference and abnormal values in the signal after the correction. Next, a signal segmentation algorithm is designed to select subcarriers that have the most obvious change in motion and segment the signal. Finally, we extract features from the segments and construct a classifier for detection and recognition. The scheme designed in this paper does not require any equipment to be carried, nor does it need to modify the related hardware equipment to detect the sleep state.

2 System Model and Framework

2.1 System Overview

In order to simulate the breathing state that may occur during sleep, the human is required to breathe normally in the test area to simulate the breathing state; stop breathing to simulate apnea; leave the test area to simulate unmanned state.

As shown in Fig. 1, the sleep state detection process based on WiFi is mainly divided into two stages. In the offline stage, CSI of unmanned state, human breathing state and apnea state are collected respectively. Then, we eliminate phase errors and noise interference in the environment, and select the optimal subcarrier. Finally, the features of the three scenes are extracted, and a detection classifier is constructed and generated. In the online phase, we collect CSI, preprocess signals, and extract signal features in the same way. The classifier built in the offline phase is used to complete the sleep state detection.

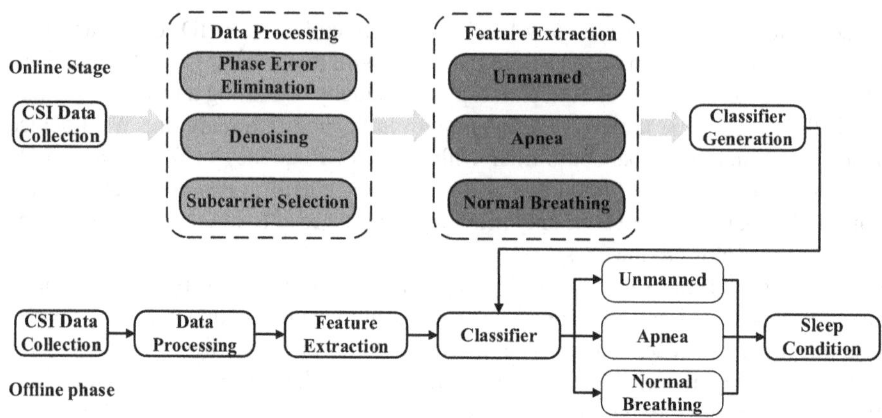

Fig. 1. A system architecture of a sleep apnea monitoring method based on home WiFi.

2.2 Channel State Information Overview

The advent of orthogonal frequency division multiplexing (OFDM) technology helps us extract CSI from the transmitting end to the receiving end of the wireless signal. To characterize multipath propagation, wireless channels are usually modeled with channel impulse response (CIR). It can be expressed as:

$$h(t) = \sum_{l=1}^{L} \alpha_l \delta(t - \tau_l) \tag{1}$$

where L is the total number of propagation paths; $\delta(t)$ is the Dirichlet function; α_l and τ_l are the amplitude attenuation and time delay of the i-th path, respectively. Since multipath transmission shows frequency selective fading in the frequency domain, it can also be characterized by channel frequency response (CFR) $H(f)$. CFR and CIR are Fourier transforms:

$$H(f) = \mathrm{FT}[h(t)] = \sum_{l=1}^{L} \alpha_l e^{-j2\pi f \tau_l} \tag{2}$$

where f is the frequency. In the time domain, the accepted signal $y(t)$ is the convolution of the transmitted signal $s(t)$ and $h(t)$:

$$y(t) = s(t) \otimes h(t) \tag{3}$$

Correspondingly, in the frequency domain, the received signal spectrum $Y(f)$ is the product of the transmitted signal spectrum $S(f)$ and $H(f)$:

$$Y(f) = S(f) \cdot H(f) \tag{4}$$

CSI is the sampled version of CFR. Assuming that there are K subcarriers on one antenna and M packets are received, the CSI can be expressed as a matrix:

$$\mathbf{CSI}_{K\times M} = \begin{bmatrix} csi_{1,1} & csi_{1,2} & \cdots & csi_{1,M} \\ \vdots & \vdots & \ddots & \vdots \\ csi_{K,1} & csi_{K,2} & \cdots & csi_{K,M} \end{bmatrix} \tag{5}$$

where $csi_{k,m}(k \in [1, K],\ m \in [1, M])$ represents the sum of all paths of the k-th subcarrier in the m-th data packet.

3 Data Processing

3.1 Phase Error Elimination

In actual operation, the actual received phase consists of the true phase value and the offset value:

$$\hat{\phi}_k = \phi_k + \frac{2\pi}{K}k\eta + \rho \tag{6}$$

where $\hat{\phi}_k$ and ϕ_k represent measured phase and actual phase of the k-th subcarrier, respectively. η represents the phase offset and ρ represents constant error. In this paper, the phase error is eliminated by linear fitting method [12], and the result comparison is shown in Fig. 2. It can be seen that the phase after linear calibration is more stable than the original phase. Although the processed phase is not completely equal to the true phase, it is very close to the true value, and the error can be relatively ignored.

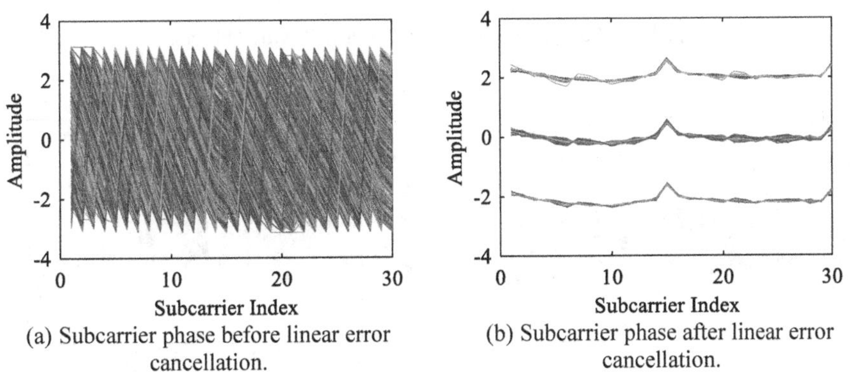

(a) Subcarrier phase before linear error cancellation.

(b) Subcarrier phase after linear error cancellation.

Fig. 2. The comparison result before and after the subcarrier phase linearity error is eliminated.

3.2 Noise Cancellation

In the measured environment, due to the multipath effect in the room and the factors of the device, the received signal exists various noises, resulting in the required useful signals being submerged in the noise. Thus, the process of amplitude processing of CSI in this paper includes the removal of outliers and wavelet denoising. As shown in the black box in Fig. 3(a), there will always be a few outliers that deviate from the original signal trajectory. We use a filter based on the median absolute deviation to filter out the sample values outside of $[\mu - 3 \cdot \sigma, \mu + 3 \cdot \sigma]$, where μ represents the mean, and σ represents the standard deviation. Then, we replace them by the median of the data, as shown in Fig. 3(b).

After removing the influence caused by outliers, the wavelet denoising method is used to process the CSI amplitude. In this paper, 'db3' is used as a wavelet basis to decompose the signal in 5 layers. The comparison results before and after denoising are shown in Fig. 4. Before denoising, the signal is submerged in noise, and after denoising, the waveform becomes smooth and can reflect the changing state of the channel.

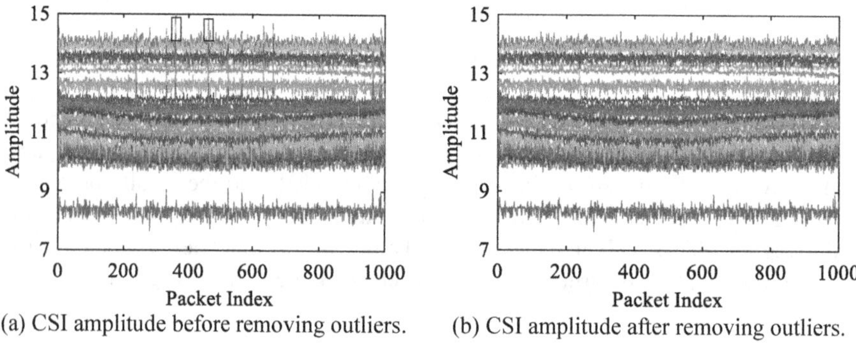

(a) CSI amplitude before removing outliers. (b) CSI amplitude after removing outliers.

Fig. 3. CSI amplitude before and after removing outliers.

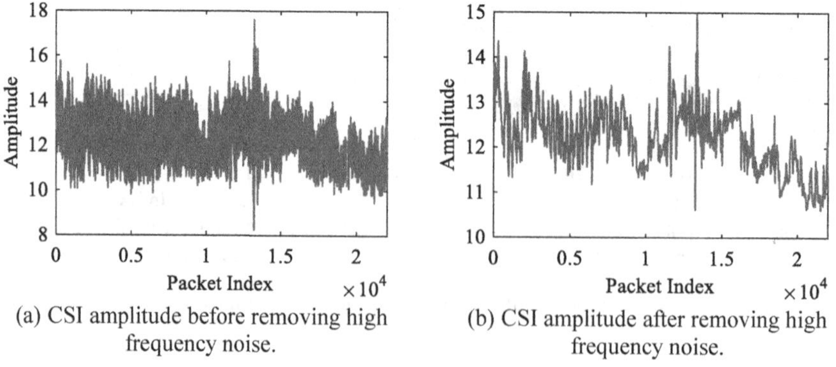

(a) CSI amplitude before removing high frequency noise. (b) CSI amplitude after removing high frequency noise.

Fig. 4. Comparison of CSI amplitude before and after denoising.

3.3 Subcarrier Selection Method

Due to the different frequency of the subcarriers, the sensitivity to changes caused by human breathing, apnea and unmanned states is also different. Therefore, we need to filter out subcarriers that do not change significantly. In this paper, we use the change of CSI amplitude to quantify the sensitivity of subcarriers to the sleep state. Assuming that the signal length is M, we calculate the variance of the CSI signal of the k-th subcarrier V_k:

$$V_k = \frac{1}{M-1}\sum_{m=1}^{M}(csi_{k,m} - \frac{1}{M}\sum_{m=1}^{M} csi_{k,m})^2 \tag{7}$$

4 Feature Extraction

4.1 Signal Segmentation Method

Since the collected data may contain the subject being in apnea or the subject leaving the test area, etc., for monitoring the sleep state accurately, before performing feature extraction, these fragments need to be segmented from the data, and identify and classify them. Therefore, this paper uses a sliding window-based method [13] to segment the signal. We assume that the window length is N and the signal length is M, and calculate the variance V_n of the CSI signal difference between two adjacent windows:

$$\begin{aligned} V_n = \frac{1}{N-1}\sum_{m=1}^{N}((csi_{n,m} - csi_{n-1,m}) \\ - \frac{1}{N}\sum_{m=1}^{N}(csi_{n,m} - csi_{n-1,m}))^2 \end{aligned} \tag{8}$$

where n indicates the number index of windows. Then, We normalize V_n to get V_n':

$$V_i' = \frac{V_i - \min\{V_i\}}{\max\{V_i\} - \min\{V_i\}}, \quad (1 \le i \le n) \tag{9}$$

Firstly, we set the start flag as "False". When it is "False", compare each V_i' with the threshold σ. If $V_i' \ge \lambda$, we set the start time $\mathrm{T}_{begin} = (i-1)\cdot N$ and change the start flag to "True". When the start flag is "True", if $V_i' < \lambda$, we set the middle Node V_{test}' as $\omega \cdot V_i' + (1-\omega)\cdot V_{i+1}'$. To make a judgment to V_{test}', if $V_{test}' < \beta \cdot V_{i+1}'$, we set the end time node $\mathrm{T}_{end} = i \cdot N$. After traversing all V_n', we can get the start and end time node of all the fragments. In this paper, we set the threshold λ as 0.65, and the weighted parameters ω and β as 0.85 and 3, respectively.

4.2 Feature Extraction

In this paper, we use the difference in phase difference between antennas in the presence of human breathing, apnea, and absence to detect the sleep state of the human. We calculate the phase difference D between the antennas, and the mean, variance, range and quartile moment of it:

$$\begin{cases} E_D = \frac{1}{M}\sum_{m=1}^{M} D(m) \\ V_D = \frac{1}{M-1}\sum_{m=1}^{M} (D(m) - E_D)^2 \\ R_D = \max(D) - \min(D),\ D' = sort(D), \\ Q_D = D'((M+1)\cdot 0.75) - D'((M+1)\cdot 0.25) \end{cases} \tag{10}$$

where $sort(\cdot)$ means sort from small to large. Assuming that the number of antennas, subcarriers and collected samples are X, K and Y, respectively, the input feature dimension of the classifier is $2X\cdot(X-1)\cdot K \times Y$. In this paper, back-propagation (BP) neural network is used to learn and map the feature set.

5 Experimental Results

In this paper, we utilize the csitools based on Linux and use the form of one transmitter and three receivers. The transmitting end uses a directional antenna to send data, and the receiving end uses three omnidirectional antennas to receive data. We use two mini hosts equipped with Intel 5300 wireless network card as transmitter and receiver and select channel 149 with a center frequency of 5.749 GHz to send and receive data. The packet sending rate is set to 500 per second. The offline data collected in the experiment are in three states: unmanned, apnea, and human breathing. 450 sets of data were collected respectively to extract features to form the training data set.

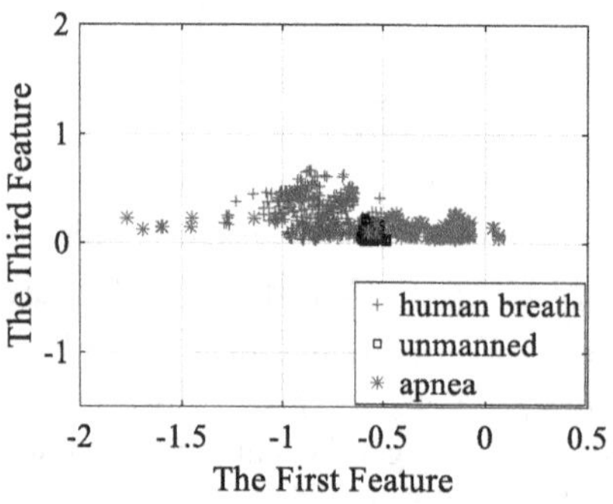

(a) Scatterplot of the distribution of the first and third features.

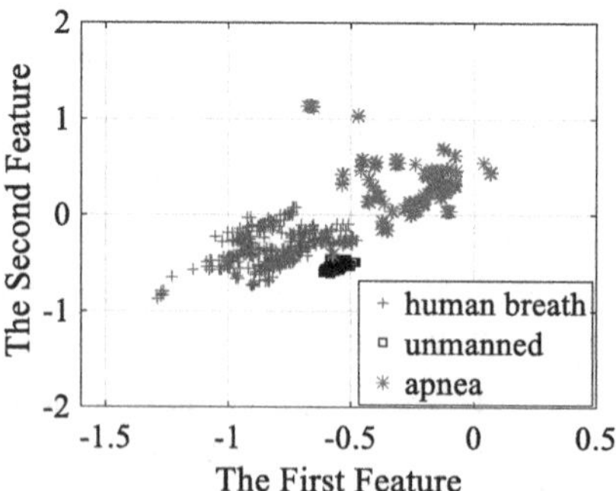

(c) Scatterplot of the distribution of the first and second features.

Fig. 5. Scatterplot of feature distribution.

For feature extraction, we observe the performance of the extracted signal feature through the feature distribution map, and randomly take two of the features to draw a feature distribution scatter plot, as shown in Fig. 5. As can be seen from Fig. 5(a), there is a small amount of confusion between the first feature and the third feature due to the similarity between apnea and unmanned state. But from Fig. 5(b), the first feature and the second feature are almost free of mixed grains and has a clear distribution area, which can better distinguish the three states.

The confusion matrix obtained after the classification is shown in Fig. 6. Then, we calculate its Precision and precision Recall:

$$\begin{cases} \text{Precision} = \dfrac{TP}{TP + FP} \\ \text{Recall} = \dfrac{TP}{TP + FN} \end{cases} \tag{11}$$

	human breath	apnea	unmanned
human breath	99.1%	0.7%	0.2%
apnea	4.0%	95.6%	0.4%
unmanned	0.0%	0.4%	99.6%

Fig. 6. Classification confusion matrix of three states.

Combining Fig. 6 and (11), the Precision and Recall values for presence of breath, apnea, and unmanned states are 96.12%, 98.85%, 99.33%, 99.11%, 95.56%, 99.56%, respectively. Then, we use macro F1-Score as a measurement indicator to weight it on average:

$$macro\,F_1{-}score = 2\times \frac{\frac{1}{3}\cdot\sum_{j=1}^{3}\text{Precision}(j)\times\frac{1}{3}\cdot\sum_{j=1}^{3}\text{Recall}(j)}{\frac{1}{3}\cdot\sum_{j=1}^{3}\text{Precision}(j)+\frac{1}{3}\cdot\sum_{j=1}^{3}\text{Recall}(j)} \tag{12}$$

The calculated Macro F1-Score is about 0.98. For the accuracy and recall rate in the macro F1-Score comprehensive model, the greater the value is, the higher the quality of the classification model is. Compared with the common CSI-based breathing detection system, this article fully considers the effect caused by getting up and turning over during actual sleep, and abstracts effective data segment without interferences produced by human's movement. The detection accuracy rate can reach more than 95.6%. Compared with the existing breath detection scheme, the scheme proposed in this paper has more practical application value.

6 Conclusion

In this paper, we propose a WiFi-based sleep apnea state detection system that does not require the subject to wear any additional equipment or modify the hardware facilities, which can complete the sleep state only using commercial WiFi equipment. Firstly, we construct the signal model and preprocess the received signal. Then, we design a segmentation algorithm to select the optimal subcarrier and segments the signal. Finally, the features of segments are extracted, and a classifier is constructed to recognize breath, apnea and unmanned state in the sleep state. Experimental results show that this approach can not only eliminate the noise CSI information, but also achieve a recognition accuracy rate of more than 95.6%. How to recognize and detect sleep status for multiple people will become our future work.

Acknowledgment. This work was supported by the National Natural Science Foundation of China (61771083, 61704015), Science and Technology Research Project of Chongqing Education Commission (KJQN201800625), and Chongqing Natural Science Foundation Project (cstc2019jcyj-msxmX0635).

References

1. http://www.toodaylab.com/44685/
2. http://www.fitbit.com/
3. Shambroom, J.R., Fábregas, S.E., Johnstone, J.: Validation of an automated wireless system to monitor sleep in healthy adults. J. Sleep Res. **21**(2), 221–230 (2012)
4. Nguyen, P., Transue, S., Choi, M.H.: WiKiSpiro: non-contact respiration volume monitoring during sleep. In: The Eighth Wireless of the Students, by the Students, and for the Students Workshop. ACM, pp. 27–29 (2016)
5. Nguyen, P., Zhang, X., Halbower, A.: Continuous and fine-grained breathing volume monitoring from afar using wireless signals. In: IEEE INFOCOM 2016-IEEE Conference on Computer Communications, pp. 10–14. IEEE (2016)
6. Adib, F., Mao, H., Kabelac, Z.: Smart homes that monitor breathing and heart rate. In: Proceedings of the ACM Conference on Human Factors in Computing Systems. ACM, pp. 837–846 (2015)
7. Hou, Y., Wang, Y., Zheng, Y.: TagBreathe: monitor Breathing with Commodity RFID Systems. In: 2017 IEEE 37th International Conference on Distributed Computing Systems (ICDCS), pp. 969–981. IEEE (2017)
8. Abdelnasser, H., Harras, K.A., Youssef, M.: Ubibreathe: a ubiquitous noninvasive wifi-based breathing estimator. In: China Proceedings of the IEEE MobiHoc, no. 15, pp. 277–286 (2015)
9. Liu, J., Wang, Y., Chen, Y.: Tracking vital signs during sleep leveraging off-the-shelf wifi. In: ACM International Symposium. ACM, pp. 267–276 (2015)
10. Wang, X., Yang, C., Mao, S.: PhaseBeat: exploiting CSI phase data for vital sign monitoring with commodity WiFi devices. In: Proceedings of the 2017 IEEE International Conference on Distributed Computing Systems (ICDCS), pp. 1230–1239 (2017)
11. Chen, C., Han, Y., Chen, Y.: TR-BREATH: time-reversal breathing rate estimation and detection. IEEE Trans. Biomed. Eng. **65**(3), 489–501 (2017)

12. Li, F., Xu, C., Liu, Y.: Mo-sleep: Unobtrusive sleep and movement monitoring via WiFi signal. In: 2016 IEEE 35th International Performance Computing and Communications Conference, Las Vegas, pp. 173–180 (2016)
13. Wu, X., Chu, Z., Yang, P.: TW-See: human activity recognition through the wall with commodity wifi devices. IEEE Trans. Veh. Technol. **68**(1), 306–319 (2019)

Energy-Efficient DAC Scheme Based on Unit Capacitor Switching for SAR ADCs

Liangbo Xie(✉) and Yan Ren

School of Communication and Information Engineering, Chongqing University of Posts and Telecommunications, Chongqing, China
xielb@cqupt.edu.cn

Abstract. With the development of Internet of Things (IoTs), the number of sensor nodes is growing rapidly. These sensors are usually passive or supplied by batteries and are usually a mixed-signal circuit. Analog to digital converter (ADC) is a core element in the sensor, and the power consumption of occupies a considerable part of the whole sensor. SAR ADC is a good candidate for the sensor due to its good energy-efficiency, medium resolution and speed. As the key part of SAR ADC, digital-to-analog converter (DAC) dominates the power consumption of the SAR ADC when dynamic comparator is employed. In order to improve the energy efficiency of the DAC, this paper proposes energy-efficient DAC scheme based on unit capacitor switching. By employing a capacitor-splitting structure and introducing a third voltage reference V_q equal to a quarter of the voltage reference V_{ref}, the unit capacitor can be employed to generate the last bit, which in turn reduces the DAC area. Simulation results show that the proposed scheme reduces the switching energy by 99.03% and the DAC area by 87.5% compared to the conventional SAR ADC structure, which achieves good energy-efficiency and area-efficiency.

Keywords: SAR ADC · DAC switching scheme · Energy-efficiency · Unit capacitor switching · Area-efficiency

1 Introduction

With the development of Internet of Things (IoTs), sensors are wildly deployed in order to meet requirements of information acquisition. In order to extend the life of sensors, the power consumption is very stringent, especially for implantable, portable and wearable devices. As the key component of these sensors, analog-to-digital converter (ADC) consumes a large amount power of the sensors. Compared to flash ADC, sigma-delta ADC and pipeline ADC, successive approximation register (SAR) features low power, medium speed, low complexity and medium resolution, and is a good candidate for those low power applications.

In SAR ADC, capacitive DAC (CDAC) dominates the power consumption and the area of the SAR ADC, and the power consumption and the area of DAC increase with the resolution of ADC. In recent years, various DAC scheme have been developed to improve

X. Wu et al. (Eds.): QShine 2020, LNICST 381, pp. 48–56, 2021.
https://doi.org/10.1007/978-3-030-77569-8_4

the performance of the DAC [1–9]. In [1], by using a set and down method, the switching energy is reduced by 81.26% compared to the conventional structure. In [2], the V_{cm}-based scheme reduces the switching energy by 87.54% compared with the conventional scheme. Although schemes in [1, 2] greatly improve the energy-efficiency, the DAC area is still too large, which occupies a large die area. In order to improve the energy efficiency and reduce the DAC area, schemes in [3–8] proposes different methods to improve the performances. Compared with the conventional schemes, the low frequency dependence switching scheme in [3], the tri-level scheme in [4], the Vcm-based monotonic switching scheme in [5], the switching scheme in [6], the scheme with high accuracy in [7], the scheme in [8] and the low common-mode voltage variation scheme in [9] reduce the switching energy by 95.34%, 96.89%, 97.66%, 98.84%, 98.44%, 98.84% and 98.45%, respectively. And these schemes in [3–8] all achieve an area reduction of 75% compared to the conventional scheme. In order to further reduce the DAC area, a V_{aq}-based tri-level switching scheme is developed [10], and the DAC area is only 12.5% of the conventional scheme, which reduces the DAC area by half compared with schemes in [3–9]. However, the switching energy is only comparable to the tri-level scheme in [4]. In order to further improve the energy efficiency and reduce the DAC area, this paper proposes a DAC scheme based on unit capacitor switching, which improves the energy efficiency by 99.03% and reduces the DAC area by 87.5% compared to the conventional scheme.

2 Proposed DAC Switching Scheme

The proposed architecture for a 10-bit SAR ADC is shown in Fig. 1. The SAR ADC is mainly divided into three parts: SAR logic, DAC and comparator. Here, a capacitor-splitting DAC structure is adopted as in [7], and the most significant bit (MSB) capacitor is divided into small LSB parts, which is the same as the least significant bit (LSB) capacitors. The DAC consists of four sub-arrays, i.e. two MSB sub-arrays and two LSB sub-arrays. Different from other schemes in [2–9], the proposed architecture introduces a new reference voltage V_q, which is a quarter of the reference voltage V_{ref}.

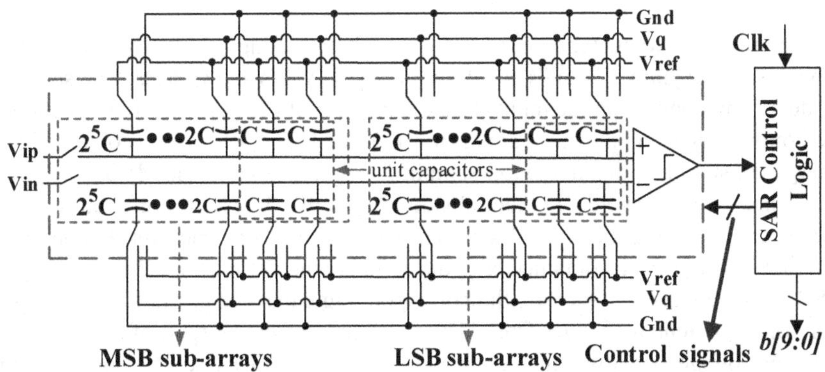

Fig. 1. Proposed SAR ADC structure

2.1 Switching Procedure of the Proposed Scheme

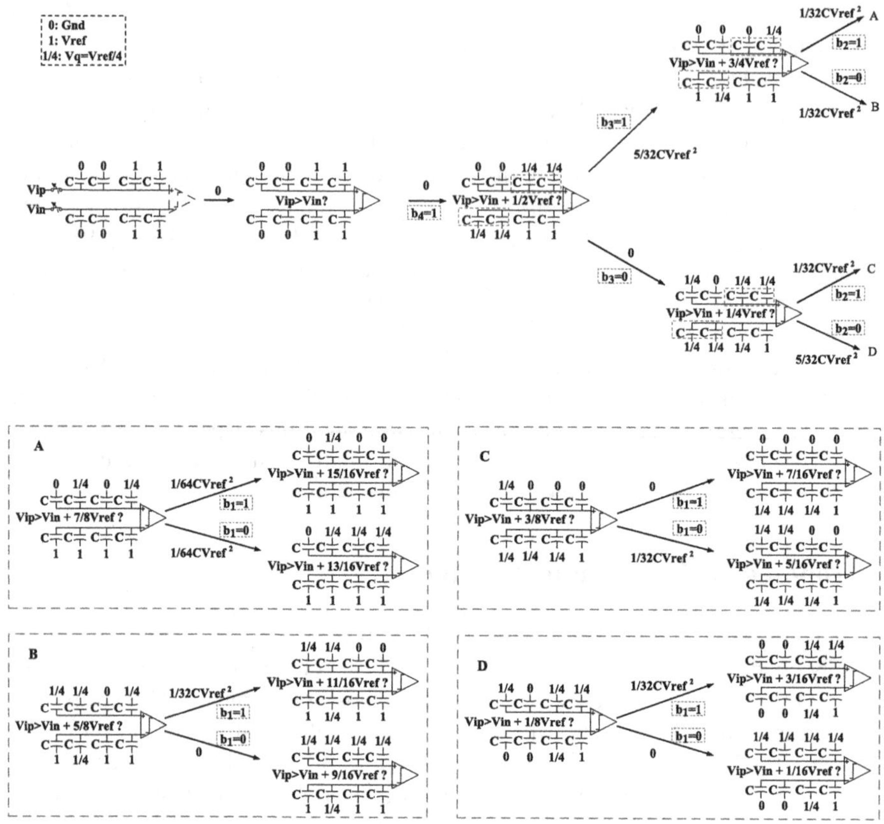

Fig. 2. The DAC switching procedure for a 5-bit SAR ADC model

The complete DAC switching procedure is explained using a 5-bit ADC in Fig. 2. As the conversion is symmetric, only the case for $V_{ip} > V_{in}$ is shown here. The sampling is conducted by connecting the inputs to the top plates of the DAC while the sequences of the bottom plates on both sides are set to [0, 0, 1, 1]. The MSB (b_4) is obtained with no energy consumption due to the top-sampling technique. When the MSB is acquired, the sequence of higher voltage potential side is set to [0, 0, 1/4, 1/4] and the sequence on the other side changes to [1/4, 1/4, 1, 1]. No energy is consumed for this operation and the MSB-1 (b_3) can be obtained after the comparison. During the MSB-2 (b_2) cycle, if MSB-1 (b_3) equals to logic '1', the corresponding capacitor of the LSB array with higher voltage potential changes from '1/4' to '0' and the counterpart of the LSB array with lower voltage potential changes from '1/4' to '1'; otherwise, the corresponding capacitor of the LSB array with higher voltage potential changes from '1' to '1/4' and the counterpart of the LSB array with lower voltage potential changes from '0' to '1/4'. This operation can be applied to the generations of MSB-2 to LSB-2 for an N-bit ADC.

The last two bits (LSB-1 and LSB) are obtained by reusing unit capacitors. The capacitors connection change depends on the comparison results of MSB, LSB-3 and LSB-2, which is shown clearly in Table 1. In order to clearly explain the LSB-1 and LSB generation, unit capacitors are divided into unit capacitors and dummy capacitors as Fig. 3 shows. And C_{d_DACPM}, C_{u_DACPM}, C_{d_DACPL}, C_{u_DACPL}, C_{d_DACNM}, C_{u_DACNM}, C_{d_DACNL} and C_{u_DACNL} in Table 1 represent the dummy capacitor in MSB sub-array of P-side, the unit capacitor in MSB sub-array of P-side, the dummy capacitor in LSB sub-array of P-side, the unit capacitor in LSB sub-array of P-side, the dummy capacitor in MSB sub-array of N-side, the unit capacitor in MSB sub-array of N-side, the dummy capacitor in LSB sub-array of N-side and the unit capacitor in LSB sub-array of N-side, respectively.

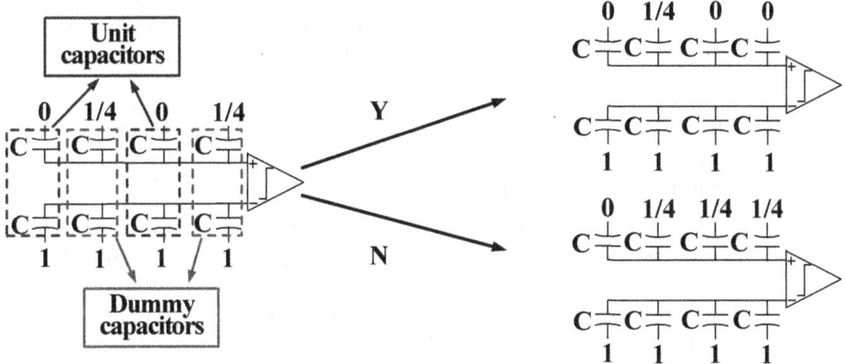

Fig. 3. Illustration of the unit capacitor switching scheme

Case 1: If MSB = 1, LSB-3 = 1 and LSB-2 = 1, for LSB-1 cycle, the dummy capacitor of the MSB-array with the higher voltage potential changes from '0' to '1/4', and the dummy capacitor of the MSB-array with the lower voltage potential changes from '1/4' to '1', this operation achieves an absolute voltage change $1/8V_{ref}$. After the comparison, the LSB-1 is obtained. If LSB-1 = 1, the dummy capacitor of the LSB-array with the higher voltage potential changes from '1/4' to '0'; otherwise, the unit capacitor of the LSB-array with the higher voltage potential changes from '0' to '1/4'. After the DAC settles, the LSB can be obtained after the comparator finished the comparison.

Case 2: If MSB = 1, LSB-3 = 1 and LSB-2 = 0, for LSB-1 cycle, the dummy capacitor and the unit capacitor of the MSB-array with the lower voltage potential both change from '0' to '1/4', and an absolute voltage change $1/8V_{ref}$ can be obtained. The capacitor connection change to obtain the LSB is just the same as Case 1.

Case 3: If MSB = 1, LSB-3 = 0 and LSB-2 = 1, for LSB-1 cycle, the dummy capacitor and the unit capacitor of the MSB-array with the higher voltage potential both change from '1/4' to '0'. For the LSB cycle, if LSB-1 = 1, the unit capacitor of the MSB-array with the higher voltage potential is reconnected from '1/4' to '0'; otherwise, the dummy capacitor of the MSB-array with the lower voltage potential is reconnected from '0' to '1/4'.

Case 4: If MSB = 1, LSB-3 = 0 and LSB-2 = 0, for LSB-1 cycle, the dummy capacitor and the unit capacitor of the MSB-array with the higher voltage potential both change from '1/4' to '0'. For the LSB cycle, if LSB-1 = 1, the unit capacitor of the MSB-array with the higher voltage potential is reconnected from '1/4' to '0'; otherwise, the dummy capacitor of the MSB-array with the lower voltage potential is reconnected from '0' to '1/4'.

Table 1. Capacitor operations of the last two bits

	MSB = 1 LSB-3 = 1 LSB-2 = 1	MSB = 1 LSB-3 = 1 LSB-2 = 0	MSB = 1 LSB-3 = 0 LSB-2 = 1	MSB = 1 LSB-3 = 0 LSB-2 = 0
LSB-1 cycle	C_{d_DACPM}: 0 to 1/4 C_{d_DACNM}: 1/4 to 1	C_{u_DACPM}: 0 to 1/4 C_{d_DACPM}: 0 to 1/4	C_{u_DACPL}: 1/4 to 0 C_{d_DAPL}: 1/4 to 0	C_{u_DACPM}: 1/4 to 0 C_{d_DACPM}: 1/4 to 0
LSB cycle	LSB-1 = 1, C_{d_DACPL}: 1/4 to 0; LSB-1 = 0, C_{u_DACPL}: 0 to 1/4	LSB-1 = 1, C_{d_DACPL}: 1/4 to 0; LSB-1 = 0, C_{u_DACPL}: 0 to 1/4	LSB-1 = 1, C_{d_DACPM}: 1/4 to 0; LSB-1 = 0, C_{u_DACPM}: 0 to 1/4	LSB-1 = 1, C_{d_DACPM}: 1/4 to 0; LSB-1 = 0, C_{u_DACPM}: 0 to 1/4

2.2 The Impact of the Accuracy of V_q

As Sect. 2.1 described, the MSB is acquired after the sampling and no capacitor connection change is performed. Thus, the generation of the MSB has no relationship with the third reference voltage V_q. During the phase of the MSB-1, all capacitors of the LSB sub-array with higher voltage potential changes from '1' to '1/4' and those of the MSB sub-array with lower voltage potential changes from '0' to '1/4'. Supposing the variation of V_q is ΔV and MSB is '1', then during the phase of the MSB-1, the following equations can be obtained

$$V_{DACP}(MSB1) = V_{ip} - \frac{1}{2}[V_{ref} - (V_q + \Delta V)] \tag{1}$$

$$V_{DACN}(MSB1) = V_{in} - \frac{1}{2}(V_q + \Delta V) \tag{2}$$

$$V_{DAC}(MSB1) = V_{ip} - V_{ip} - \frac{1}{2}V_{ref} \tag{3}$$

where $V_{DACP}(MSB1)$, $V_{DACN}(MSB1)$ and $V_{DAC}(MSB1)$ are the voltage of positive-side DAC array, the voltage of negative DAC array and the differential voltage of the DAC, respectively. From Eq. (3), it can be seen that the variation of V_q has no impact on the DAC voltage. The generation of MSB-2 to LSB-2 is similar to MSB-1, where the operations are complementary, which means the inaccurate transition of one side is compensated by the other side. Thus, the variation of V_q does not affect the transitions. However, the generation of the last two-bits only involves capacitor(s) of one side, thus the inaccurate transition due to the variation of V_q can not be compensated. Fortunately, the voltage changes of the last two-bit are small, the effect of V_q is minimized.

3 Simulation Results

3.1 Switching Energy

In order to evaluate the switching scheme of the proposed scheme, behavioral models of SAR ADC with the proposed DAC switching scheme and other published schemes are built using Matlab. During simulations, a 10-bit SAR ADC is employed, and the negative value of the switching energy was treated as zero in calculation as in [6]. The reset energy can be eliminated by the technique in [11], and it is not considered in the simulation. Figure 4 illustrates the switching energy versus the digital code of different schemes. Table 2 summarizes the main performances of different DAC switching schemes. The switching energy and the unit capacitor number of the proposed unit capacitor switching based scheme are 13.24 CV_{ref}^2 and 256, which improves the energy efficiency by 99.03% and reduces the DAC area by 87.5% compared to the conventional one. Compared to other schemes, the proposed scheme is the most energy-efficient one and the DAC area is the smallest one.

3.2 Linearity

As analysis in Sect. 3.1, the switching energy is related to the value of the unit capacitor. The unit capacitor should be as small as possible when considering the switching energy. However, due to the process variation, the practical value of a capacitor usually deviates from its nominal value, and the capacitor mismatch limits the smallest value of the unit capacitor adopted in the design procedure. Assume that the unit capacitor follows a Gaussian distribution with a mean value of C_u and a standard deviation of σ_u. Each capacitor in the binary-weighted capacitor array is consisted of the unit capacitor connected in parallel.

In order to evaluate the linearity of the switching scheme, Monte Carlo simulations are carried out using Matlab. During simulations, the unit capacitor is regarded as a Gaussian random variable with standard deviation of 1% ($\sigma_u/C_u = 0.01$). Figure 5 shows simulation results of DNL (Differential-Non-linearity) and INL (Integral-Nonlinearity) of 500 Monte Carlo runs of a 10-bit DAC with the proposed switching scheme. The root-mean-square (RMS) values of DNL and INL are 0.259LSB and 0.242LSB, respectively. As the first comparison cycle is mismatch free, the worst DNL occurs at $1/4V_{FS}$ and $3/4V_{FS}$, where V_{FS} stands for full scale signal.

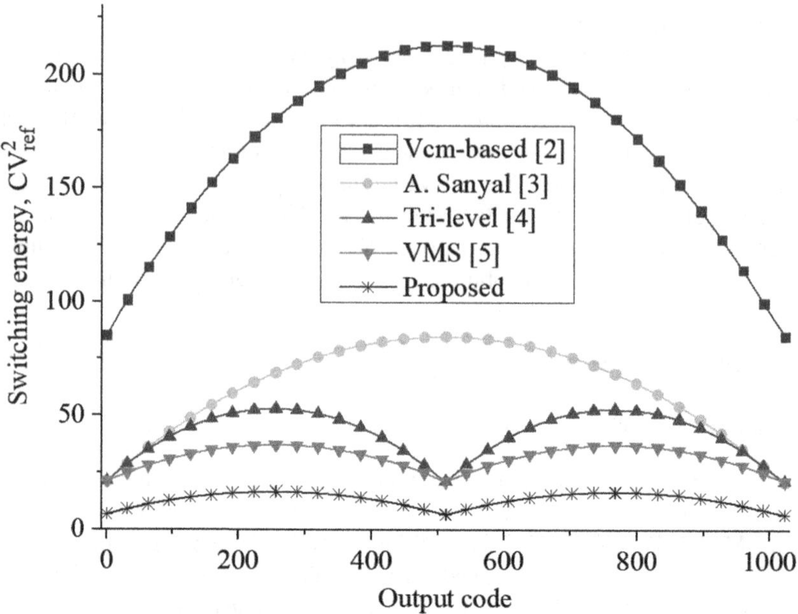

Fig. 4. Output code against switching energy

Table 2. Performance comparisons of different schemes for 10-bit SAR ADC

Switching scheme	Switching energy (CV_{ref}^2)	Energy savings	Total unit capacitor number	Area reduction
Conventional	1363.3	Ref.	2048	Ref.
Monotonic [1]	255.5	81.26%	1024	50%
V_{cm}-based [2]	170.17	87.54%	1024	50%
A. Sanyal [3]	63.56	95.34%	512	75%
Tri-level [4]	42.42	96.89%	512	75%
VMS [5]	31.88	97.66%	512	75%
Tong [6]	15.88	98.84%	512	75%
Xie [7]	21.2	98.44%	512	75%
Zhu [9]	15.88	98.84%	512	75%
Zhao [10]	48.03	96.48%	256	87.5%
This work	13.24	99.03%	256	87.5%

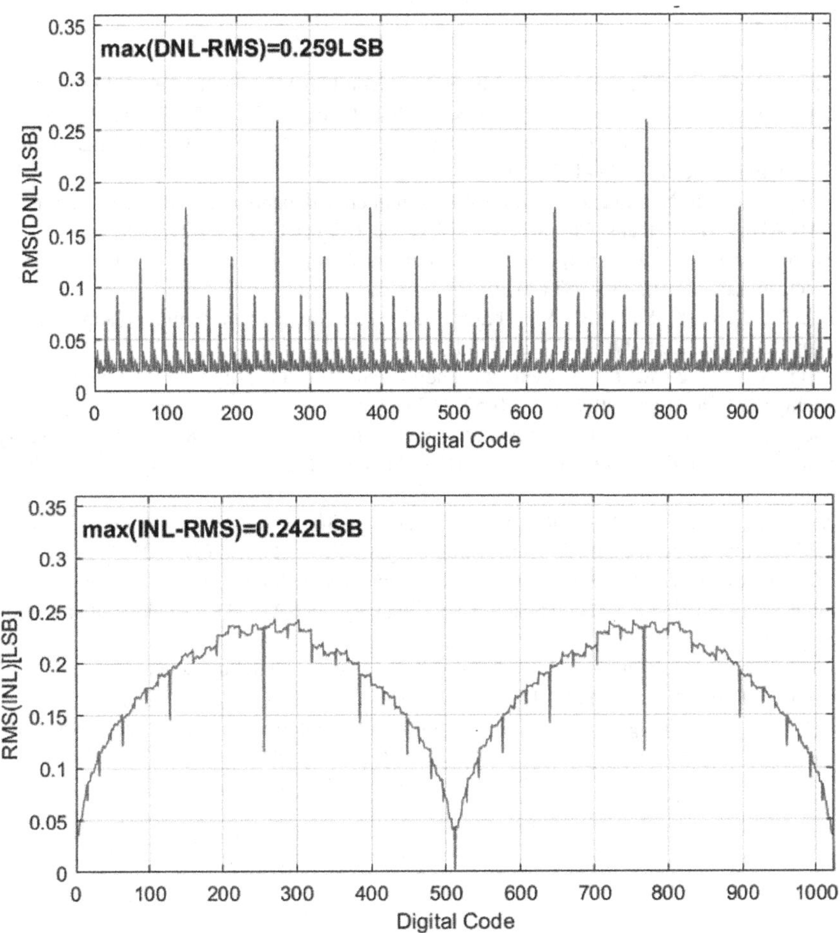

Fig. 5. The standard deviation of DNL and INL of the proposed scheme

4 Conclusion

This paper presents a low energy and small DAC area switching scheme for SAR ADC. By employing a capacitor splitting structure and introducing a third reference voltage V_q equals to 1/4 V_{ref}, the proposed DAC scheme can reuse the unit capacitors to reduce switching energy and the DAC area. Compared to the conventional structure, the proposed structure reduces the switching energy by 99.03% and the DAC area by 87.5%, which achieves both energy efficiency and area reduction.

Acknowledgements. This work was supported partly by the National Natural Science Foundation of China (No. 61704015), and General program of Chongqing Natural Science Foundation (special program for the fundamental and frontier research) (No. cstc2019jcyj-msxmX0108).

References

1. Liu, C.C., Chang, S.J., Huang, G.Y., Lin, Y.Z.: A 10-bit 50-MS/s SAR ADC with a monotonic capacitor switching procedure. IEEE J. Solid-State Circuits **45**(4), 731–740 (2010)
2. Zhu, Y., et al.: A 10-bit 100-MS/s reference-free SAR ADC in 90 nm CMOS. IEEE J. Solid-State Circuits **45**(6), 1111–1121 (2010)
3. Sanyal, A., Sun, N.: An energy-efficient low frequency dependence switching technique for SAR ADCs. IEEE Trans. Circuits Syst. II Express Briefs **61**(5), 294–298 (2014)
4. Yuan, C., Lam, Y.: Low-energy and area-efficient tri-level switching scheme for SAR ADC. Electron. Lett. **48**(9), 482–483 (2012)
5. Zhu, Z., Xiao, Y., Song, X.: Vcm-based monotonic capacitor switching scheme for SAR ADC. Electron. Lett. **49**(5), 327–329 (2013)
6. Tong, X., Zhang, Y.: 98.8% switching energy reduction in SAR ADC for bioelectronics application. Electron. Lett. **51**(14), 1052–1054 (2015)
7. Xie, L., Su, J., Wang, Y., Liu, J., Wen, G.: Switching scheme with 98.4% switching energy reduction and high accuracy for SAR ADCs. Analog Integr. Circuits Signal Process. **90**(3), 681–686 (2016)
8. Zhu, Z., Liang, Y.: A 0.6-V 38-nW 9.4-ENOB 20-kS/s SAR ADC in 0.18μm CMOS for medical implant devices. IEEE Trans. Circuits Syst. I Regul. Papers **62**(9), 2167–2176 (2015)
9. Li, J., Li, X., Huang, L., Wu, J.: An energy-efficient switching scheme with low common-mode voltage variation and no-capacitor-splitting DAC for SAR ADC. Analog Integr. Circuits Signal Process. **104**(1), 93–101 (2020)
10. Zhao, J., Mei, N., Zhang, Z., Meng, L.: Vaq-based tri-level switching scheme for SAR ADC. Electron. Lett. **54**(2), 66–68 (2018)
11. Xie, L., Nie, W., Yang, X., Zhou, M.: A group reset method for energy-efficient SAR ADC switching schemes. Analog Integr. Circuits Signal Process. **96**(1), 183–187 (2018)

The SDN-Governed Ad Hoc Swarm for Mobile Surveillance of Meteorological Facilities

Xi Chen[1,3], Tao Wu[2(✉)], and Ying Tan[1]

[1] The Key Laboratory for Computer Systems of State Ethnic Affairs Commission, Southwest Minzu University, Chengdu, China

[2] School of Computer Science, Chengdu University of Information Technology, Chengdu, China
wut@cuit.edu.cn

[3] School of Information and Communication Engineering, University of Electronic Science and Technology of China, Chengdu, China

Abstract. Accurate meteorological observation relies heavily on the proper and precise working of meteorological facilities. Nevertheless, a big portion of meteorological facilities are deployed in outdoor environments hardly within the reach of convenient monitoring and reliable networking infrastructure, hence the surveillance challenge. Given such an infrastructure-less/-poor environment (deserts, oceans, etc.) for meteorological facilities, Ad Hoc nodes are possible candidates for mobile surveillance. However, pure Ad Hoc networking without a logical centric node can barely provide consistent collaboration between mobile nodes during monitoring. This paper proposes a tunnelled overlay structure that bridges the Ad Hoc protocol stack and the SDN (Software-Defined Networking) protocol stack based on network virtualization techniques, so that robust distributed Ad Hoc mobile nodes are grouped in the form of an SDN-governed swarm to conduct the mobile surveillance task with joint efforts under the consistent control of the centric SDN controller. In addition, mobile nodes are equipped with TensorFlow-based image recognition feature, capable of transmitting the recognized results of harmful creatures that might cause facility damages, to suggest proper protection measures, control/avoid facility loss, etc. Experiments on the prototype SDN-governed Ad Hoc swarm are carried out in a real-world university campus meteorological station to demonstrate its feasibility and functionalities.

Keywords: SDN (Software-Defined Networking) · Ad Hoc · Image recognition · Mobile surveillance

Supported by National Key Research and Development Program of China (2018YFC1507005), China Postdoctoral Science Foundation (2018M643448), Sichuan Science and Technology Program (2019YFG0110, 2020YFG0189), Fundamental Research Funds for the Central Universities, Southwest Minzu University (2020NQN18), and SEAC Southeast Asia Research Center at Southwest Minzu University (SE2019Y07).

X. Wu et al. (Eds.): QShine 2020, LNICST 381, pp. 57–75, 2021.
https://doi.org/10.1007/978-3-030-77569-8_5

1 Introduction

Meteorological facilities play a key role in various economic aspects, such as weather forecast, agricultural production, tourism, etc. The accuracy of meteorological observation relies heavily on the proper and precise working of meteorological facilities, thus requiring close surveillance of these devices and equipments. However, a big portion of these facilities are usually deployed in outdoor environments hardly within the reach of convenient surveillance and reliable network infrastructure, some of which even have to face with severe environmental conditions. For instance, for those remotely located meteorological facilities in, e.g., deserts, steep ocean cliffs, etc., they might encounter extreme heat, erosion, wind damage, etc. Given such a complex and infrastructure-less/-poor environment and diverse facilities to monitor, simply introducing fixed sensors and cameras might not fully meet the surveillance requirements of outdoor meteorological facilities. For example in a meteorological station as shown in Fig. 1, meteorological and auxiliary facilities might not be perfectly monitored due to that their positions are out of the range of fixed cameras, or obscured by unexpected objects such as growing trees, falling rocks, (see Q1 in Fig. 1, i.e., an obscured humidity sensor) etc. Facilities might also be damaged by wild animals (Q2), e.g., birds, or severe environmental conditions, e.g., strong wind. Besides, meteorological (e.g., temperature sensors, humidity sensors, solar sensors, wind monitors, rain monitors) and auxiliary facilities (e.g., cameras) also need to be periodically or regularly checked/monitored in order to make sure they work properly. For such scenarios, a group of mobile nodes (such as unmanned aerial vehicles or wheeled vehicles in Fig. 1) can be commanded to take a patrol of meteorological facilities and accordingly adapt on-site formations to avoid obstacles, to track harmful creatures, etc., and timely transfer surveillance or damage report to administrators so that facility loss can be controlled or even avoided asap.

Mobile nodes organized and networked through Ad Hoc networking [7] in a flexible and on-demand fashion are possible candidates for mobile surveillance of meteorological facilities deployed in fields. The mobility, flexibility, and robustness in volatile networked environment are desired benefits. However, several issues must be addressed.

- Consistent control over the whole mobile swarm. Mobile nodes in an Ad Hoc swarm work in a distributed and self-determined fashion. Their moving or sensing behaviors might not comply with the requirements of the global surveillance task if no consistent control is given.
- Collaborative on-site formation adaptation. Obstacles might block the surveillance angle. Mobile nodes under the control of a common "head" are able to adjust positions accordingly and collaboratively to form various formations such as triangles, rectangles, etc., for better monitoring views.
- Image recognition capability. Surveillance by only images would require excessive human intervention to identify harmful creatures or damage status if any. If pictures or videos taken by mobile nodes can be timely recognized on-site

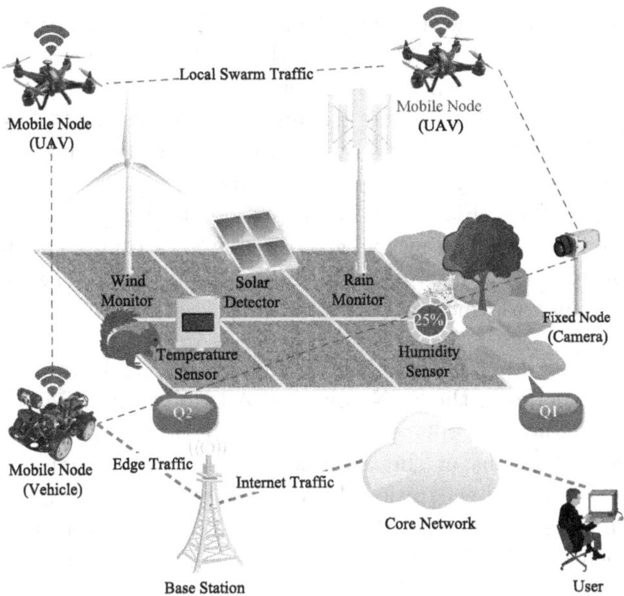

Fig. 1. An outdoor meteorological station under mobile surveillance

and the results of which can be meaningfully transmitted to administrators, appropriate measures are able to be taken as soon as possible.

To summary, pure Ad Hoc networking, which is essentially a distributed structure without central control, can barely afford consistent collaboration between mobile nodes during monitoring, hence the challenge in consistent and streamlined collaboration during mobile surveillance. One possible solution is to apply the centralized SDN (Software-Defined Networking) [15] on top of the distributed Ad Hoc underlay if its mobility, flexibility, and robustness are to be sustained. SDN is widely appliced in DCN (Data center Networks) [21,26,30], WAN (Wide Area Networks) [24,28], network virtualization [37,39], network resources optimization [5,27], QoS provisioning [3,4,6], service function chaining [25,29,32,34], etc. Works are also seen in the literature that integrate SDN into distributed network underlays, to improve security [12,14], to conduct access control and flow scheduling [33], to refine data aggregation [11], to facilitate machine learning integration [22], etc. Nevertheless, how to combine the benefits from the distributed Ad Hoc networking and the centralized SDN, setup the direct control from a uniform controlling interface, and apply the overlaid and hybrid network structure in real-world mobile surveillances of valuable assets are still seldom studied. In this paper, we propose the SDN-governed Ad Hoc swarm, an approach to construct a centralized SDN overlay on top of the distributed Ad Hoc underlay for mobile surveillance of meteorological facilities. The contribution of this paper includes as follows:

- A tunnelled overlay structure based on VxLAN (Virtual eXtensible Local Area Network) [18] that bridges the Ad Hoc protocol stack and SDN protocol stack to implement uniform control over distributed underlay.
- The extended OpenFlow [19] to accommodate self-defined physical actions so that mobility and sensibility of the Ad Hoc swarm can be consistently controlled without involving hardware details.
- Combined with TensorFlow-based image recognition, a prototype of low-cost mobile surveillance Ad Hoc swarm governed by SDN is implemented on Raspberry Pi platform. Extensive experiments of this prototype swarm were conducted in a real-world meteorological station.

The rest of this paper is organized as follows: Sect. 2 introduces the architecture of mobile nodes in the SDN-governed Ad Hoc swarm. Section 3 specifies how the SDN overlay is built on top of the Ad Hoc underlay, and the centralized control by means of the extended OpenFlow. Section 4 specifies the TensorFlow-based image recognition deployed on the swarm. Section 5 conducts various experiments to test the SDN-governed Ad Hoc swarm prototype in a real meteorological station. Related works are summarized in Sect. 6. Finally, this paper is concluded in Sect. 7.

2 Mobile Node Architecture

The architecture of the mobile node is shown in Fig. 2, which is roughly divided into 3 layers. The HW (hardware) layer is based on the Raspberry Pi platform, mainly providing wheeled mobility, camera monitoring and various sensors such as temperature sensors, infrared detectors, etc. Note that other mobility platform such as unmanned aerial vehicle (UAV) can also be considered if affordable. OS (Operating System) layer provides software functionalities like networking, computing, storage, etc. It is able to provide on the operating system several wireless communication protocols, such as WiFi, ZigBee, LoRa, NB-IoT, etc., as well as Ad Hoc routing protocols, such as AODV, DSDV, DSR, OLSR, etc., to enable peer-to-peer distributed control, packed with operating systems or manually installed as needed. Since high data rate transmission is needed in our application scenario to support camera surveillance and photo sharing for image recognition, we adopt the higher-rate WiFi as the fundamental wireless communication mechanism between mobile nodes. Low-power and lower data-rate communications, such as ZigBee for shorter range and LoRa/NB-IoT for longer range can be considered if corresponding chips are available.

The key extension for the centralized control over the distributed Ad Hoc underlay is the SDN layer. Every node is equipped with the SDN data plane, e.g., Open vSwitch (OVS) [23]. Meanwhile, swarm head might be re-assigned to other mobile nodes due to damage, high workload, low battery, etc., hence every node must be eligible to be elected as the swarm head to control and manage other mobile nodes so that the centralized SDN overlay can be continuously maintained. Thus, the SDN control plane must be deployed on every node as well, such as Floodlight, ONOS [1], OpenDaylight [20], etc. In this paper, we adopt OVS as the SDN switch and lightweight Floodlight as the SDN controller. SDN controller (i.e., Floodlight) is activated once a mobile node is elected/designated as swarm head while SDN switch (i.e., OVS) is always standing by on every node to accept control by swarm head through OpenFlow.

Fig. 2. Mobile node architecture

3 Ad Hoc Swarm Networking

3.1 Ad Hoc Underlay Networking

Every mobile node needs to participate in a two-stage networking to enable the centralized SDN control over the Ad Hoc swarm, i.e., the Ad Hoc underlay networking and SDN overlay networking. And, the Ad Hoc underlay does not have fixed networking infrastructure. Therefore, the wireless NIC (network interface card) of the mobile node should be configured to work in the ad hoc mode. This can be done by editing /etc./network/interfaces. For example, the following configuration (Listing 1.1) sets the IP address of the physical wireless NIC wlan0 as 10.0.0.1. It works in the ad hoc mode as specified in the configuration, so that infrastructure such as access points (AP) are not needed for mutual communication between mobile nodes. Meanwhile, all participating mobile nodes in the same swarm must be configured with exactly the same ESSID (Extended Service Set ID), for example "my-swarm" in Listing 1.1. Network mask "netmask" can be adapted according to the scale of the swarm to be constructed. Take the 10.0.0.0 IP subnet as an example. The IP addresses in such a subnet are by default class A addresses, whose network masks default to 255.0.0.0 (with 8 bits for network IDs and 24 bits for host IDs). It provides an address space capable of hosting almost 2^{24} nodes, which is more than needed in our scenario where an Ad Hoc swarm usually consists of several or tens of collaborative mobile nodes. Therefore, CIDR (classless inter-domain routing) can be used to partition such big address spaces by providing different network masks. For example, 10.0.0.0

subnet with an altered network mask 255.255.255.0 (also written as 10.0.0.0/24) gives a 2^8 space whereas 10.0.0.0/28 gives an even smaller space with 2^4 mobile nodes. Other mobile nodes have similar settings except for different IP addresses. Mobile nodes can ping each other using physical IP addresses (i.e., 10.0.0.x) and receive ICMP replies normally at this point of time, even though no APs are present. The following configuration demonstrates how wireless NICs should be configured when WiFi is adopted as the wireless communication for a distributed Ad Hoc swarm. The above configuration can be scripted and executed to enable automatic Ad Hoc underlay networking.

```
auto wlan0
allow-hotplug wlan0
iface wlan0 inet static
address 10.0.0.1
netmask 255.255.255.0
network 10.0.0.0
broadcast 10.0.0.255
wireless-essid my-swarm
wireless-mode ad-hoc
wireless-channel 3
```

Listing 1.1. The Ad Hoc Configuration

3.2 SDN Overlay Networking

Tunnelled Overlay Structure. Ad Hoc protocol stack and SDN protocol stack are two independent stacks. To combine flexibility from Ad Hoc and manageability from SDN, the two stacks must be somehow bridged. The key to deploying the SDN overlay above the Ad Hoc underlay to offer consistent and streamlined control is the deployment of OVS, the software-ized SDN switch. OVS consists of several modules working in both user space and kernel space of a Linux system. Core modules include ovsdb-server, ovs-vswitchd, ovs kernel module in kernel space, etc. ovsdb-server keep records of configurations. ovs-vswitchd communicates with SDN controllers through OpenFlow. ovs-vswitchd also connects with ovs kernel module through netlink and supports multiple datapaths, i.e., the virtual bridges that forward data. When ovs kernel module receives traffic, the actual forwarding is done by datapaths. Datapaths can be created on demand by OVS, and equipped with multiple vports (virtual ports on a datapath) for traffic ingress and egress. When directing packets to another vport upon traffic arrival, a vport matches packet fields with OpenFlow-installed flow table entries, and forwards matched packets or inquire the SDN controller for miss-matched ones. In other words, traffic forwarding by OVS is essentially instructed by (the flow table entries installed by) OpenFlow. The primary workflow of OVS can be found in Fig. 3.

According to the workflow of OVS mentioned above, to control the traffic flows running through the Ad Hoc underlay, wireless NICs of mobile nodes

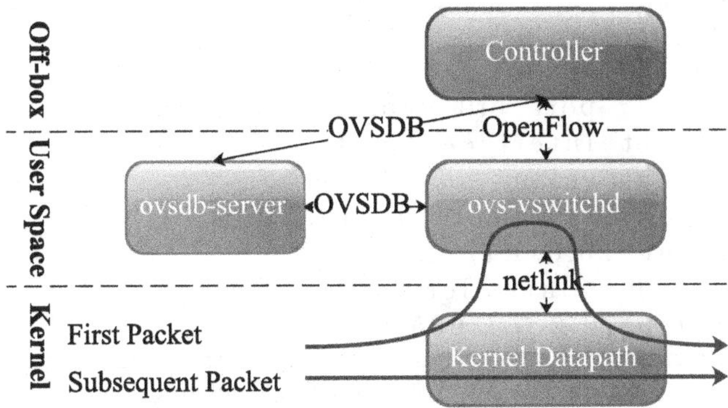

Fig. 3. OVS workflow

must be bound to datapaths created by OVS, so that traffic flows run into OVS modules, leading to an direct control by OpenFlow. Nevertheless the physical wireless NICs have already participated in the Ad Hoc underlay networking in the previous phase, thus simply bounding wireless NICs to datapaths fails the Ad Hoc underlay communication. Alternatively, we can create extra datapaths to encapsulate physical wireless NICs, so that the inner physical wireless NICs inside datapaths are controlled indirectly by OpenFlow. In this way, the SDN overlay becomes possible. In every mobile node, we create an OVS datapath named br0. At this point of time, every node has two forwarding devices, the physical wireless NIC wlan0, participating in the Ad Hoc underlay networking, and the virtual datapath br0, to participate in the SDN overlay controlling. To encapsulate wlan0 without compromising the Ad Hoc underlay communication, it requires end-to-end tunneling over the physical communication channels by means of datapath and vport. To achieve this, VxLAN is adopted for pairwise tunneling. The ovs-vsctl utitity provided by OVS offers VxLAN support. For every other mobile node, br0 adds a vport to tunnel all the way to the wlan0 of that node. Besides, br0 is assigned an independent IP address other than that of its own wlan0. Pseudo code for mobile node with wlan0 assigned 10.0.0.1 (n1 for short) tunnelling another node with wlan0 assigned 10.0.0.2 (n2) looks like Listing 1.2. In the above settings, it not only tunnels n1 and n2 using VxLAN, it also instructs the OVS datapath br0 to be controlled by the SDN controller process residing on 10.0.0.1:6653 through OpenFlow. Note 6653 is the official transport number for OpenFlow. The SDN controller IP address 10.0.0.1 indicates both SDN data plane and control plane are activated in n1 and n1 controls itself. Similar settings must be configured at the other side of the tunnel, i.e., n2, with symmetric changes on remote IP address as 10.0.0.1 and 20.0.0.2/24 for n2 br0. In this way, n1 controls both n1 and n2, forming the SDN overlay.

```
ovs-vsctl del-br br0
ovs-vsctl add-br br0
ovs-vsctl add-port br0 vport1
ovs-vsctl set interface vport1
          type=vxlan
          options:remote_ip=10.0.0.2
ifconfig br0 20.0.0.1/24 up
ovs-vsctl set-controller br0
          tcp:10.0.0.1:6653
```

Listing 1.2. VxLAN Tunneling Configuration

Imagine there are m nodes in an Ad Hoc swarm. It requires $m(m-1) \sim O(m^2)$ tunnels for full collaboration in the Ad Hoc underlay. Therefore, the control and collaboration at runtime is quite complex given larger m, demanding sophisticated algorithms. Nevertheless, if an SDN overlay is constructed, it only imposes $m-1$ OpenFlow channels for full control and centralized collaboration. The administrator controls only the SDN controller through northbound interface, and it gives instructions on behalf of the administrator through OpenFlow, leading to much simpler runtime management. Figure 4 demonstrates the architecture of the SDN-controlled Ad Hoc swarm ($m = 4$) where the upmost node with purple shadow is the SDN controller. The above configuration can be scripted and executed to enable automatic SDN overlay networking. Meanwhile, the selection of the SDN controller (i.e., the swarm head) can be done manually for a small scale swarm. Alternatively, derivatives of cluster head election algorithm such as LEACH [31] widely used in ZigBee can be tuned and used for such a purpose.

Centralized Controlling. We can see from the mobile node architecture that the control and management of the Ad Hoc swarm can be simply achieved by invoking the SDN controller northbound interface residing on the swarm head. In this way, the control and management of the swarm is simplified as the interaction with the swarm head that sends unified OpenFlow directives to other mobile nodes. Therefore, OpenFlow can now be regarded as the middleware to unify heterogenous hardware platforms, offering centralized control and management over peer-to-peer distributed Ad Hoc underlay. This "divide and conquer" approach scales when there are a large amount of mobile nodes under control. In this paper, we go one step further to extend OpenFlow from pure data forwarding actions to physical actions including moving, monitoring, and sensing.

OpenFlow is originally designed to instruct data forwarding on switches. Switches match the incoming packets against flow table, and execute corresponding OpenFlow actions for matched packets. For those miss-matched packets, switches send OpenFlow packet-in PDUs to the SDN controller to inquire how these packets should be processed. The SDN controller replies with packet-out to instruct forwarding actions, and switches also install corresponding flow table entries so that similar packets are to be locally processed the same way without further inquiries. OpenFlow supports several forwarding-related actions (see Listing 1.3, line 3–9) in version 1.3 which is a widely deployed version.

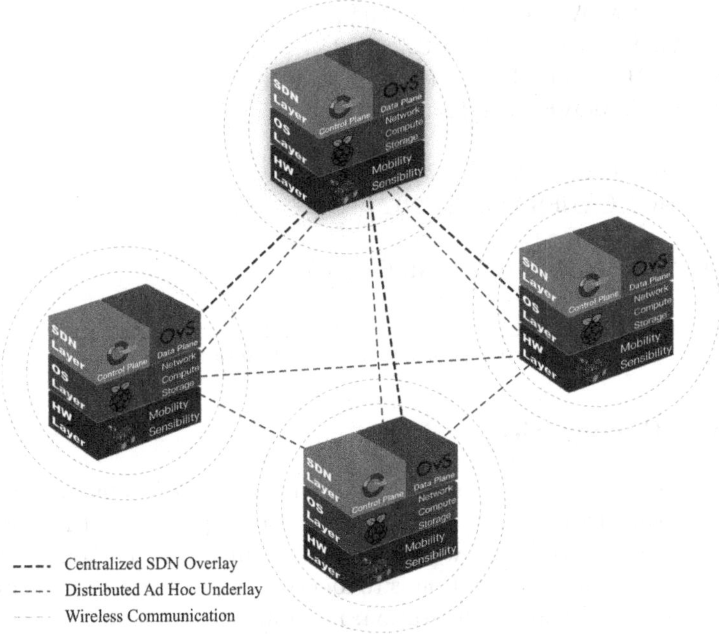

Fig. 4. SDN-governed Ad Hoc Swarm

Since mobile nodes involves moving, sensing, monitoring, etc., to fully implement the control and management of mobile node using OpenFlow, OpenFlow must be extended with corresponding actions. For example in Listing 1.3, action {OFPACT_MOVE_FORWARD, 101} has been added to instruct the movement of the mobile node to move forward; action {OFACT_CAMERA_ON, 201} has been added to instruct the mobile node to turn on the camera and conduct live surveillance; action {OFACT_IMAGE_RECOGNIZE, 201} for image recognition (see Sect. 4). These actions are to be parsed locally on mobile nodes upon arrival. And subsequently, hardware driver APIs are to be invoked internally, transparent to upper layer applications or administrators, to drive corresponding equipments on board of these mobile nodes.

```
static const struct ofpact_map of12[]=
{
  {OFPACT_OUTPUT, 0},
  {OFPACT_SET_MPLS_TTL, 15},
  {OFPACT_DEC_MPLS_TTL, 16},
  {OFPACT_PUSH_VLAN, 17},
  {OFPACT_STRIP_VLAN, 18},
  {OFPACT_PUSH_MPLS, 19},
  {OFPACT_POP_MPLS, 20},
  ...
```

```
    {OFPACT_MOVE_FORWARD, 101},
    {OFPACT_MOVE_BACKWARD, 102},
    {OFPACT_MOVE_LEFT, 103},
    {OFPACT_MOVE_RIGHT, 104},
    ...
    {OFPACT_CAMERA_ON, 201},
    {OFPACT_CAMERA_OFF, 202},
    ...
    {OFPACT_IMAGE_RECOGNIZE, 301},
}
```

Listing 1.3. OpenFlow 1.3 Actions

4 Image Recognition

Image recognition plays a key role in mobile surveillance of meteorological facilities. We developed the image recognition feature deployed on the Raspberry Pi platform based on TensorFlow and OpenCV framework. The purpose of the TensorFlow-based image recognition is to identify several harmful creatures seen around the target meteorological station, i.e., rats, sparrows, dogs, eagles, etc., and send not only the taken images but also the recognized results once damage occurs to the administrator in a timely fashion, to alarm possible creature-caused damages to meteorological facilities (e.g., rat-bites) at early stage and/or to suggest protection measures.

We selected the ssd_mobilenet_v1 network and the pre-trained ssd_mobilenet_v1_coco model to conduct the transfer learning, to accelerate the training. Mobilenet is a lightweight neural network model proposed by Google in 2017, which is suitable for embedded mobile nodes. There are 28 layers in total, each of which has a BatchNorm layer and a ReLU layer. Depth separable convolution is used in mobilenet to divide the standard convolution kernels into depth convolution kernels and 1×1 point-wise convolution kernels, so that the computation load can be reduced for mobile nodes. SSD (single shot multibox detector) is one of the most popular classification frameworks. Its main task is to conduct classification after features extraction from neural networks. The main idea of SSD is to carry out intensive sampling uniformly at different positions in the figure, and adopt different scales and aspect-ratios during sampling to carry out classification and regression. SSD uses convolution to extract the detection from the multi-scale feature map, so that objects of different scales can be simultaneously detected, that is, bigger scales for smaller objects detection and vise versa. ssd_mobilenet adds 8 convolution layers after the last layer of mobilenet, and extracts 6 of them for detection and the rest for convolution kernels for coordinate regression. As of model training, the original ssd_mobilenet_v1_coco model has a 90-dimensional output layer that we substituted with a 4-dimensional (representing rats, sparrows, dogs, and eagles) softmax layer to generate the probabilities indicating which creature the identified object belongs to. The above process applies in the static image recognition. If image recognition must be applied in video streaming such as camera surveillance, OpenCV framework needs to be deployed along with TensorFlow to extract frames from video streaming and conduct the serial image recognition for each frame.

At this point of time, we have constructed an Ad Hoc swarm under the centralized control of the SDN controller activated on the swarm head. Since every mobile node is equipped with OVS, it can be uniformly and compliantly instructed to move, adapt on-site formation, monitor, recognize objects on the taken photos, etc., by means of our extended OpenFlow messages that contain self-defined physical actions.

5 Evaluation

5.1 Swarm Networking and Centralized Controlling

The prototype of the SDN-controlled Ad Hoc swarm is implemented based on the Raspberry Pi platform. Raspbian operating system was installed on very Raspberry Pi. Floodlight was deployed as the SDN controller, and OVS is also installed as the SDN data plane, to comprehensively build the SDN layer over every mobile node. For the administrator, northbound REST API provided by Floodlight can be invoked to remotely control the swarm head (i.e., the SDN controller), and the head delivers OpenFlow messages that contain actions to instruct how the Ad Hoc swarm should function and collaborate. For indoor experiments in this section, three such mobile nodes (P1–P3, as shown in Fig. 5a) were participating in the networking, among which, one node was working as the swarm head, which was also an ordinary Ad Hoc swarm node under control itself. The logical SDN overlay topology was discovered by the Floodlight controller and shown in Fig. 5b.

(a) Underlay Ad Hoc Swarm (b) Overlay SDN Topology

Fig. 5. SDN-controlled Ad Hoc swarm

The SDN overlay worked in the 20.0.0.0 IP segment while the Ad Hoc underlay worked in the 10.0.0.0 IP segment. Administrators can push flow entries through Floodlight's northbound API (relative URL is /wm/staticflowpusher/json) to instruct data forwarding, demonstrated by Fig. 6. A flow entry that drops all data from source IP 20.0.0.1 (i.e., P1) to destination IP 20.0.0.4 (i.e., P3) in this experiment was pushed by the administrator, as shown in Fig. 6a. We can see from Fig. 6c that subsequent pingings complained about the destination host unreachable. Dynamic controlling over data forwarding of the Ad Hoc underlay was also implemented, as shown in Fig. 6b where a flow entry that deletes the previous traffic-dropping instruction was pushed by the administrator. We can see from Fig. 6d, the pinging was re-activated. Notice that the pinging operation directly uses the SDN overlay IP addresses (20.0.0.x), instead of those of the Ad Hoc underlay (10.0.0.x). This indicates that the forwarding of the Ad Hoc underlay is fully controlled by the SDN overlay, without any underlying hardware/configuration/communication details involved. Therefore, the administrator interacts with only the Ad Hoc swarm head to implement full control over the whole Ad Hoc swarm through the SDN layer deployed on every node.

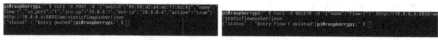

```
pi@raspberrypi:~ $ curl -X POST -d '{"switch":"00:00:d2:a4:ec:11:6c:42","name":"flow-1","in_port":"1","src-ip":"20.0.0.1","dst-ip":"20.0.0.4","active":"true"}' http://10.0.0.4:8080/wm/staticflowpusher/json
{"status" : "Entry pushed"}pi@raspberrypi:~ $
```

(a) The flow entry that drops traffic between P1 and P3

```
pi@raspberrypi:~ $ curl -X DELETE -d '{"name":"flow-1"}' http://10.0.0.4:8080/wm/staticflowpusher/json
{"status" : "Entry flow-1 deleted"}pi@raspberrypi:~ $
```

(b) Deleting the traffic-dropping flow entry

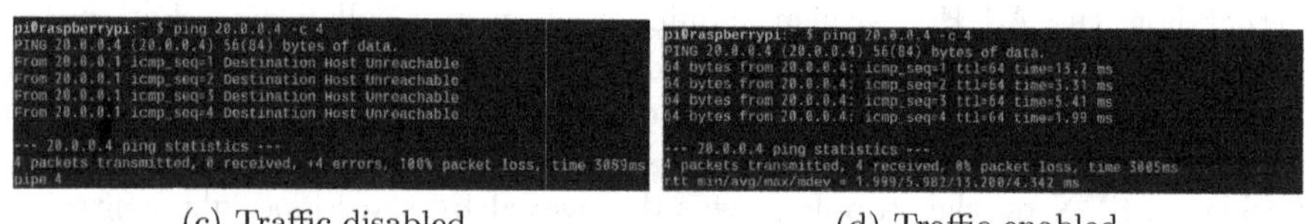

(c) Traffic disabled (d) Traffic enabled

Fig. 6. Flow entries pushing

Self-defined physical actions of the extended OpenFlow, as we have introduced in Sect. 3.2, can also be pushed from the SDN controller located on the swarm head using Packet-out PDUs, like ordinary OpenFlow actions for data forwarding. A flow entry containing action {OFACT_CAMERA_ON, 201} was pushed, and the Ad Hoc swarm node that received it turned on its camera to conduct live surveillance, as shown in Fig. 7a. Another flow entry containing {OFPACT_MOVE_FORWARD, 101} action was pushed, and the Ad Hoc swarm node that received it moved in different directions for keyboard strokes w (forward), s (backward), a (left) and d (right), as shown in Fig. 7b. These physical actions are particularly useful for collaborative mobile surveillance over valuable assets or facilities, including meteorological facilities remotely deployed, in case of being unexpectedly obscured by obstacles such growing tree branches, damaged by wild animals, etc., as will be shown in outdoor experiments in Sect. 5.2.

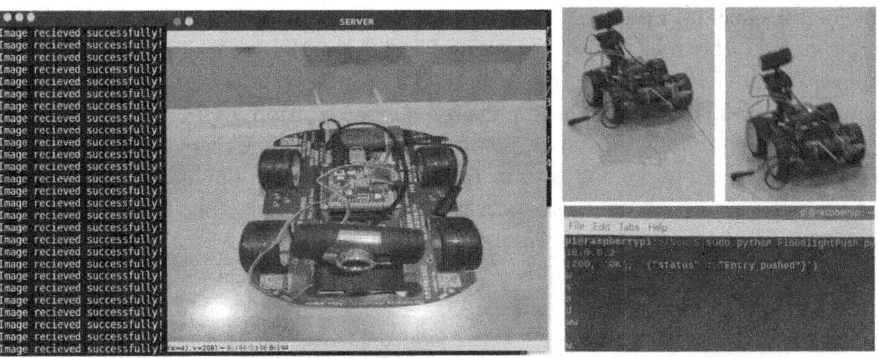

(a) SDN-controlled Camera Surveillance (b) SDN-controlled Moving

Fig. 7. Self-defined physical actions of the extended OpenFlow

5.2 Mobile Surveillance and Image Recognition

The prototype of the SDN-governed Ad Hoc swarm, augmented with wheels, cameras, sensors, etc., was also deployed in the meteorological station of Chengdu University of Information Technology (CUIT) for outdoor mobile surveillance of meteorological facilities. CUIT features its meteorology, and has a professional meteorological station located in its campus, as shown in Fig. 8a. The size of the main station is about 88 m × 41 m and that of the secondary radar station is about 20 m × 21 m. Some facilities, especially the radar, are obscured by nearby trees, hard to be effectively monitored by the fixed camera, as we can see from Fig. 8b. An SDN-governed Ad Hoc swarm consisting of three wheeled mobile nodes was deployed for evaluation. A flow entry containing action Camera_On was pushed by the SDN controller (i.e., the swarm head), and the node that received it turned on its camera to conduct live surveillance and image recognition as needed. To bypass obstacles and get better monitoring angles, the swarm head also instructed mobile nodes to gradually move around the radar and other meteorological facilities while conducting live surveillance, by sending OpenFlow messages that contained movement actions in different directions. Figure 8c (1)–(3) shows this mobile monitoring process that gradually avoided bushes and grass, and revealed a clear view of the meteorological radar. Besides, mobile surveillance makes it more feasible to conduct multi-angle monitoring compared with fixed camera shooting. For example, Fig. 8c (4) shows a reverse view of the monitored radar, taken by anther mobile node in the triangular on-site formation commanded by the swarm head. In addition, the node-to-node wireless coverage in the field is about 100 m, given the current mobile node design, which is sufficient for the mutual communication between nodes, and the surveillance of the meteorological station.

Image recognition based on TensorFlow was also tested in this experiment due to that small-sized harmful animals are seen around the meteorological facilities, which might cause damage. Figure 8d–8e demonstrate the recognition and marking of birds in a picture. The marking function, which draws a highlighted frame around the recognized object and indicates the recognition confidence, requires TensorFlow's object detection API as well as a more complex offline training process. It says in these pictures the recognition confidence was about 89% in Fig. 8d and 99% in Fig. 8e, respectively. The average recognition precision was above 90% for larger objects in a picture. The delay for image recognition and marking was 0.7 s on average even though the computing platform is the comparatively low-performance mobile Raspberry Pi.

(a) The Meteoreology Station (b) The Swarm and the Blocked Radar

(c) Mobile Surveillance around the Radar

(d) Bird Recognition (e) Bird Recognition

Fig. 8. Mobile surveillance of a meteorological radar

6 Related Works

Efforts have been made to extend SDN from wired networking towards the wireless networking. OpenRoads [35] merged SDN with WiFi/WiMAX technologies to deploy in university campus networks to facilitate wireless network innovation. It provides wireless extensions to the original OpenFlow. OpenRoads consists of three layers, namely the flow table layer, which coordinates OpenFlow and WiFi/WiMAX; the slicing layer, for FlowVisor-based network slicing; and the controller layer implemented by NOX [10]. In particular, OpenRoads demonstrated how wireless handover can be effectively conducted by means of SDN in wireless environments. In reference [8], OpenFlow is applied to WMN (Wireless Mesh Network), and an SDN solution based on virtualization technology and cross-layer flow table rules for WMN is proposed to enable unified control. Similarly, applications of SDN in wireless networks are also reported in references [9,36]. These works focus on SDN-equipped wireless networks, lacking the support for ad hoc networking and uniform control, hence difficulty in coordinating Ad Hoc nodes to conduct the consistent and coordinated mobile surveillance.

Recent years have also seen the efforts in integrating SDN into Ad Hoc networking. Reference [16] proposed and envisioned the software-defined VANET architecture and various services. Architecturally, it includes a remote centralized SDN controller located in a telecommunications datacenter, base stations, fixed RSUs (Road Side Unit), and mobile vehicles. Vehicles and RSUs constitute the data plane, whose data transmission is controlled by the SDN controller. Base stations and the remote controller conduct remote communication via wireless communications, such as LTE or WiMAX. This work gave some preliminary envisions on the software-defined VANET. Reference [38] proposed to control the OVS entities deployed on Ad Hoc nodes by the SDN controller, as we do in this paper. It maintains the interoperability with other SDN by applying best-practice SDN methodologies in implementation. However, the main weakness of this work is its lack of consideration of mobility management commonly seen in Ad Hoc networking. Reference [17] proposed to bridge SDN domains with west-east interfaces in a peer-to-peer manner, which offered new possibilities for interconnectivity and collaboration between software-defined MANETs.

SDN has also been introduced into the monitoring of valuable assets in recent years. Reference [2] extended the SDN southbound interface that is dedicated to devices discovery in a way that messages for adaptive device status monitoring are piggybacked in the LLDP (Link Layer Discovery Protocol), and proposed to apply this adaptive monitoring mechanism in energy Internet. Reference [14] proposed to monitor and improve the use of water resources by means of secure SDN techniques. Reference [13] adopted a similar approach to our work in that hardware-specifics are encapsulated using software and exposed as unified interfaces for easy invocation and composition, i.e., the concept of Software-Defined Device. Our work defers in that OpenFlow is used as the middleware to unify data forwarding and physical actions, thus the enhancement of interoperability and compatibility with wired SDN networks.

7 Conclusions

Ad Hoc networking provides good survivability and flexibility in non-static network topologies, thus a good choice for on-demand mobile surveillance for meteorological facilities deployed in fields where communication and monitoring infrastructures might be in shortage. However, the distributed structure prevents it from being consistently controlled hence the difficulty in coordination in mobile surveillance. This paper proposes to ingrate SDN methodologies into Ad Hoc networking to construct a collaborative and controlled swarm, so that moving, sensing, image recognition, etc., are uniformly controlled by the centralized SDN controller by means of extended OpenFlow. This overlaid and hybrid structure simplifies the global control over an Ad Hoc swarm in that only the Ad Hoc swarm head that activates the SDN controller is involved for the manipulation of the whole swarm. Extensive experiments in a real-world meteorological station demonstrate the feasibility. In our future work, topology dynamics such as swarm fusion and fission due to the distance variability will be studied to offer finer-grained control over the distributed underlay.

References

1. Berde, P., et al.: ONOS: towards an open, distributed SDN OS. In: Proceedings of the Third Workshop on Hot Topics in Software Defined Networking, HotSDN 2014, pp. 1–6. ACM, New York (2014). https://doi.org/10.1145/2620728.2620744
2. Chen, X., Chen, Y., Sangaiah, A.K., Luo, S., Yu, H.: MonLink: piggybackStatus monitoring over LLDP in software-defined energy internet. Energies **12**(6), 1147 (2019). https://doi.org/10.3390/en12061147, https://www.mdpi.com/1996-1073/12/6/1147
3. Chen, X., et al.: Reinforcement learning based QoS/QoE-aware service function chaining in software-driven 5g slices. Trans. Emerg. Telecommun. Technol **29**(11), e3477 (2018).https://doi.org/10.1002/ett.3477, https://onlinelibrary.wiley.com/doi/abs/10.1002/ett.3477
4. Chen, X., Wu, J., Wu, T.: The Top-K QoS-aware paths discovery for source routing in SDN. KSII Trans. Internet Inf. Syst. **12**(6), 2534–2553 (2018). https://doi.org/10.3837/tiis.2018.06.006
5. Chen, X., Wu, T.: Towards the semantic web based northbound interface for SDN resource management. In: IEEE 11th International Conference on Semantic Computing (ICSC), San Diego, CA, United states, pp. 40–47 (2017). https://doi.org/10.1109/ICSC.2017.27
6. Chen, X., Wu, T., Xie, L.: The declarative and reusable path composition for semantic web-driven SDN. IEICE Trans. Commun. **E101.B**(3), 816–824 (2018). https://doi.org/10.1587/transcom.2017EBP3233
7. Corson, D.S.M., Macker, J.P.: Mobile Ad hoc Networking (MANET): Routing Protocol Performance Issues and Evaluation Considerations. No. 2501 in Request for Comments, RFC Editor (1999). https://doi.org/10.17487/RFC2501, https://rfc-editor.org/rfc/rfc2501.txt, published: RFC 2501
8. Dely, P., Kassler, A., Bayer, N.: OpenFlow for wireless mesh networks. In: 2011 Proceedings of 20th International Conference on Computer Communications and Networks (ICCCN), pp. 1–6, July 2011. https://doi.org/10.1109/ICCCN.2011.6006100

9. Detti, A., Pisa, C., Salsano, S., Blefari-Melazzi, N.: Wireless mesh software defined networks (wmSDN). In: 2013 IEEE 9th International Conference on Wireless and Mobile Computing, Networking and Communications (WiMob), pp. 89–95, October 2013. https://doi.org/10.1109/WiMOB.2013.6673345
10. Gude, N., et al.: NOX: towards an operating system for networks. SIGCOMM Comput. Commun. Rev. **38**(3), 105–110 (2008). https://doi.org/10.1145/1384609.1384625
11. Hakiri, A., Berthou, P., Gokhale, A., Abdellatif, S.: Publish/subscribe-enabled software defined networking for efficient and scalable IoT communications. IEEE Commun. Mag. **53**(9), 48–54 (2015). https://doi.org/10.1109/MCOM.2015.7263372
12. Han, Z., Li, X., Huang, K., Feng, Z.: A software defined network-based security assessment framework for CloudIoT. IEEE Internet Things J. **5**(3), 1424–1434 (2018). https://doi.org/10.1109/JIOT.2018.2801944
13. Hu, P., Ning, H., Chen, L., Daneshmand, M.: An open Internet of Things system architecture based on software-defined device. IEEE Internet Things J. 2583–2592 (2018). https://doi.org/10.1109/JIOT.2018.2872028
14. Kamienski, C., et al.: SWAMP: smart water management platform overview and security challenges. In: 2018 48th Annual IEEE/IFIP International Conference on Dependable Systems and Networks Workshops (DSN-W), pp. 49–50, June 2018. https://doi.org/10.1109/DSN-W.2018.00024
15. Kreutz, D., Ramos, F.M.V., Esteves Verissimo, P., Esteve Rothenberg, C., Azodolmolky, S., Uhlig, S.: Software-defined networking: a comprehensive survey. In: Proceedings of the IEEE, vol. 103, pp. 14–76 (2015). https://doi.org/10.1109/JPROC.2014.2371999
16. Ku, I., Lu, Y., Gerla, M., Gomes, R.L., Ongaro, F., Cerqueira, E.: Towards software-defined VANET: architecture and services. In: 2014 13th Annual Mediterranean Ad Hoc Networking Workshop (MED-HOC-NET), pp. 103–110, June 2014. https://doi.org/10.1109/MedHocNet.2014.6849111
17. Lin, P., Bi, J., Chen, Z., Wang, Y., Hu, H., Xu, A.: WE-bridge: west-east bridge for SDN inter-domain network peering. In: 2014 IEEE Conference on Computer Communications Workshops (INFOCOM WKSHPS), pp. 111–112, April 2014. https://doi.org/10.1109/INFCOMW.2014.6849180
18. Mahalingam, M., et al.: Virtual eXtensible Local Area Network (VXLAN): a Framework for Overlaying Virtualized Layer 2 Networks over Layer 3 Networks (2014). https://doi.org/10.17487/RFC7348, https://www.rfc-editor.org/info/rfc7348
19. Mckeown, N., et al.: OpenFlow: enabling innovation in campus networks. ACM SIGCOMM Comput. Commun. Rev. **38**(2), 69–74 (2008). https://doi.org/10.1145/1355734.1355746
20. Medved, J., Varga, R., Tkacik, A., Gray, K.: OpenDaylight: towards a model-driven SDN controller architecture. In: Proceeding of IEEE International Symposium on a World of Wireless, Mobile and Multimedia Networks 2014, pp. 1–6 (2014). https://doi.org/10.1109/WoWMoM.2014.6918985
21. Miao, W., et al.: SDN-enabled OPS with QoS guarantee for reconfigurable virtual data center networks. IEEE/OSA J. Opt. Commun. Netw. **7**(7), 634–643 (2015). https://doi.org/10.1364/JOCN.7.000634
22. Mohammadi, M., Al-Fuqaha, A., Guizani, M., Oh, J.S.: Semisupervised deep reinforcement learning in support of IoT and smart city services. IEEE Internet Things J. **5**(2), 624–635 (2018). https://doi.org/10.1109/JIOT.2017.2712560

23. Pfaff, B., et al.: The design and implementation of open vSwitch. In: 12th USENIX Symposium on Networked Systems Design and Implementation (NSDI), vol. 40, pp. 117–130. USENIX Association, Oakland, CA (2015)
24. Soliman, M., Nandy, B., Lambadaris, I., Ashwood-Smith, P.: Exploring source routed forwarding in SDN-based WANs. In: IEEE International Conference on Communications (ICC), pp. 3070–3075, June 2014. https://doi.org/10.1109/ICC.2014.6883792
25. Sun, G., Li, Y., Liao, D., Chang, V.: Service function chain orchestration across multiple domains: a full mesh aggregation approach. IEEE Trans. Netw. Serv. Manage. **15**(3), 1175–1191 (2018). https://doi.org/10.1109/TNSM.2018.2861717
26. Sun, G., Liao, D., Zhao, D., Xu, Z., Yu, H.: Live migration for multiple correlated virtual machines in cloud-based data centers. IEEE Trans. Serv. Comput. **11**(2), 279–291 (2018). https://doi.org/10.1109/TSC.2015.2477825
27. Sun, G., Zhang, F., Liao, D., Yu, H., Du, X., Guizani, M.: Optimal energy trading for plug-in hybrid electric vehicles based on fog computing. IEEE Internet Things J. **6**(2), 2309–2324 (2019). https://doi.org/10.1109/JIOT.2019.2906186
28. Sun, G., Chang, V., Yang, G., Liao, D.: The cost-efficient deployment of replica servers in virtual content distribution networks for data fusion. Inf. Sci. **432**, 495–515 (2018). https://doi.org/10.1016/j.ins.2017.08.021
29. Sun, G., Li, Y., Yu, H., Vasilakos, A.V., Du, X., Guizani, M.: Energy-efficient and traffic-aware service function chaining orchestration in multi-domain networks. Fut. Gener. Comput. Syst. **91**, 347–360 (2019).https://doi.org/10.1016/j.future.2018.09.037,http://www.sciencedirect.com/science/article/pii/S0167739X1831848X
30. Sun, G., Liao, D., Zhao, D., Sun, Z., Chang, V.: Towards provisioning hybrid virtual networks in federated cloud data centers. Fut. Gener. Comput. Syst. **87**, 457–469 (2018). https://doi.org/10.1016/j.future.2017.09.065
31. Tyagi, S., Kumar, N.: A systematic review on clustering and routing techniques based upon LEACH protocol for wireless sensor networks. J. Netw. Comput. Appl. **36**(2), 623–645 (2013).https://doi.org/10.1016/j.jnca.2012.12.001,http://www.sciencedirect.com/science/article/pii/S1084804512002482
32. Wen, T., Yu, H., Du, X.: Performance guarantee aware orchestration for service function chains with elastic demands. In: 2017 IEEE Conference on Network Function Virtualization and Software Defined Networks (NFV-SDN), pp. 1–4, November 2017. https://doi.org/10.1109/NFV-SDN.2017.8169854
33. Wu, D., Arkhipov, D.I., Asmare, E., Qin, Z., McCann, J.A.: UbiFlow: mobility management in urban-scale software defined IoT. In: 2015 IEEE Conference on Computer Communications (INFOCOM). pp. 208–216, April 2015. https://doi.org/10.1109/INFOCOM.2015.7218384
34. Yang, K., Zhang, H., Hong, P.: Energy-aware service function placement for service function chaining in data centers. In: 2016 IEEE Global Communications Conference (GLOBECOM), pp. 1–6 (Dec 2016). https://doi.org/10.1109/GLOCOM.2016.7841805
35. Yap, K.K., et al.: OpenRoads: empowering research in mobile networks. SIGCOMM Comput. Commun. Rev. **40**(1), 125–126 (2010).https://doi.org/10.1145/1672308.1672331
36. Yap, K.K., et al.: Blueprint for introducing innovation into wireless mobile networks. In: Proceedings of the Second ACM SIGCOMM Workshop on Virtualized Infrastructure Systems and Architectures, VISA 2010. pp. 25–32. ACM, New York (2010). https://doi.org/10.1145/1851399.1851404

37. Yu, H., Qiao, C., Wang, J., Li, L., Anand, V., Wu, B.: Regional failure-resilient virtual infrastructure mapping in a federated computing and networking system. IEEE/OSA J. Opt. Commun. Netw. **6**(11), 997–1007 (2014). https://doi.org/10.1364/JOCN.6.000997
38. Yu, H.C., Quer, G., Rao, R.R.: Wireless SDN mobile ad hoc network: from theory to practice. In: 2017 IEEE International Conference on Communications (ICC), pp. 1–7, May 2017. https://doi.org/10.1109/ICC.2017.7996340
39. Yu, H., Wen, T., Di, H., Anand, V., Li, L.: Cost efficient virtual network mapping across multiple domains with joint intra-domain and inter-domain mapping. Opt. Switch. Netw. **14**(PART 3), 233–240 (2014). https://doi.org/10.1016/j.osn.2014.05.020

Research on Optimizing the Location and Capacity of Electric Vehicle Charging Stations

Lingling Yang[1], Jiali Chen[1], Wenzao Li[1,2(✉)], and Zhan Wen[1]

[1] College of Communication Engineering, Chengdu University of Information Technology, Chengdu, China
yanglledu@sina.com, lwz@cuit.edu.cn

[2] School of Computing Science, Simon Fraser University, Burnaby, Canada

Abstract. Charging stations deployment is an important problem in Electric Vehicle (EV) networks. The distribution of EV is complicated in urban environments. Therefore, reasonable location deployment will avail to reduce construction costs and improve user experience. Aim to this, this paper comprehensively considers the cost of charging stations and the charging costs of EVs. Studied the charging station location, charging station capacity and the optimization algorithms for charging station location, and proposed a method for estimating the optimal location and optimal capacity allocation of EV charging stations. Firstly, this paper uses the Voronoi diagram to divide the service range of the charging stations, then uses the differential evolution algorithm combined with the particle swarm optimization algorithm (DEIPSO) to solve the charging station location model, and finally consider the residence time of EV in the charging station, use queuing theory to solve the charging station capacity allocation model. The experimental results shows that DEIPSO can better jump out of the local optimum and achieve the global optimum; the proposed model can plan the charging station on the basis of fully considering the total charging costs of charging stations and EVs.

Keywords: EV · Charging stations · Location · Capacity allocation · DEIPSO algorithm

1 Introduction

1.1 Background and Motivation

At present, the related research of EV charging stations at home and abroad mainly focuses on the following aspects: (1) Research on Location Model of Charging Station. Mehmet, C.C. et al. [1] believe that the optimal deployment location of the charging station is closely related to the number of EVs and traffic density in the planning area, and propose to use data mining methods to estimate the optimal location of the charging station; Johannes, S. et al. [2] solve the problem of user's charging difficulties, considering the user's charging habits and propose a method for selecting the location of the charging

X. Wu et al. (Eds.): QShine 2020, LNICST 381, pp. 76–90, 2021.
https://doi.org/10.1007/978-3-030-77569-8_6

station; Bi, R. et al. [3] considered the influence of different vehicle owners' charging behavior and established multiple models for comparative analysis. The results showed that the vehicle owners' charging behavior will have a greater impact on the simulation results, so on this basis, optimize the location of the charging station; Phonrattanasak, P. [4] under the constraints of the distribution network and traffic restrictions, the planning model of charging station is established by considering the total cost of fast charging station and the total loss of distribution network. (2) Research on the charging station capacity allocation model. To meet the charging needs of EVs, the vehicles do not consider the problem of waiting in line when charging. Some research scholars determine the number of chargers in charging stations by calculating the maximum charging demand within the service range of charging stations [5–7]; by considering the queuing problem of EVs during charging, some researchers have proposed to use the method of queuing theory [8–10] to establish the capacity allocation model of charging stations. (3) Research on the algorithm for solving the location model. Mehar, S. et al. [11] added a new operator to the traditional genetic algorithm to estimate the optimal location of the charging station. The improved genetic algorithm can prevent the algorithm from prematurely converging and improve the efficiency of the algorithm; Han, F.J. [12] combined Voronoi diagram with traditional particle swarm optimization algorithm to improve the optimization effect and optimization speed of the algorithm; because the traditional particle swarm optimization algorithm is easy to fall into the local optimum, some scholars [5, 6, 13] combined Voronoi diagram with improved particle swarm optimization algorithm to further improve the optimization speed of the algorithm.

1.2 Challenges and Our Solution

EVs have become an important part of the new energy development strategy, and is the development direction of new energy vehicles. The construction of EV charging stations is a prerequisite for the development and popularization of EVs. Due to the relatively high construction cost of charging stations and the large amount of land and power resources they occupy, the development of charging stations is relatively slow. The layout of the charging stations that have been constructed is unreasonable in space and has a great blindness. The unreasonable deployment of charging stations will affect the urban transportation network planning, increase the driving costs of vehicles, and make it difficult for operators to make profits or even lose money. Whether the planning of EV charging stations is reasonable will directly affect the number of EVs used and the improvement of service levels.

The above literature has important guiding significance for the planning of charging stations, but them ignores that the location of EV charging stations will affect many aspects, which in turn will not be conducive to the development of charging stations, and different location optimization algorithms solving the location problem of the charging station is also a key part, the quality of the selection algorithm is directly related to the accuracy of the final optimization result. Therefore, in the location and capacity model of the charging station, this paper comprehensively considers the actual influencing factors in real life, and considers the number of EVs, the charging behavior of users, the cost of building stations and operating costs of charging stations in different regions, and proposed the location and volume model of EVs charging station. On the basis of the traditional particle swarm optimization algorithm is easy to fall into the local optimal, proposing a differential evolution particle swarm optimization algorithm (DEIPSO) to verify the feasibility of the model.

1.3 Paper Structure

The rest of the paper is organized as follows: Sect. 2 introduces the prediction model of EV charging demand points in the planning area. Section 3 introduces the mathematical model of EV charging station location. Section 4 introduces the mathematical model of EV charging station capacity allocation and objective function. Section 5 introduces the constraints in this model. Section 6 introduces the DEIPSO algorithm. Section 7 introduces the simulation scenarios of the experiment in this paper. Section 8 introduces the simulation results of the model in this paper. Section 9 summarizes the paper.

2 Division of Charging Demand Points

Charging demand is the number of EVs that need to go to the charging station for charging in a certain area and time. The charging demand is closely related to the traffic density. In areas with charging demand, the traffic density near the area will also increase. Therefore, the service capacity of the charging station should match the traffic density of the corresponding area to meet the charging needs of EVs as much as possible. EVs are located in most areas of the city, and vehicles will go to the nearest charging station to charge without the guidance of charging. Therefore, in this paper, the planning area is divided into several smaller areas [14], and each small area is a charging demand cell. Nie, Y. et al. [15] proposed that in any area, traffic flow is conserved for a period of time. According to this conclusion, it can be assumed that the number of EVs in each small area remains unchanged, so for the convenience of calculation, the geometric center of each cell is regarded as a charging load point, according to the number of EVs at each charging load point, calculating the charging demand of each cell.

3 Mathematical Model of EV Charging Station Location

The construction of the charging station not only needs to consider the construction cost of the charging station, but also needs to consider the driving cost of EVs. This paper mainly considers the fixed construction cost of the charging station, the annual operating cost and the charging satisfaction of EV users, and establishes a mathematical model for the location of EV charging stations, as shown in Eq. (1).

$$TotalCost = CSC + \frac{1}{\varphi\left(v_j, d_{ij}\right)} \tag{1}$$

In Eq. (1), *TotalCost* is the total cost; *CSC* is the construction cost of the charging station; $\varphi(v_j, d_{ij})$ is the EV user satisfaction.

The construction cost of the charging station is composed of the fixed construction cost and the annual operating cost of the charging station, as shown in Eq. (2).

$$CSC = \sum\nolimits_{i \in CS} f_{cs}(N_i)R_Z + u_{cs}(N_i) \tag{2}$$

The fixed construction cost f_{cs} of the charging station is shown in Eq. (3).

$$f_{cs}(N_i) = W_i + q_i N_i + m_i \tag{3}$$

In Eq. (3), W_i is the fixed investment cost of each charging station; q_i is the construction investment cost related to the charger in the charging station; m_i is the investment cost related to the transformer in the charging station; N_i is the number of charging piles in charging station i.

By reading a large number of references and combining the simulation environment of this paper, the annual operating cost of the charging station is shown in Eq. (4).

$$u_{cs}(N_i) = 0.1 f_{cs}(N_i) \tag{4}$$

R_Z is the discount factor of the charging station as shown in Eq. (5).

$$R_Z = \frac{\left(rr(1+rr)^{ms}\right)}{(1+rr)^{ms-1}} \tag{5}$$

In Eq. (5), rr is the discount rate and ms is the depreciation period of the charging station.

EV user satisfaction: EV user satisfaction indicates the evaluation of the charging station by the EV users at the charging station, as shown in Eq. (6).

$$\varphi\left(v_j, d_{ij}\right) = \frac{1}{VTC + VTE + CSL} \tag{6}$$

In Eq. (6), VTC is the cost of the travel time for EVs to reach the charging station; VTE is the cost of energy consumption for EVs to reach the charging station; CSL is the cost of waiting time for EVs at charging station, and the size of the function value indicates the satisfaction of the EV user with the charging station.

In this paper, consider the non-linear coefficient of urban roads to calculate the distance traveled by EVs, as shown in Eq. (7).

$$D_{ij} = \lambda_{ij} * \gamma_{ij} * d_{ij} \tag{7}$$

In Eq. (7), λ_{ij} is the non-linear coefficient of the urban road from the demand point j to the charging station i; γ_{ij} is the reentry coefficient of the EV journey from the demand point j to the charging station i; d_{ij} is the linear distance from the demand point j to the charging station i.

$$\lambda_{ij} = \frac{d_{tij}}{d_{ij}} \tag{8}$$

The minimum value of λ_{ij} is 1, and the smaller the λ_{ij}, the more convenient the traffic between the two points.

The time-consuming cost of an EVs in road is shown in Eq. (9).

$$VTC = \frac{365\beta_{time}N_c\left(\sum_{i\in CS}\sum_{j\in J_{CD}}pn_jD_{ij}\right)}{v_j} \tag{9}$$

In Eq. (9), β_{time} is the time cost of EVs; p is the daily fast charging probability of EVs; n_j is the number of EVs at demand point j; v_j is the average driving speed of EVs; N_c is the number of daily charging of EVs, as shown in Eq. (10).

$$N_{\mathrm{c}} = \frac{E_{1km} * k}{\text{battery}} \tag{10}$$

In Eq. (10), E_{1km} is the energy consumption of EVs; k is the daily mileage of EVs; *battery* is the battery capacity of EVs.

The cost of energy consumption for EVs to reach the charging station is shown in Eq. (11).

$$VTE = 365mN_c\left(\sum_{i\in CS}\sum_{j\in CD}n_jD_{ij}E_{1km}\right) \tag{11}$$

In Eq. (11), m is the electricity price in the planned area.

4 Mathematical Model of EV Charging Station Capacity Allocation

When an EV is charging at a charging station, if the charging station does not have an idle charging pile, it needs to wait in line for service. Queuing theory is through statistical research on the arrival and service time of service objects, to obtain statistical laws of quantitative indicators such as waiting time, queue length, and length of busy period, and then to improve the structure of the service system or reorganize the service objects according to these laws. So that the service system can meet the needs of the service target, but also can make the organization's expenses the most economical or some indicators are optimal. The planning of the number of charging piles for EV charging stations is to meet the charging needs of EVs and to optimize the economics of the charging station. Therefore, this paper uses the queuing theory multi-service desk model (M/M/S) to establish charging stations capacity allocation model. In the queuing system of the charging station, the arrival time of EVs follows the negative exponential distribution with the parameter λ, and the service time of each service desk is independent of each other, with the negative exponential distribution with the parameter μ. The average queue length L_s of the EV at the charging station is shown in Eq. (12).

$$L_s = \frac{P_0\rho^{N_i}\rho_{N_i}}{N_i!\left(1-\rho_{N_i}\right)^2} + \rho \tag{12}$$

In Eq. (12), P_0 is the probability that all charging piles in the charging station are idle, as shown in Eq. (13).

$$P_0 = \left[\sum_{n=0}^{N_i-1}\frac{\rho^n}{n!} + \frac{\rho^{N_i}}{N_i!\left(1-\rho_{N_i}\right)}\right]^{-1} \tag{13}$$

In Eq. (13), n is the number of EVs.

$$\rho_{N_i} = \frac{\rho}{N_i} = \frac{\lambda}{N_i \mu} \tag{14}$$

The residence time of the EV at the charging station is shown in Eq. (15).

$$W_s = \frac{L_s}{\lambda} \tag{15}$$

The cost of the waiting time of an EV at a charging station is shown in Eq. (16).

$$CSL = 365\beta_{time}N_c\left(\sum\nolimits_{i\in CS}\sum\nolimits_{j\in CD} n_j\left(W_s - \frac{1}{\mu}\right)\right) \tag{16}$$

The objective function of this paper is shown in Eq. (17).

$$Cost = min(TotalCost) \tag{17}$$

In Eq. 17, Cost is the lowest cost considering the cost of the charging station (construction cost, annual operating cost) and the cost of the EV (road travel time cost, energy consumption cost, waiting time cost at charging station).

5 Model Constraints

The number of charging piles in each charging station is constrained by Eq. (18).

$$N_{i,min} \leq N_i \leq N_{i,max} \tag{18}$$

In Eq. (18), $N_{i,min}$ is the minimum number of charging piles included in the charging station; $N_{i,max}$ is the maximum number of charging piles included in the charging station.

The distance constraint between charging stations is shown in Eq. (19).

$$D_{min} \leq D_{ij} \leq 2 * D_{min} \tag{19}$$

In Eq. (19), D_{min} is the minimum distance between two charging stations.

The distance constraint from the charging demand point to the charging station is shown in Eq. (20).

$$max(D_{ij}) \leq D_{max} \tag{20}$$

In Eq. (20), D_{max} is the maximum service radius of the charging station.

In order to avoid the long queue of EVs at the charging station and ensure the stability of the queuing system, the arrival rate of EVs must be less than the product of the service rate of the charging station and the number of charging piles, as shown in Eq. (21).

$$\lambda \leq \mu N_i \tag{21}$$

The residence time limit of EVs at charging stations is shown in Eq. (22).

$$W_s \leq W_{s-max} \tag{22}$$

In Eq. (22), W_{s-max} is the maximum residence time of the EV at the charging station ($W_{s-max} = 40$ min).

6 Location Algorithm Analysis

6.1 Voronoi Diagram

Voronoi diagram is composed of a set of continuous polygons formed by a set of vertical bisectors connecting two adjacent points. In the Voronoi diagram, the distance from any point within a polygon to the control points that constitute the polygon is less than the distance to the control points of other polygons. In this paper, assuming that the coordinates of the charging station are control points, using these control points to draw a Voronoi diagram, the service area of the charging station can be divided.

6.2 Improve Particle Swarm Optimization

The particle swarm optimization searches the search space in parallel through a group of initialized groups, and realizes the evolution of the population through the competition and cooperation between individuals in the population. Fewer parameters need to be set, and the operation is simple.

In particle swarm optimization, each particle represents a potential solution to the problem, and the fitness of the particle is judged by the fitness function. The initial value of the particle swarm is a group of random particles, and the optimal solution is found according to the iteration. In each iteration, the particle updates its position and velocity according to the individual optimal value and the global optimal value. The particle velocity update formula is shown in Eq. (23).

$$V_t = \omega * V_t + c_1 r_1 (P_{best} - x_t) + c_2 r_2 (g_{best} - x_t) \tag{23}$$

In Eq. (23), ω is the inertial weight; c_1 c_2 is the learning factor; r_1 r_2 is the random number in the range [0,1]; P_{best} is the individual optimal value of the particle; g_{best} is the global optimal value of the particle. The speed update formula consists of 3 parts, $\omega * V_t$ is the inertial part, the motion inertia of the reaction particle; $c_1 r_1 (P_{best} - x_t)$ is the cognitive part, and the reaction particle has a tendency to update its own history optimally; $c_2 r_2 (g_{best} - x_t)$ is the social part, and the reaction particle has a tendency to update the historical optimal value in the directed group.

When the learning factor c_1 is greater than c_2, particles pay attention to their historical position; when c_1 is less than c_2, particles pay more attention to social information. We can find that in the early stage of the particle movement, the particle needs to update to its historical optimal value; in the later stage of the movement, the particle needs to pay more attention to update to the group optimal value. Therefore, in this paper, the improved particle swarm algorithm (IPSO) is used. In IPSO, asymmetric arccosine strategy is used to set the learning factor, as shown in Eqs. (24) and (25).

$$c_1 = c_{1end} + (c_{1start} - c_{1end}) * (1 - \mathrm{acos}(-2^*\mathrm{n}/(\mathrm{N}+1) + 1)/\pi) \tag{24}$$

$$\mathrm{c}_2 = \mathrm{c}_{2\mathrm{end}} + (\mathrm{c}_{2\mathrm{start}} - \mathrm{c}_{2\mathrm{end}})^*(1 - \mathrm{acos}(-2^*\mathrm{n}/(\mathrm{N}+1) + 1)/\pi) \tag{25}$$

In Eq. (24) and (25), $c_{1start} = 2.75$, $c_{1end} = 0.5$, $c_{2start} = 1.25$, $c_{2end} = 2.25$.

Equation (26) shows the particle position update formula.

$$X_t = X_t + V_t \tag{26}$$

In this paper, the particle out-of-bounds problem is solved by restricting the particle's active area. The particle's active area *Particle_Area* is shown in Eq. (27).

$$Particle\_Area \in \left[x_{min}, x_{max}\right] \cup \left[y_{min}, y_{max}\right] \tag{27}$$

In Eq. (27), $x_{min}, x_{max}, y_{min}, y_{max}$ are the smallest horizontal axis coordinate point, the largest horizontal axis coordinate point, the smallest vertical axis coordinate point, and the largest vertical axis coordinate point in the simulation environment.

6.3 Differential Evolution Algorithm

Differential evolution (DE) algorithm is a group-based adaptive global optimization algorithm. The core part of the algorithm includes mutation, hybridization and selection operations. The mutation operation of the differential evolution algorithm is shown in Eq. (28).

$$V_{di} = X_{dr1} + F_0(X_{dr2} - X_{dr3}) \tag{28}$$

In Eq. (28), X_d is the initial population; $r_1 r_2 r_3$ is a random value ($r_1 r_2 r_3 \epsilon [1, N]$), and $r_1 \neq r_2 \neq r_3 \neq i$; N is the population size. The hybridization operation is shown in Eq. (29).

$$U_{dij} = \begin{cases} V_{dij} rand \leq CR \, or \, randi(1, Tn) = j \\ X_{dij} rand > CR \, or \, randi(1, Tn) \neq j \end{cases} \tag{29}$$

In Eq. (29), *CR* is the mutation probability, $j \in [1, Tn]$. The selection operation is shown in Eq. (30).

$$X_d = \begin{cases} U_d \, fit(U_d) < fit(X_d) \\ X_d \, fit(U_d) \geq fit(X_d) \end{cases} \tag{30}$$

In Eq. (30), $fit(U_d)$ is the fitness of the population U_d; $fit(X_d)$ is the fitness of the population X_d.

6.4 Differential Evolution Improve Particle Swarm Optimization

In this paper, by combining DE algorithm and IPSO to improve the global optimization ability of the solution, the steps are as follows.

(1) According to the charging demand in the planning area and the service capacity of the charging station, estimate the number range of charging stations in the planning area $Tn \in [N_{min}, N_{max}]$.

(2) Use the DE algorithm to plan the charging station, set the range of the charging station site *CSposition*, $CSposition \in [x_{min}, x_{max}] \cup [y_{min}, y_{max}]$, and randomly generate the charging station site *Xd* within this range, $Xd = [(x_1, y_1), (x_2, y_2), \ldots, (x_N, y_N)]$, N is the population size.
(3) Set the initial dimension of the DE algorithm, that is, the initial number of charging stations $Tn = N_{min}$.
(4) According to Eqs. (28), (29), and (30), mutation operation, hybridization operation, and selection operation are performed respectively. After each operation, a new charging station position must be generated reasonably according to the constraints.
(5) Set the maximum number of iterations of the algorithm. When the algorithm reaches the maximum number of iterations, the algorithm stops. At this time, the position set *DEX* of the charging station and the corresponding adaptation value *DEF* are counted.
(6) The IPSO is used to plan the charging station, the initial station site range is *CSposition*, and the initial charging station site is $X, X = X_d$.
(7) Set the initial dimension of the IPSO, that is, the number of charging stations $Tn = N_{min}$.
(8) Calculate the local optimal solution and the global optimal solution, update the particle speed according to Eq. (23), and update the particle position according to Eq. (26), and calculate the local optimal solution and global optimal solution of the updated population.
(9) Set the maximum number of iterations of the algorithm. When the algorithm reaches the maximum number of iterations, the algorithm stops. At this time, the charging station position set *IPSOX* and the corresponding adaptation value *IPSOF* are counted.
(10) Combining *DEF* and *IPSOF* to obtain a new fitness value combination *NewFit*, $NewFit = [DEF, IPSOF]$, the dimension of *NewFit* is $2N$.
(11) Sort *NewFit* in ascending order to obtain the new population's fitness value *NewF*, and take the first N fitness values for the charging station location *NewP*.
(12) Using *NewP* as the initial site, repeat steps (7), (8), (9). Obtain the charging station location set *IPSOX* and the corresponding fitness value *IPSOF*. By selecting the minimum value of *IPSOF*, the minimum total cost and the optimal deployment position of the charging station can be obtained.
(13) $Tn = Tn + 1$, repeat steps (4)–(12), until $Tn > N_{max}$, the algorithm stops.
(14) Count the total cost corresponding to the number of different charging stations, and select the number and location of charging stations corresponding to the optimal cost.

7 Simulation Scenario

The experiment process of this paper uses MATLAB software to simulate. In order to increase the scalability of the model, the algorithm realization module, environmental parameter module, and parameter calculation module are independently established in the program. This paper sets the simulation parameters based on the reference [14] simulation environment, as shown in Table 1.

Table 1. Simulation parameters.

Parameter	Value
Number of charging demand points (n)	34
Fixed investment (W_i)	2 millions
Investment-related to the unit price of the charging piles in the charging station (q)	0.35 millions
The investment cost related to the transformer in the charging station (e_i)	0.2 millions
Discount rate (rr)	0.08
Charging station depreciation period (ms)	20 years
Average driving speed of EVs (v)	30km/h
Non-linear coefficient of the urban road (λ_{ij})	1.2
Minimum number of charging piles in the charging station ($N_{i,min}$)	3
Maximum number of charging piles in the charging station ($N_{i,max}$)	30

Algorithm parameters settings are shown in Table 2.

Table 2. Algorithm parameters.

Parameter	Value
Number of particles (N)	20
Maximum number of iterations ($MaxIter$)	100
Inertial weights (W_s and W_e)	0.9 and 0.4
Scaling factor (F)	0.4
Mutation probability (CR)	0.6

The location of the charging demand points is shown in Fig. 1.

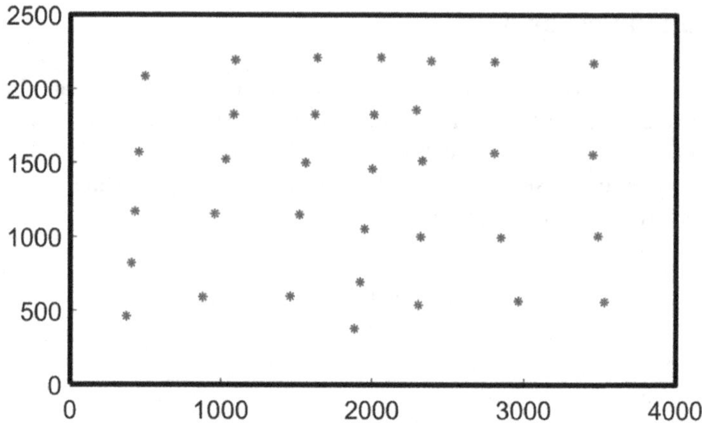

Fig. 1. Distribution of charging demand points

8 Simulation Results

The convergence curves of IPSO, DE, and DEIPSO are shown in Fig. 2.

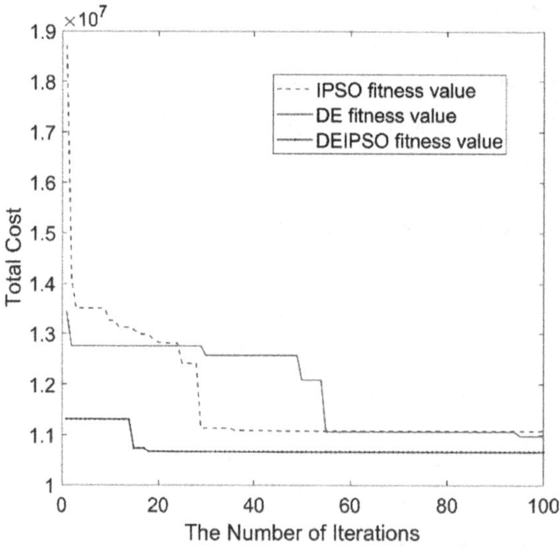

Fig. 2. Convergence curve

In Fig. 2, we can find that compared with IPSO and DE, DEIPSO can converge earlier and can jump out of the local optimum.

The changes in the cost of charging stations and the cost of driving EVs with the number of charging stations deployed are shown in Fig. 3.

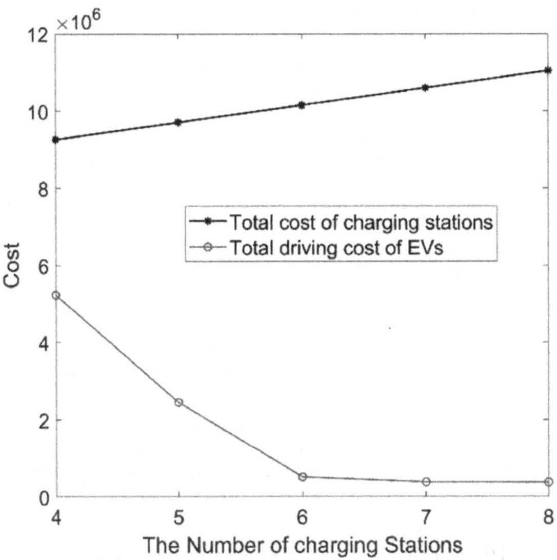

Fig. 3. Different costs vary with the number of charging stations deployed

In Fig. 3, we can find that as the charging station increases, the cost of the charging stations also increases, because the construction cost of the charging station is positively related to the number of charging stations deployed. The driving cost of EVs decreases as the number of charging stations increases, the reason is that as the number of charging stations increases, it will be easier for EVs to find the nearest charging station and reduce the distance traveled. Therefore, when selecting the number of charging stations to be deployed, it is necessary to comprehensively consider the cost of charging stations and the driving cost of EVs, as shown in Fig. 4.

In Fig. 4, We can find that when the cost of charging stations and the cost of EVs are considered comprehensively, the total cost of deploying six charging stations is the lowest. This is because when the number of charging stations is six, the total cost of charging station cost and EVs cost is the lowest.

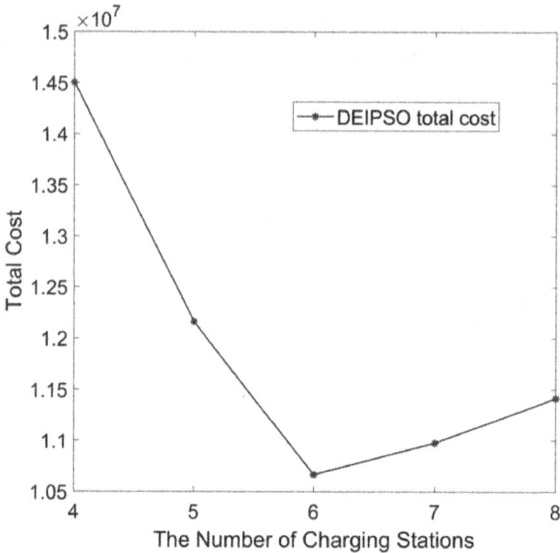

Fig. 4. The total cost varies with the number of charging stations deployed

The total cost of IPSO, DE, and DEIPSO changes with the number of charging stations deployed is shown in Table 3.

Table 3. Total cost changes.

Algorithm	IPSO	DE	DEIPSO
Total cost of deploying 4 charging stations (ten million)	1.5332	1.4833	1.4503
Total cost of deploying 5 charging stations (ten million)	1.2164	1.3645	1.2162
Total cost of deploying 6 charging stations (ten million)	1.2152	1.1311	1.0668
Total cost of deploying 7 charging stations (ten million)	1.1080	1.0979	1.0978
Total cost of deploying 8 charging stations (ten million)	1.4694	1.1501	1.1417

In Table 3, we can find that no matter how many charging stations are deployed, the total cost of using the DEIPSO algorithm is lower than the IPSO and DE algorithms. This is because the DE algorithm increases the diversity of the IPSO population and reduces the risk of falling into a local optimum.

The location distribution of charging stations in the planned area is shown in Fig. 5.

When six charging stations are deployed, the service indicators of the charging piles in the charging station are shown in Table 4.

In Table 4, we can find that the average residence time of EVs at the charging station is within 40min.

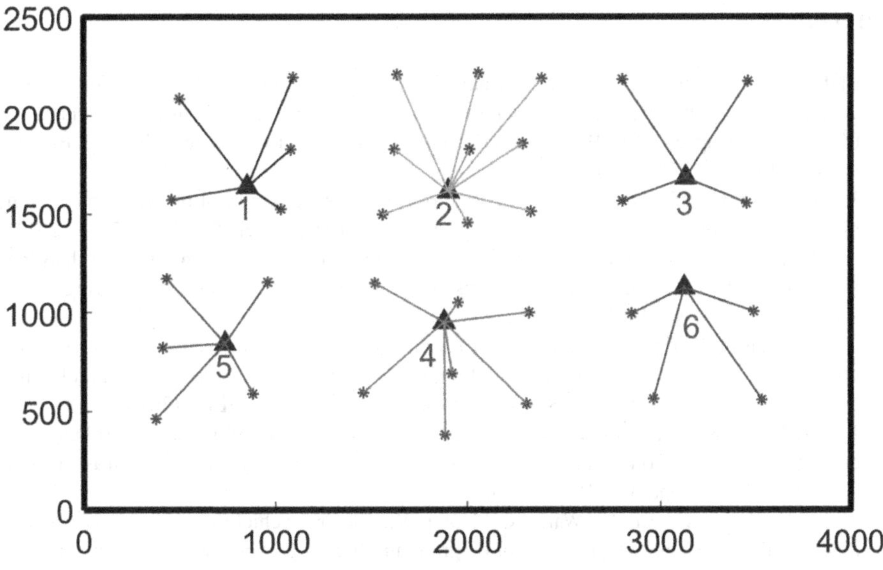

Fig. 5. Location of charging station deployment

Table 4. Service indicators of the charging piles in the charging station.

Charging station serial number	1	2	3	4	5	6
Number of charging piles	18	21	22	22	13	23
Residence time (min)	39.984	38.874	37.794	37.146	36.516	39.756

9 Conclusion

This paper comprehensively considers the cost of EVs charging and the cost of charging stations construction, and proposes a method of location and capacity for EV charging stations. The model of EV charging station location, the model of EV charging station capacity allocation, and the optimization algorithm of the location of the charging station are studied respectively. The DEIPSO algorithm is used to verify the feasibility of the site selection model for the site selection of charging stations; the queuing theory of this paper can be used to reasonably plan the number of piles in each charging station; by comparing the DE, IPSO, and DEIPSO algorithms, it is found that the DEIPSO algorithm can achieve a certain degree Jump out of the local optimal solution, finding the global optimal solution, and better deploy the charging station location.

Acknowledgement. This work supported by Sichuan Science and Technology Program 2019ZDZX0005.

References

1. Mehmet, C.C., Merve, Y., Arif, G., Hasan, K.: Estimation of optimal locations for electric vehicle charging stations. In: 2017 IEEE International Conference on Environment and Electrical Engineering and 2017 IEEE Industrial and Commercial Power Systems Europe (EEEIC / I&CPS Europe), Milan (2017)
2. Johannes, S., Matthias, E.: Finding suitable locations for charging stations-implementation of customers' preferences in an allocation problem. Int. Electric Veh. Conf. **12**, 17–19 (2014)
3. Bi, R., Xiao, J.J., Viswanathan, V., Knoll, A.: Influence of charging behaviour given charging station placement at existing petrol stations and residential car park locations in Singapore. Procedia Comput. Sci. **80**, 335–344 (2016)
4. Phonrattanasak, P.: Optimal placement of EV fast charging stations considering the impact on electrical distribution and traffic condition. In: 2014 International Conference and Utility Exhibition on Green Energy for Sustainable Development (ICUE). IEEE (2014)
5. Jiao, D.J., Su, X.L., Yan, X.X., Wang, W.C.: A location and capacity planning scheme for charging station based on Voronoi diagram and catfish particle swarm optimization algorithm. Autom. Technol. Appl. **37**(03), 5–10 (2018)
6. Ma, X.F., Wang, H., Li, Y., Wang, C., Hong, X.: Electric vehicle charging station planning based on variable weight Voronoi diagram and hybrid particle swarm optimization. J. Electrotech. **32**(19), 160–169 (2017)
7. Chen, J.P., Ai, Q., Xiao, F.: Electric vehicle charging station planning based on user travel needs. Electric Power Autom. Equipment **36**(06), 34–39 (2016)
8. Zheng, C.Q.: Research on the most optimal location of urban electric vehicle charging facilities. Nanchang University (2016)
9. Yang, H.M., Deng, Y.J., Qiu, J., Li, M., Lai, M.Y,. Dong, Z.Y.: Electric vehicle route selection and charging navigation strategy based on crowd sensing. IEEE Trans. Ind. Inform. (2017)
10. Jiao, D.J.: Research on the optimization of location and volume of electric vehicle charging stations. Shanxi University (2018)
11. Mehar, S., Senouci, S.M.: An optimization location scheme for electric charging stations. In: International Conference on Smart Communications in Network Technologies. Paris, France (2013)
12. Han, F.J.: Research on Optimizing the Location and Capacity of Electric Vehicle Charging Station. Nanchang University (2015)
13. Wang, H.: Research on optimal layout of electric vehicle charging stations based on hybrid discrete particle swarm optimization. North China Electric Power University (Beijing) (2017)
14. Liu, H.: Location and capacity optimization of electric vehicle charging stations based on particle swarm genetic hybrid algorithm. Xi'an University of Technology (2016)
15. Nie, Y., Ghamami, M.: A corridor-centric approach to planning electric vehicle charging infrastructure. Transp. Res. Part B **57**(57), 172–190 (2013)

Research on Semantic Vision SLAM Towards Dynamic Environment

Nanyang Bai(✉), Tianji Ma, Wentao Shi, and Lutao Wang

Chengdu University of Information Technology, Chengdu, China
wanglt@cuit.edu.cn

Abstract. Simultaneous localization and mapping (SLAM) is considered to be the basic ability of intelligent mobile robots. In the past few decades, thanks to community's continuous and in-depth research on SLAM algorithms, the current SLAM algorithms have achieved good performance. But there are still some problems. For example, most SLAM algorithms have the assumption of a static environment, but in real life, most of the environment contains moving objects, so how to deal with the moving objects in the environment requires careful consideration. What's more, traditional geometric maps cannot specific environmental semantic information for mobile robots, so how to make robots truly understand the surrounding environment to complete some advanced tasks is also a difficult problem. In this paper, we design a scheme to improve the accuracy and robustness of SLAM in a dynamic environment. And we realize the perception of semantic information of objects in the environment through the object detection algorithm of deep learning neural network.

Keywords: SLAM · Semantic recognition · Semantic map · Dynamic target detection

1 Introduction

In the past years, with the development of AR (Augments Reality), UAV (Unmanned Aerial Vehicle), and UGV (Unmanned Ground Vehicle), visual SLAM has been extensively investigated. The mainstream vision sensors are divided into monocular cameras, stereo cameras, and RGB-D cameras. The monocular camera's simple solution has advantages in terms of size, power, and cost. However, there are also some problems, such as the inability to observe the scale and state initialization. By using more complex equipment, such as stereo cameras or RGB-D cameras, these problems could be solved, and the robustness of the visual SLAM system is also greatly improved.

Thanks to the SLAM system's continuous research by the research group, the visual SLAM system framework has been quite mature. It usually consists of several essential parts, such as feature extraction front-end, state estimation back-end, and closed-loop detection. Additionally, some advanced SLAM algorithms have achieved satisfactory performance, such as ORBSLAM2 [1], LSD-SLAM [2].

X. Wu et al. (Eds.): QShine 2020, LNICST 381, pp. 91–102, 2021.
https://doi.org/10.1007/978-3-030-77569-8_7

However, some issues remain unsolved. For example, these algorithms all assume the strong constraint of the static environment. When there are dynamic objects in the environment, its robustness and accuracy will be significantly decreased. Besides, these algorithms only provide geometric maps. It cannot provide support for advanced tasks such as intelligent obstacle avoidance.

In a dynamic environment, the SLAM algorithm's robustness will be significantly affected whether those algorithms are based on the feature method or the direct method because dynamic objects in the environment will corrupt the state estimation quality and lead to system failure. For example, dynamic objects in the environment may deceive the feature association in the visual SLAM algorithm. Therefore, to improve the entire system's stability, it is especially important to deal with dynamic objects in the environment.

The typical SLAM method only provides a geometric map composed of points and planes, which does not contain the surrounding environment's semantic information. Compared with geometric maps, semantic maps have the advantages of intuitive visualization and effective human–machine-environments interaction. According to the summary by Cadena et al. [3] We have now entered the third stage of SLAM research, videlicet, a stage of robust perception: the realization of robust performance, high-level understanding, resource perception, and task-driven perception represents the theme of this era.

This paper focuses on reducing the impact of dynamic objects in vision SLAM by combining the Mask R-CNN network with epipolar geometry. Simultaneously, the semantic information is bound to the octree map to obtain the semantic graph. Provide conditions for robots to achieve advanced tasks such as intelligent navigation.

In the rest of this paper, the structure is as follows. Section 2 provides an overview of semantic SLAM and SLAM in dynamic environments. Then Sect. 3 presents the framework of this whole SLAM system in detail. And we discuss how to detect dynamic objects and produce semantic maps. Subsequently, Sect. 4 provides the results of our program and ORB-SLAM2 on the TUM RGB-D dataset. Finally, in Sect. 5, we give a brief conclusion and discussion about our work.

2 Related Work

2.1 Semantic SLAM

Generally speaking, semantic SLAM is to use a neural network to provide road sign information for traditional SLAM solutions. The semantic SLAM system consists of two essential components: a semantic extractor and a modern V-SLAM framework. The semantic information is mainly extracted and derived from two processes. They are object detection and semantic segmentation [4].

Object detection is recognized as an essential branch of CV, whose development can be roughly divided into handcraft feature-based machine learning stage (2001–2013) and learning feature-based deep learning (2013-present). The former is extremely dependent on handcraft features of images [5–9]. It also requires many computing resources. In recent years, due to the introduction of deep learning and graphics processing units,

Object detection's accuracy and efficiency have been greatly improved in theory and practice. Therefore, more and more SLAM adds semantic modules into the system.

The earliest semantic map was proposed by Pham et al. [10] They used SSD, which has a fast detection speed, and ORB-SLAM2, which can achieve real-time positioning and promote each other. Then through dividing the depth map, object detection, and finally, output a semantic map with semantic information. Pronobis et al. [11, 12] proposed an online system using lasers and cameras to construct a semantic map of the environment. McCormac et al. proposed a dense three-dimensional semantic mapping method SemanticFusion [13] using convolutional neural networks. By combining CNNs with a dense SLAM solution ElasticFusion [14]. It ensures the dense long-term consistency of indoor positioning and eliminates the multi-circle scanning trajectory's cumulative error. It integrates the semantic prediction probability of CNNs from the multi-view points into the map to obtain a three-dimensional dense semantic map.

2.2 Dynamic Segmentation

In the SLAM community, relevant information extracted from static objects is considered stable and effective, while information extracted from moving objects is known to decline the algorithm's performance. For dynamic objects in the environment, advanced SLAM systems either treat them as outliers and eliminate them in different ways. Either use a separate target tracking module to track it.

One of the earliest works about SLAM in dynamic environments is presented by Hahnel et al. [15] use an Expectation–Maximisation (EM) algorithm to update the probabilistic estimate about which measurements correspond to a static/dynamic object and remove them from the estimation when they correspond to a dynamic object. Alcantarilla et al. [16] introduce dense scene flow for dynamic objects detection and show improved localization and mapping results by removing "erroneous" measurements on dynamic objects from the estimation. Tan et al. [17] propose an online keyframe update that reliably detects changed features in terms of appearance and structure and discards them if necessary. Kundu et al. [18] extend egomotion estimation with MBSfM [19] techniques similar to estimate the SE (3) trajectories of the third-party motions in a scene, but they constrain all the motions to the horizontal plane.

3 System Description

3.1 SLAM

In practical applications, the accuracy of attitude estimation and harsh environments' reliability are the critical factors for evaluating autonomous robots. ORB-SLAM2, as a relatively lightweight SLAM system, has an excellent performance in a static environment. We added a dynamic object detection module and a semantic map module for it. As shown in Fig. 1, we have designed five threads to control the SLAM system's five main modules. For the part of the dynamic environment, we place a real-time semantic segmentation network in a separate sub-thread and filter out the scene's dynamic targets by combining semantic segmentation and moving consistency checking methods. This

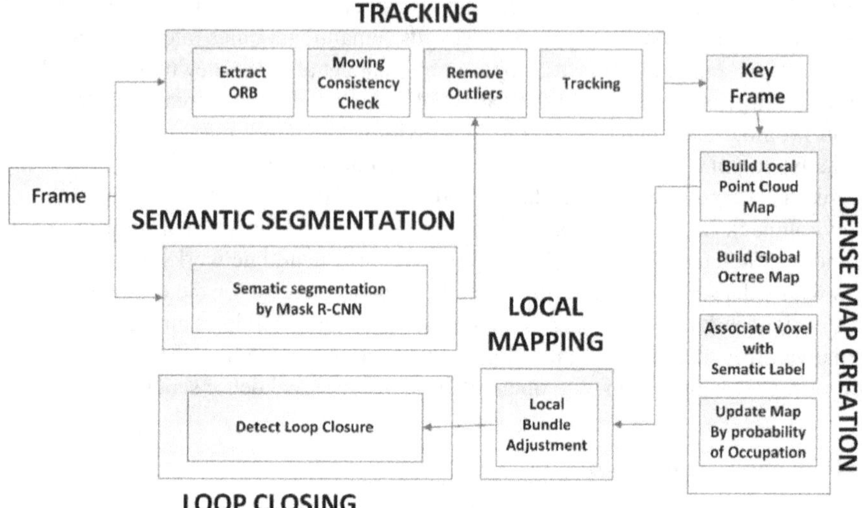

Fig. 1. The framework of our system. Our work mainly focuses on semantic extraction and dynamic detection of input images.

improves the robustness and accuracy of the SLAM system in dynamic scenarios. For the semantic map part, we also designed a separate thread to build the octree map. The semantic map is realized in the form of binding with semantic tags.

3.2 Semantic Segmentation

Semantic segmentation is an integral part of image processing and image understanding in machine vision technology. Semantic segmentation is to classify each pixel in the image, determine each point's category (such as background, person, or car), and divide the area.

To obtain the semantic information and potential moving objects in the environment, we use Mask R-CNN [20] to perform semantic segmentation on the image to obtain the objects in the environment and their semantic information. Mask R-CNN is a region-based semantic segmentation method that uses selective search to extract many target proposals. It then calculates the CNN features of each proposal. Finally, class-specific linear support vector machines are used to classify each region. Compared with other CNN networks, Mask R-CNN has higher speed and accuracy.

Like Fig. 2, it can subdivide the dynamic or movable objects in the image (such as people, animals, bicycles, cars, motorcycles, planes, buses, trains, trucks, boats, etc.). Most of the objects in our lives are included in their recognizable range.

3.3 Moving Consistency Check

In the Mask R-CNN network, we detect and process potential moving objects in the image, but these potential moving objects are not necessarily in a real state of motion.

Fig. 2. The result of object detection by Mask R-CNN in COCO dataset

They may also be static, such as a car parked on the side of the road. If we remove all the features of the image located on the potential moving objects, although the negative impact of motion on the accuracy can be avoided, we will also lose a large number of effective features. This may cause the tracking of the SLAM system to be lost due to the lack of matching features. Therefore, it is particularly important to use the movement consistency to detect the true state of all potential dynamic objects. Geometric constraints (such as epipolar lines, triangulation, basic matrix estimation or reprojection error equations) are effective ways to determine the state of feature motion [21].

In this experiment, we use the feature of epipolar constraint to distinguish the dynamic and static features of the object. As shown in Fig. 3, the static feature satisfies the standard constraints of the epipolar geometry (Fig. 3 (a)). If the tracked feature is too far from the polar line, it is likely to be a dynamic feature (Fig. 3 (b)).

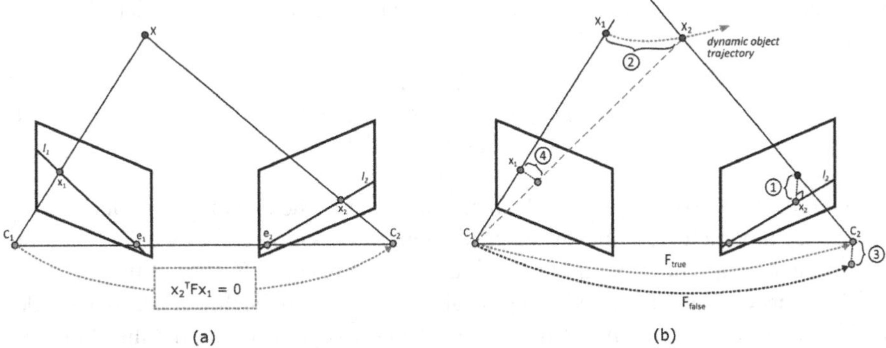

Fig. 3. (a) The transformation form point x_1 to point x_2 be define by epipolar constraint in static scenes. (b) Violation of geometric constraints in a dynamic environment: (1) The tracked feature is too far from the epipolar line; (2) Back-projected rays from the tracked features do not meet; (3) The dynamic feature has an impact on the accuracy of the basic matrix estimation; (4) The reprojected feature is too far away from the observed feature

Figure 3(a) shows that the feature point transformation in the static background satisfies the epipolar geometric constraint. When the point x_1 changes to the point x_2 in the static scene, the epipolar constraint is:

$$x_2^T F x_1 = 0 \tag{1}$$

To filter out the feature points of motion more effectively, our solution process is as follows: first, we extract the feature points of the previous frame. Secondly, calculate the corresponding displacements of these points in the current frame by using the LK optical flow method of pyramid layering. Then, we estimate the fundamental matrix by the RANSAC algorithm through the previous frame's feature points and the corresponding points of the current frame. Next, use the fundamental matrix to calculate the epipolar line in the current frame. Finally, the matching point's distance to its corresponding epipolar line is calculated, and the motion state of the feature point is judged by comparing it with the preset threshold.

The feature points of the previous frame are projected into the current frame through the fundamental matrix. Let P_1 and P_2 denote the matching points in the previous frame and the current frame, respectively, and their homogeneous coordinate forms are:

$$P_1(u_1, v_1, 1), P_2(u_2, v_2, 1) \tag{2}$$

Among them, u and v respectively represent the position of the point in the image. According to the principle of epipolar geometry, we can get the epipolar line l_1 through the fundamental matrix F and the point P_1. The expression is:

$$l_1 = \begin{bmatrix} X \\ Y \\ Z \end{bmatrix} = FP_1 = F\begin{bmatrix} u_1 \\ v_1 \\ 1 \end{bmatrix} \tag{3}$$

X, Y, Z in the expression represent the line vectors of the epipolar line. Moreover, the distance from the matching point to its corresponding epipolar line is determined as follows:

$$D = \frac{\left|P_2^T FP_1\right|}{\sqrt{||X^2|| + ||Y^2||}} \tag{4}$$

D represents the distance from the matching point to the epipolar line. Our moving consistency module determines whether the point is a motion point by calculating the distance and comparing it with our previous preset threshold. Finally, if the matching point's distance to its corresponding epipolar line is less than the threshold, we consider the feature point is static. In contrast, if the distance is greater than the threshold, we consider the matching point to be moving.

3.4 Remove Outliers

In the Mask R-CNN network, we detect and extract potential moving targets in the image. Nevertheless, these potential moving targets are not certainly in a moving state, such as a car parked on the side of the road. If all the feature points in the potential moving target are eliminated indiscriminately, the SLAM system may lose track due to the lack of features. Therefore, it is particularly important to combine the moving consistency detection results with semantic segmentation to judge the movement state detection's potential moving targets.

Thanks to the semantic segmentation network, we can quickly obtain the complete contours of potential moving targets. We judge the target's movement state by the number of moving feature points in the potential moving targets' contour. Suppose the number of moving feature points is bigger than the threshold. We regard the target as a moving object. We will delete all the feature points in the target's contour, then use the remaining feature points for pose estimation. In this way, we can accurately eliminate the outliers that will affect the attitude estimation, thereby improving the system's accuracy.

3.5 Semantic Map

The maps used in the SLAM system are divided into point cloud map and octree map. The advantage of the point cloud map is that it can be efficiently generated directly from RGB-D images and does not require additional processing. However, the point cloud map is usually large and carry too much useless information, such as shadows and wrinkles. Simultaneously, the point cloud map cannot handle dynamic objects because the point cloud map can only add points during the construction process. The octree map is more flexible and updatable than the point cloud map. Furthermore, the octree map can be stored more efficiently and is easy to navigate.

We maintain the octree map through the octree map thread in the system. This thread will combine the keyframes obtained in the tracking thread with the segmentation results obtained in the semantic segmentation thread. We use the transformation matrix and depth image in the keyframe to create a local point cloud map. The local point cloud map is convenient for the system to perform local BA operations. And we convert and store the local point cloud map in the global octree map. The semantic information is merged into the octree map by binding the octree map's voxels to a specific color. We assign every semantic label to each different color. For example, red represents people, and blue represents cars, etc. In this way, the semantic information in the map can be updated efficiently.

In General, what is saved in the map should be the static background in the environment, so dynamic objects should not exist on the map. We can use semantic segmentation results to filter out dynamic objects. However, the accuracy of semantic segmentation is limited. In complex situations, for example, objects overlap each other, the semantic segmentation results may be incomplete or even wrong. We use a probability model to evaluate the possibility of a single voxel being occupied quantitatively to solve this problem. Let P denote the probability that a voxel n is occupied from time z_1 to z_t, and its expression is:

$$\mathrm{P}(n|z_{1:t}) = \left[1 + \frac{1-P(n|z_t)}{P(n|z_t)} \frac{1-P(n|z_{1:t-1})}{P(n|z_{1:t-1})} \frac{P(n)}{1-P(n)}\right]^{-1} \quad (5)$$

It can be seen that the value of this formula depends on the prior probability of P(n), $\mathrm{P}(n|z_{1:t-1})$, and $\mathrm{P}(n|z_t)$ at time z_t. By using the log-odds notation [22], it can be expressed as:

$$\mathrm{L}(n|z_{1:t}) = \mathrm{L}(n|z_{1:t-1}) + \mathrm{L}(n|z_t) \quad (6)$$

$$\mathrm{L}(n) = \log\left[\frac{P(n)}{1-P(n)}\right] \quad (7)$$

$L(n|z_{1:t})$ represents the log-odds score of voxel n from the start time z_1 to time z_t. When the voxel is repeatedly observed to be occupied, the voxel's log-odds score will increase. Otherwise, it will decrease. The occupied probability P of a voxel can be calculated by inverse logit transform. The state of the voxel is judged by comparing the probability P with our predefined threshold. When the probability P is greater than the threshold, we consider that the voxel is stably occupied. In other words, this voxel belongs to a static object. Through this method, we can handle the map construction problem in a dynamic environment well.

4 Experimental Results

In this experiment, we used the TUM RGB-D dataset as the test data and compared our solution with the original ORB-SLAM2. The TUM RGB-D data set [23] consists of 39 kinds of sequences recorded in different indoor scenes at full frame rate (30 Hz) using Microsoft Kinect sensors. The data set provides RGB-D images and real trajectories. In this dataset's walking sequence, there will be a large number of scenes of people moving. These moving people will significantly reduce the robustness and accuracy of the SLAM algorithm. This data set is highly dynamic. Therefore, it is challenging for the SLAM algorithm.

Figure 4(a) shows the evaluation of the absolute pose error (APE) between our measured value and the real value. This data evaluates the overall consistency of our measurement data with the real trajectory. Figure 4(b) Represents the real-time attitude error between our measured value and the real trajectory. The lower the color temperature of the track color, the closer our measured value is to the real value. The higher the color temperature of the track color, the greater the deviation between our measured value and the real value. Figure 5(a) shows the evaluation of the relative pose error (RPE) between our measured value and the real value. This data evaluates the translation and rotation drift between our scheme and the real trajectory. Figure 5(b) Represents the real-time translation drift error and rotation drift error between our measured value and the real trajectory. The lower the color temperature of the track color, the smaller the error.

Figure 6(a) shows the difference in our scheme, ORB-SLAM2, and the real trajectory. It reflects the system's integral accuracy. Figure 6(b) shows the deviation of our scheme and ORB-SLAM2 in the x, y, and z directions compared to the real trajectory. Figure 6 (c) shows the error between the two methods and the real trajectory attitude. Moreover, Tables 1, 2 and 3 show the comparison of rotation deviations and translation deviations in different schemes. It is clear that our solution is more robust and accurate in a dynamic environment than the original ORB-SLAM2.

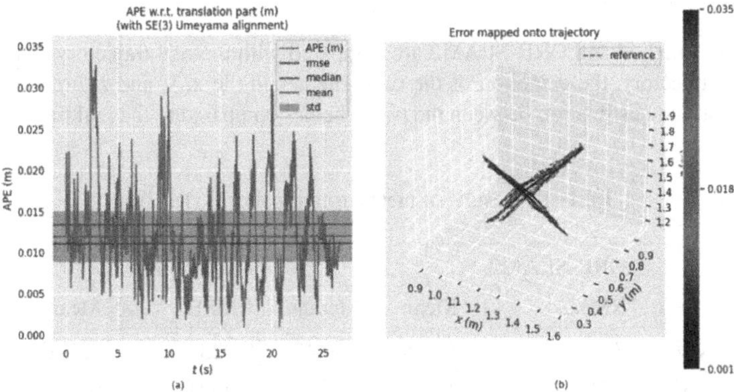

Fig. 4. (a) The absolute pose error (APE) between the measured value and the trajectory includes root mean square error (RMSE), median, and mean. (b) The deviation of the measurement data on the track position

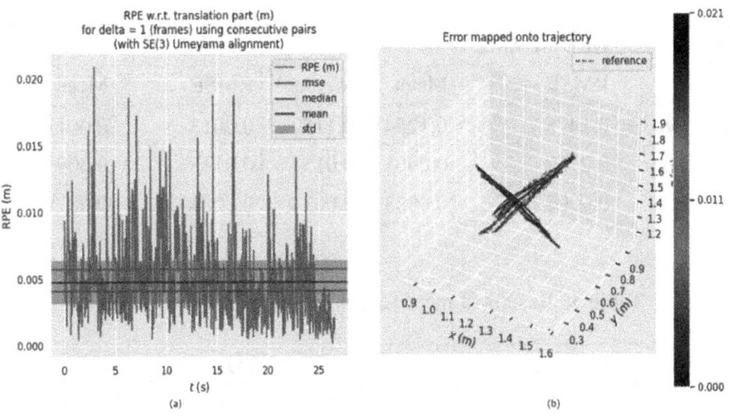

Fig. 5. (a) The relative pose error (RPE) between the measured value and the trajectory includes root mean square error (RMSE), median, and mean. (b) The deviation of the measurement data on the track position

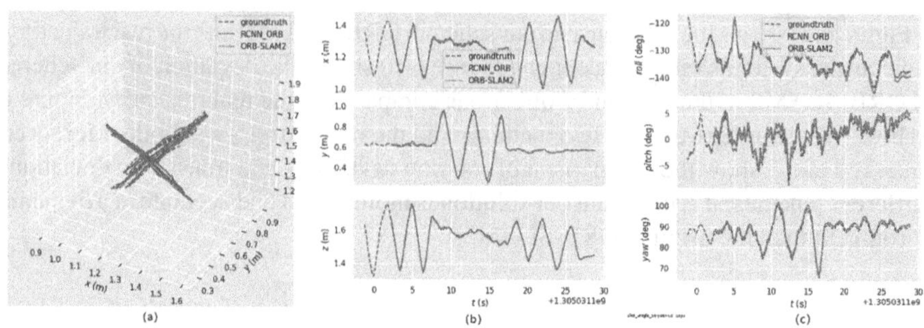

Fig. 6. (a) Our method and ORB-SLAM2 are compared with the real trajectory. (b) Compared with the real trajectory, the deviation of the two schemes in the x, y, and z directions. (c) The difference in attitude estimation between the two schemes compared to the real trajectory

Table 1. Results of metric rotational drift (RPE)

Sequences	ORB-SLAM2			Our Scheme		
	RMSE	Mean	Median	RMSE	Mean	Median
fr3_walking_xyz	7.7424	5.8754	4.5440	0.9234	0.5836	0.4197
fr3_walking_static	3.8754	1.5744	0.4571	0.2975	0.2415	0.2276
fr3_sitting_static	0.2887	0.2559	0.2495	0.2775	0.2417	0.2355

Table 2. Results of metric translational drift (RPE)

Sequences	ORB-SLAM2			Our Scheme		
	RMSE	Mean	Median	RMSE	Mean	Median
fr3_walking_xyz	0.1475	0.1254	0.1152	0.0013	0.0012	0.0012
fr3_walking_static	0.2167	0.0901	0.0154	0.0115	0.0091	0.0084
fr3_sitting_static	0.0095	0.0083	0.0074	0.0084	0.0078	0.0072

Table 3. Results of metric absolute trajectory error (ATE)

Sequences	ORB-SLAM2			Our Scheme		
	RMSE	Mean	Median	RMSE	Mean	Median
fr3_walking_xyz	0.7541	0.6484	0.5864	0.0247	0.0194	0.0172
fr3_walking_static	0.4052	0.3574	0.3028	0.0085	0.0075	0.0067
fr3_sitting_static	0.0087	0.0075	0.0062	0.0065	0.0054	0.0048

5 Conclusion

In this paper, we discussed some of the SLAM problems, firstly the accuracy problems in dynamic environments, secondly the limitations of geometric maps. We have proposed a scheme to solve these problems. We use the semantic segmentation technology of the deep learning network to capture potential dynamic objects in the image and use geometric constraints to interpret the potential moving objects' real state. SLAM's accuracy in a dynamic environment is improved by removing feature points from moving objects. Simultaneously, we extracted semantic information from the segmented image and combined it with traditional geometric maps to realize SLAM's perception of environmental information. After testing, our solution has significantly improved positioning accuracy and robustness compared to ORB-SALM2 in a dynamic environment. However, there are still some problems with our system. For example, static objects at the edge of images may be misunderstood as dynamic objects because they disappear from the next frame's edge. It may lead to a decrease in the system's accuracy, which is also the direction of our efforts in the future.

References

1. Mur-Artal, R., Tardós, J.D.: ORB-SLAM2: an open-source SLAM system for monocular, stereo and RGB-D cameras. IEEE Trans. Robot. **33**, 1255–1262 (2017)
2. Engel, J., Schöps, T., Cremers, D.: LSD-SLAM: large-scale direct monocular SLAM. In: Fleet, D., Pajdla, T., Schiele, B., Tuytelaars, T. (eds.) ECCV 2014. LNCS, vol. 8690, pp. 834–849. Springer, Cham (2014). https://doi.org/10.1007/978-3-319-10605-2_54
3. Cadena, C., Carlone, L., Carrillo, H.: Past, present, and future of simultaneous localization and mapping: toward the robust-perception age. IEEE Trans. Robot. **32**(6), 1309–1332 (2016)
4. Xia, L., Cui, J., Shen, R.: A survey of image semantics-based visual simultaneous localization and mapping: application-oriented solutions to autonomous navigation of mobile robots. Int. J. Adv. Robotic Syst. **17**(3), 172988142091918 (2020)
5. Viola: Robust real-time object detection. Int. J. Comput. Vis. **57**(2), 87 (2001)
6. Dalal, N., Triggs, B.: Histograms of oriented gradients for human detection. In: CVPR, San Diego, CA, USA, pp. 20–25 (2005)
7. Felzenszwalb, F., Girshick, R., McAllester, D.: Object detection with discriminatively trained part-based models. IEEE Trans. Patt. Anal. Mach. Intell. **32**, 9 (2009)
8. Girshick, R.: From rigid templates to grammars: object detection with structured models. University of Chicago, Chicago (2012)
9. Lin, T.Y., Doll, P., Girshick, R.: Feature pyramid networks for object detection, pp. 936–944. CVPR, Honolulu, HI, USA (2017)
10. Sunderhauf, N., Pham, T.T., Latif, Y., Milford, M., Reid, I.: Meaningful maps with object-oriented semantic mapping. In: 2017 IEEE/RSJ International Conference on Intelligent Robots and Systems (IROS) (2017)
11. Vasudevan, S., Gchter, S., Nguyen, V.: Cognitive maps for mobile robots—an object based approach. Robot. Auton. Syst. **55**, 359–371 (2017)
12. Pronobis, A., Jensfelt, P.: Large-scale semantic mapping and reasoning with heterogeneous modalities. In: IEEE International Conference on Robotics & Automation, pp. 3515–3522 (2012)
13. Mccormac, J., Handa, A., Davison, A., Leutenegger, S.: Semanticfusion: dense 3D semantic mapping with convolutional neural networks, pp. 4628–4635 (2016)

14. Whelan, T., Salas-Moreno, R.F., Glocker, B., Davison, A.J., Leutenegger, S.: Elasticfusion: real-time dense slam and light source estimation. Int. J. Robot. Res. **35**(14), 1697–1716 (2016)
15. Hhnel, D., Triebel, R., Burgard, W.: Map building with mobile robots in dynamic environments. IEEE Int. Conf. Robot. Autom. **2**, 1557–1563 (2003)
16. Alcantarilla, P., Yebes, J., Almazan, A.: On combining visual SLAM and dense scene flow to increase the robustness of localization and mapping in dynamic environments. In: IEEE International Conference, pp. 1290–1297 (2012)
17. Tan, W., Liu, H., Dong, Z., Zhang, G., Bao, H.: Robust monocular SLAM in dynamic environments. In: IEEE International Symposium on Mixed & Augmented Reality (2013)
18. Kundu, A., Krishna, K., Jawahar, C.: Realtime multibody visual slam with a smoothly moving monocular camera. In: ICCV, pp. 2080–2087 (2011)
19. Wang, C., Thorpe, C., Thrun, S., Hebert, M., Durrant-Whyte, H.: Simultaneous localization, mapping and moving object tracking. IJRR **26**(9), 889–916 (2007)
20. Kaiming, H., Georgia, G., Piotr, D., Ross, G.: Mask R-CNN, pp. 1–1. IEEE Trans. Pattern Anal. Mach. Intell., PP (2017)
21. Saputra, M.R.U., Markham, A., Trigoni, N.: Visual slam and structure from motion in dynamic environments: a survey. ACM Comput. Surv. **51**(2), 1–36 (2018)
22. Hornung, A., Wurm, K.M., Bennewitz, M., Stachniss, C., Burgard, W.: Octomap: an efficient probabilistic 3D mapping framework based on octrees. Auton. Robot. **34**(3), 189–206 (2013)
23. Sturm, J., Engelhard, N., Endres, F., Burgard, W., Cremers, D.: A benchmark for the evaluation of RGB-D SLAM systems. Intelligent Robots and Systems (IROS) (2012)

IAA Spectral Estimation in the Selective Range

Yuan Chen[1] and Longting Huang[2(✉)]

[1] University of Science and Technology Beijing, Beijing, China
[2] Wuhan University of Technology, Wuhan, China
huanglt08@whut.edu.cn

Abstract. Recently, the iterative adaptive approach (IAA) has been proposed to be a high-resolution spectrum estimator. Its main idea is reformulating the nonlinear frequency estimation problem as a linear one, with parameters being updated iteratively according to weighted least squares. Since the derivation is based on grid searching in a fixed frequency range $[0, 2\pi)$, the accurate of the IAA is limited by the number of grid. In this paper, we proposing two generalized versions of IAA, which can work well in a flexible frequency range, so that the performance can be improved in the same grid points. Simulation results are included to demonstrate the superior of our proposed methods.

Keywords: Iterative adaptive approach (IAA) · Spectral estimation · Frequency domain

1 Introduction

Spectral analysis has been an important topic in science and engineering because many real-world signals are well described by the sinusoidal model. Basically, the frequency components of the observed data can be obtained by means of either parametric or nonparametric techniques [1]. In the parametric approach, the signal is assumed to satisfy a generating model with known functional form, which allows the derivation of the optimal spectral estimators. However, the performance of these methods degrades when there is a mismatch between the assumed and actual signal models. On the other hand, no assumptions are made about the data in the nonparametric approach. Among numerous non-parametric estimators developed in the literature, a conventional representative is the periodogram based on the Fourier transform, but its resolution is fundamentally limited by the available observation length. To improve the performance, several algorithms such as principal-singular-vector utilization for modal analysis (PUMA) [2], Capon [3], multiple signal classification (MUSIC) [4] have been proposed, which can provide high-resolution in the scenario of high signal-to-noise ratio (SNR) and large number of snapshots. In [5], amplitude and phase estimator (APES) was suggested to accurately estimate the power of the source

X. Wu et al. (Eds.): QShine 2020, LNICST 381, pp. 103–112, 2021.
https://doi.org/10.1007/978-3-030-77569-8_8

signal, which can resolve sources as well. Although these methods can obtain high accurate estimation in the case of high SNR or numerous snapshots, their performance degrades when only a few snapshots are available. This is because that accurate implementation of covariance matrix in these methods requires a large number of snapshots.

In [6], a super-resolution method, namely, the iterative adaptive approach (IAA), is developed, which iteratively obtaining spectrum estimate using the weighted least squares (WLS) approach. According to the Markov estimate [7], the weighing matrix in IAA is in fact the covariance matrix of observations. To ensure the high resolution, IAA updates the covariance matrix using the estimate iteratively, and hence, accurate implementation of the IAA covariance matrix requires the estimates in full frequency ranges of $[0, 2\pi)$. That is to say, IAA can only work well in the fixed full range. However, in the case that the coarse arrival ranges of sources are known *a priori*, full range estimation of IAA is redundant and suffers from high computational cost. Although fast implementation of IAA [8]–[9] has been proposed, it is still not a good choice.

In this paper, two generalized version of IAA, referred to as selective IAA I (SIAA I) and selective IAA II (SIAA II), are devised, which can be work well in a flexible frequency range. To be employed in any selective azimuth range, two implementation criteria of the covariance matrix are suggested, where only the spectrum estimate in the interested azimuth range is required. For SIAA I, we divide the full frequency range into interested one and non-interested one. Then the covariance matrix is modified utilizing the spectrum estimate in the interested range as well as the variance estimates outside the selective range that can be obtained by the selective spectrum estimate. While in SIAA II, we redefine the mathematical model of observations as the noise-free and noisy component, where the former is described by the selective azimuth range. The covariance matrix of SIAA II is then defined by the spectrum estimate and the variance of noise term.

The rest of this paper is organized as follows. In Sect. 2, a brief review of IAA algorithm is given. In Sect. 3, the main idea of both SIAA I and SIAA II are provided. Computer simulations in Sect. 4 demonstrate the accurate of the proposed methods. Finally, conclusions are drawn in Sect. 5.

2 Review of IAA

Here we just consider a 1-D uniformly sampled sequence of N samples. IAA is based on a uniform frequency grid with K points in full range $[0, 2\pi)$ and the frequency bin: $\omega_k = 2\pi\frac{k}{K}, k = 0, 1, \ldots, K-1$. Then the frequency components can be expressed as $\mathbf{A} = [\mathbf{a}(\omega_0) \quad \mathbf{a}(\omega_1) \quad \ldots \quad \mathbf{a}(\omega_k) \quad \ldots \quad \mathbf{a}(\omega_{K-1})]$ with $\mathbf{a}(\omega_k) = [1 \quad e^{j\omega} \quad \ldots \quad e^{jn\omega} \quad \ldots \quad e^{j(N-1)\omega}]^T$ standing for the steering vector. Then the data model can be written as

$$\mathbf{y} = \mathbf{A}\mathbf{x} + \mathbf{q} \tag{1}$$

where $\mathbf{y} = [y_0 \;\; y_1 \;\; \dots \;\; y_{N-1}]^T$ is observed data and $\mathbf{x} = [x_0 \;\; x_1 \;\; \dots \;\; x_{K-1}]^T$ is the amplitude corresponding to each frequency bin with x_k denoting the complex amplitude corresponds to the kth bin, and $\mathbf{q}$ is noise term.

IAA can solve (1) by minimize the cost function:

$$||\mathbf{y} - \mathbf{a}(\omega_k)x_k||^2_{Q^{-1}} \quad k = 0, 1, \dots, K-1 \tag{2}$$

where $||\mathbf{x}||_{\mathbf{Q}^{-1}} = \mathbf{x}^H \mathbf{Q}^{-1}\mathbf{x}$ and

$$\begin{aligned} \mathbf{Q} &= E\{(\mathbf{y} - \mathbf{a}(\omega_k)x_k)(\mathbf{y} - \mathbf{a}(\omega_k)x_k)^H\} \\ &= \mathbf{R} - p_k\mathbf{a}(\omega_k)\mathbf{a}^H(\omega_k) \end{aligned} \tag{3}$$

is the weighting matrix which is also IAA interference and noise covariance matrix. Where $p_k = |x_k|^2$ stands for the signal power at frequency ω_k. Introducing a definition of IAA covariance matrix which is:

$$\mathbf{R} = E\{\mathbf{y}\mathbf{y}^H\} = \mathbf{A}\mathbf{P}\mathbf{A}^H \tag{4}$$

where $\mathbf{P}$ is a diagonal matrix with diagonal elements from power vector $\mathbf{p} = [p_0 \;\; p_1 \;\; \dots \;\; p_{K-1}]^T$. Then we can minimize (2) with respect x_k yields [10]

$$x_k^{IAA} = \frac{\mathbf{a}^H(\omega_k)\mathbf{Q}^{-1}\mathbf{y}}{\mathbf{a}^H(\omega_k)\mathbf{Q}^{-1}\mathbf{a}(\omega_k)} \quad k = 0, 1, \dots, K-1 \tag{5}$$

Using matrix inverse lemma we can see:

$$\mathbf{a}^H(\omega_k)\mathbf{Q}^{-1} = \frac{\mathbf{a}^H(\omega_k)\mathbf{R}^{-1}}{1 - p_k\mathbf{a}^H(\omega_k)\mathbf{R}^{-1}\mathbf{a}(\omega_k)} \tag{6}$$

Then (5) can be simplified as:

$$x_k^{IAA} = \frac{\mathbf{a}^H(\omega_k)\mathbf{R}^{-1}\mathbf{y}}{\mathbf{a}^H(\omega_k)\mathbf{R}^{-1}\mathbf{a}(\omega_k)} \quad k = 0, 1, \dots, K-1 \tag{7}$$

(7) avoids the computation of p_k for each bin, so we usually use (7) replace (5) as solution of IAA.

3 Proposed Method

Although IAA estimate amplitude x_k one by one, we cannot just change the range of $\mathbf{A}$ because it will have a singular problem when we compute inverse of covariance matrix. Even when we solve the singular problem, the result is wrong so the covariance should utilize the information in full range.

The spectrum of an observed signal is composed of noise-free signal and noise. We just consider the additive Gaussian white noise here. If we have already known the locate range of frequency, ripples outside of this range are just noise which is Gaussian distribution. So in order to reduce the computation cost, we can just use IAA estimate amplitude in locate range. Here we suppose locate range is $[t_1, t_2)$.

Then we can decompose the original data model (1) into:

$$\begin{aligned}\mathbf{y} &= \mathbf{s} + \mathbf{q} \\ &= \mathbf{A}(\Psi + \Phi) \\ &= \mathbf{A}_s(\Psi + \Phi_1) + \mathbf{A}_{os}\Phi_2 \\ &= \mathbf{s}_{new} + \mathbf{q}_{new} \\ &= \mathbf{A}_s\mathbf{x}_s + \mathbf{A}_{os}[\mathbf{x}_{os1} \quad \mathbf{x}_{os2}]\end{aligned} \tag{8}$$

Where $\mathbf{s}$ is the noise-free signal and $\mathbf{q}$ is noise with Ψ being the amplitude vector of noise-free signal and Φ denoting the 'amplitude' vector of noise. We divide the frequency bin into two range: $\mathbf{A}_s$ in frequency location range $[t_1, t_2)$ and $\mathbf{A}_{os}$ in range other than frequency range $[0, t_1)$ and $[t_2, 2\pi)$. And the corresponding amplitude vector can also be divided into two parts: $\Psi + \Phi_1$ and Φ_2. From the decomposition we reconstruct a new noise-free signal $\mathbf{s}_{new}$ and a new noise $\mathbf{q}_{new}$. Here the new noise is also Gaussian distribution.

Then from the new definition of signal and noise, we can divide the whole range $[0, 2\pi)$ into $[0, t_1)$, $[t_1, t_2)$and $[t_2, 2\pi)$. And corresponding amplitude vectors can also be broken into $\mathbf{x}_{os1}$,$\mathbf{x}_s$ and $\mathbf{x}_{os2}$, which means the amplitude distributed in $[0, 2\pi)$ can be expressed as $[\mathbf{x}_{os1} \quad \mathbf{x}_s \quad \mathbf{x}_{os2}]$. We also introduce a steering vector $\mathbf{a}_s(\omega) = [1 \quad e^{j\omega} \quad \dots \quad e^{j(N-1)\omega}]$ so Eq. (8) can be simplified as:

$$\begin{aligned}\mathbf{y} &= \mathbf{s}_{new} + \mathbf{q}_{new} \\ &= \mathbf{A}_s\mathbf{x}_s + \mathbf{q}_{new}\end{aligned} \tag{9}$$

where $\mathbf{A}_s = [\mathbf{a}_s(\omega_0) \quad \dots \quad \mathbf{a}_s(\omega_{K-1})]$ with $\omega_k = t_1 + (t_2 - t_1)\frac{k}{K}$ and $\mathbf{x}_s = [x_{s_0} \quad x_{s_1} \dots \quad x_{s_{K-1}}]$ is the corresponding amplitude vector.

If we have known $\mathbf{y}$ and $\mathbf{A}_s$, we can estimate $\mathbf{x}_s$ by minimizing cost function:

$$||\mathbf{y} - \mathbf{a}_s(\omega_k)x_{s_k}||^2_{\mathbf{R}^{-1}} \quad k = 0, 1, \dots, K-1 \tag{10}$$

where $\mathbf{R}$ has the same definition with original IAA. And there are two method to compute $\mathbf{R}$ which will be shown later.

And the solution of Eq. (10) can be expressed:

$$x^{IAA}_{s_k} = \frac{\mathbf{a}^H_s(\omega_k)\mathbf{R}^{-1}\mathbf{y}}{\mathbf{a}^H_s(\omega_k)\mathbf{R}^{-1}\mathbf{a}_s(\omega_k)} \quad k = 0, 1, \dots, K-1 \tag{11}$$

3.1 SEIAA I

This algorithm is based on the original definition of $\mathbf{R}$ in frequency domain. The definition of $\mathbf{R}$ request us to know the amplitude corresponding to $[0, 2\pi)$. As we just want to update the amplitude $\mathbf{x}_s$ of $[t_1, t_2)$, we should use another method to reconstruct information outside of this range. Only the power of each complex amplitude is useful in estimating $\mathbf{R}$. As ripples outside of $[t_1, t_2)$ are Gaussian distribution and the values of ripples are very small comparing with the peak of signal, we can assume all magnitudes outside $[t_1, t_2)$ are equal and this is the basic idea of SEIAA.

We can use the energy conservation law and sample average method to compute new noise variance σ^2_{new}:

$$\begin{aligned}
\mathbf{y}^H\mathbf{y} &= t^2(\mathbf{x}_s^H\mathbf{x}_s + \sigma^2_{new}) \\
\sigma^2_{new} &= \frac{1}{N}||\mathbf{y} - t\mathbf{A}_s\mathbf{x}_s||_2^2 \\
\sigma^2_{new} &= \mathbf{x}_{os1}^H\mathbf{x}_{os1} + \mathbf{x}_{os2}^H\mathbf{x}_{os2} \\
\mathbf{x}_{os1}^H\mathbf{x}_{os1} &= \mathbf{x}_{os2}^H\mathbf{x}_{os2}
\end{aligned} \tag{12}$$

where t is coefficient to balance the relationship between $\mathbf{y}$ and $\mathbf{x}$ and $\mathbf{A}_s$ is shorten for the frequency bin of $[t_1, t_2)$. The detail of Eq. (12) can be shown in appendix A.

After we estimate the value of $\mathbf{x}_{os1}$ and $\mathbf{x}_{os2}$, we can reshape the power vector $\mathbf{p}$ as $\mathbf{p} = [|\mathbf{x}_{os1}|^2 \quad |\mathbf{x}_s|^2 \quad |\mathbf{x}_{os2}|^2]$ and then we can see the step of algorithm in Table 1.

Table 1. Steps of SEIAA I

Steps of Proposed Algorithm
1. Implementing the integrated frequency matrix $\mathbf{A}$ and frequency matrix $\mathbf{A}_s$;
2. Setting a initial value of $\mathbf{x}_s, \mathbf{x}_{os1}, \mathbf{x}_{os2}$, e.g., all set to 1;
3. Reshaping vector $\mathbf{p}$ and then $\mathbf{R}$ using Eq. (4);
4. Estimating $\mathbf{x}_s$ using solution (11);
5. Computing $\mathbf{x}_{os1}, \mathbf{x}_{os2}$ using Eq. (12);
6. Repeat steps 3 - 5 until $\frac{\|\|\mathbf{x}^{t+1} - \mathbf{x}^t\|\|_2}{\|\|\mathbf{x}^{t+1}\|\|_2}$ in tolerance;

3.2 SEIAA II

Combining the definition of $\mathbf{R}$ and Eq. (9) we can see:

$$\begin{aligned}
\mathbf{R} =& E\{\mathbf{y}\mathbf{y}^H\} \\
=& E\{(\mathbf{A}_s\mathbf{x} + \mathbf{q}_{new})(\mathbf{A}_s\mathbf{x} + \mathbf{q}_{new})^H\} \\
=& \mathbf{A}_s\mathbf{P}_s\mathbf{A}_s + E\{\mathbf{q}_{new}\mathbf{q}_{new}^H\} \\
=& \mathbf{A}_s\mathbf{P}_s\mathbf{A}_s + \sigma^2_{new}\mathbf{I}_N
\end{aligned} \tag{13}$$

where $\mathbf{P}_s$ is the diagonal matrix with diagonal entries from power vector $\mathbf{p}_s = [|x_{s_0}|^2, \ldots, |x_{s_{K-1}}|^2]$ and σ^2_{new} is the variance of new noise $\mathbf{q}_{new}$. σ^2_{new} is same with the Eq. (12). So Algorithm II can be shown in Table 2.

Table 2. Steps of SEIAA II

Steps of Proposed Algorithm
1. Implementing frequency matrix $\mathbf{A}_s$;
2. Setting a initial value of $\mathbf{x}_s$, e.g., all set to 1;
3. Estimating σ_{new}^2 using Eq. (12);
4. Constructing vector $\mathbf{p}_s$ and then $\mathbf{R}$ using Eq. (13);
5. Estimating $\mathbf{x}_s$ using solution (11);
6. Repeat steps 3 - 5 until $\frac{\|\mathbf{x}^{t+1}-\mathbf{x}^t\|_2}{\|\mathbf{x}^{t+1}\|_2}$ in tolerance;

4 Simulation ResultS

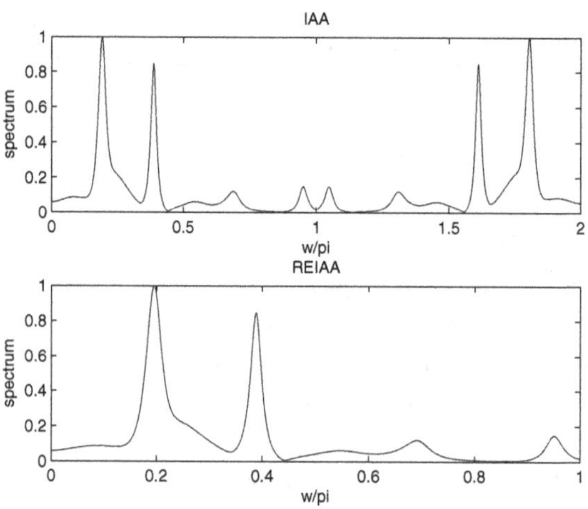

Fig. 1. Spectrum of two 1-D real tones using original IAA and REIAA

In this section, we examine the performance of the proposed algorithm. To have a directly conclusion, we set the number of grid points $K = 600$ for SEIAA and the interested range is $[0.1\pi, 0.5\pi)$, so we choose $K = 3000$ for original IAA. Meanwhile, when we estimate spectrum, SNR(signal to noise ratio) is 12 dB and te data length is $N = 40$.

Figure 1 is for a 1-D real tone. And the signal model is $y_n = \alpha_1\cos(\omega_1 n + \phi_1) + \alpha_2\cos(\omega_2 n + \phi_2) + q_n, \quad n = 0, 1, \ldots, N-1$ and parameters we set are $\alpha_1 = \alpha_2 = 1, \omega_1 = 0.2\pi, \omega_2 = 0.3\pi, \phi_1 = 0.1; \phi_2 = 0.12$. From Fig. 1 we can see, for real tones, IAA gives the spectrum which is symmetric by π. The two curves shows that the real-tone IAA (REIAA) gives the same spectrum with IAA in range $[0, \pi)$. So in order to save complexity, we use REIAA to replace IAA when estimate real tones.

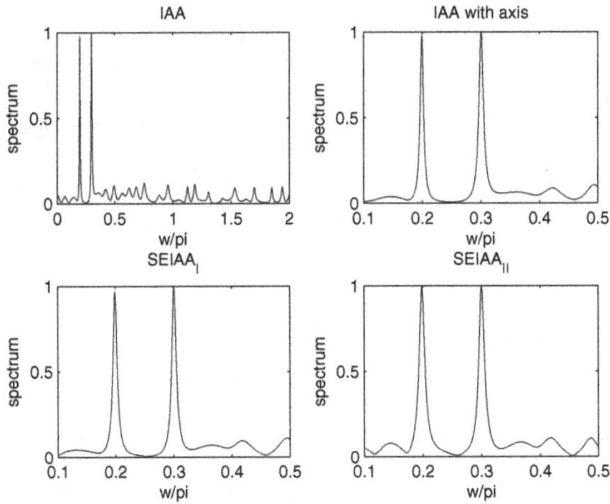

Fig. 2. Spectrum of two 1-D complex tones using original IAA and SEIAA$_I$, SEIAA$_{II}$

Table 3. Computation time of three methods

original IAA	3.9652
SEIAA I	1.2515
SEIAA II	0.2012

To simplify the problem, we use complex tones When we test SEIAA. the data model is $y_n = \alpha_1 e^{\omega_1 n+\phi_1} + \alpha_2 e^{\omega_2 n+\phi_2} + q_n, \quad , n = 0, 1, \ldots, N-1$ and the parameters we set are $\alpha_1 = \alpha_2 = 1, \phi_1 = 0.1; \phi_2 = 0.12$. Figure 2 gives four curves which is original IAA, original IAA in $[0.1\pi, 0.5\pi)$ (to compare easily),SEIAA I and SEIAA II, while the Table 3 shows the computation time of those three methods. From the curve 'IAA with axis' and 'SEIAA I', we can see the spectrum of the first two is approximately same but 'SEIAA I' runs faster than 'IAA'. For 'SEIAA II', it can resolve two peaks but it is not flat in range $[0.2\pi, 0.3\pi)$ compared with the former two methods. But 'SEIAA I' runs fastest in those three methods. So from the simulation result we can see, IAA gives the best spectrum but the highest complexity; SEIAA I gives the similar spectrum with IAA but the medium complexity in three method; SEIAA II gives a not so good spectrum but the computation time is fastest.

Figure 3 and Fig. 4 also use the same data model with Fig. 2. But here we suppose the number of grid points is same for three methods. Those two curves give us a shown that the accuracy of estimating frequency using original IAA and our methods comparing with crlb. all the test are dependent on 200 independent runs.

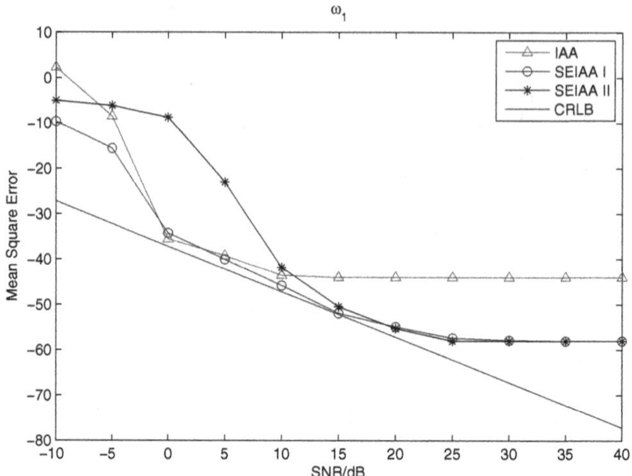

Fig. 3. Mean square error of ω_1 versus SNR

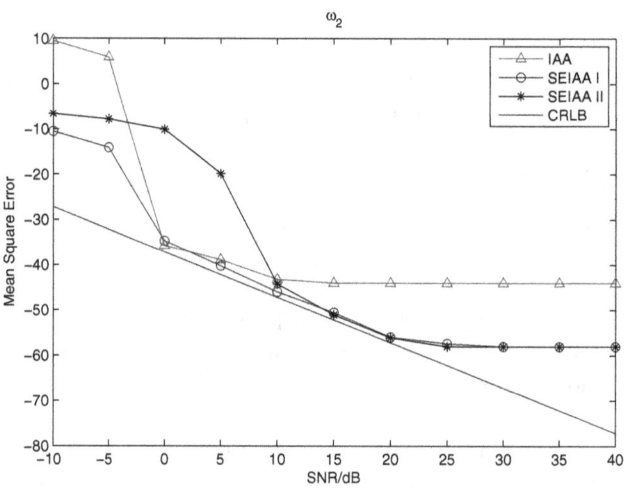

Fig. 4. Mean square error of ω_2 versus SNR

5 Conclusion

Based on the symmetric property of covariance and the symmetric property of magnitude for real tones, we simplify the IAA from $[0, 2\pi)$ to $[0, \pi)$. We also propose another two method: SEIAA I and SEIAA II to a selective range $[t_1, t_2)$ with $0 \leq t_1 < t_2 \leq 2\pi$. SEIAA I are based on assumption that all the value outside of interested range is equal and use energy conservation law to estimate those then reshape the covariance. SEIAA II reshape a new signal and new noise and use energy conservation law to compute variance of noise and then estimate covariance matrix according to its definition. And then simulation results prove the performance of SEIAA.

Funding. The work was financially supported by National Natural Science Foundation of China (Grant No. 61701021) and Fundamental Research Funds for the Central Universities (Grant No. FRF-TP-19-006A3).

Appendix: Noise Variance Relationship Between Time Domain and Frequency Domain

In IAA, If the input data is the noise $\mathbf{q}_n, \quad n = 0, \ldots, N-1$:

$$\mathbf{q} = \mathbf{A}\mathbf{x}_q \tag{14}$$

where $\mathbf{A}$ is the frequency bin in $[0, 2\pi]$ and $\mathbf{x}_q = [x_{q_0} \quad \ldots \quad x_{q_{K-1}}]$ so the variance of noise can be written:

$$\begin{aligned} \sigma^2 =& \frac{1}{N} E\{\mathbf{q}^H \mathbf{q}\} \\ =& \frac{1}{N} E\{\mathbf{x}_q^H (A^H A) \mathbf{x}_q\} \\ =& E\{|\mathbf{x}_q|^2\} \end{aligned} \tag{15}$$

where noise is i.i.d noise so $E\{\mathbf{x}_{q_k} \mathbf{x}_{q_t}\} = 0$ when $t \neq k$. From the original data model we can see and use sample average to express expectation:

$$\begin{aligned} \sigma^2 =& E\{\mathbf{q}^H \mathbf{q}\} \\ =& \frac{1}{N} ||\mathbf{y} - \mathbf{A}\mathbf{x}||_2^2 \end{aligned} \tag{16}$$

$$E\{|\mathbf{x}_q|^2\} = \sum_{k=0}^{K-1} |x_{q_k}|^2 \tag{17}$$

References

1. Stoica, P., Moses, R.: Spectral Analysis of Signals. Prentice-Hall, Hoboken (2005)
2. Qian, C., Huang, L., Sidiropoulos, N.D., So, H.C.: Enhanced PUMA for direction-of-arrival estimation and its performance analysis. IEEE Trans. Signal Process. **64**(16), 4127–4137 (2016)
3. Capon, J.: High resolution frequency-wavenumber spectrum analysis. Proc. IEEE **57**(8), 1408–1418 (1969)
4. Wang, X.P., Wang, L.Y., Li, X.M., Bi, G.A.: Nuclear norm minimization framework for DOA estimation in MIMO radar. Signal Process. **135**, 147–152 (2017)
5. Li, J., Stoica, P.: An adaptive filtering approach to spectral estimation and SAR imaging. IEEE Trans. Signal Process. **44**(6), 1469–1484 (1996)

6. Yardibi, T., Li, J., Stoica, P., Xue, M., Beggeroer, A.: Source localization and sensing: a nonparametric iterative adaptive approach based on weighted least squares. IEEE Trans. Aerosp. Electron. Syst. **46**(1), 425–443 (2010)
7. Händel, P.: Markov-based single-tone frequency estimation. IEEE Trans. Circuits Syst. Analog Digital Signal Process. **47**(10), 2857–2863 (1999)
8. Xue, M., Xu, L., Li, J.: IAA spectral estimation: Fast implementation using the Gohberg-Semencul factorization. IEEE Trans. Signal Process. **59**(7), 3251–3261 (2011)
9. Glentis, G.-O., Jakobsson, A.: Superfast approximative implementation of the IAA spectral estimate. IEEE Trans. Signal Process. **60**(1), 472–478 (2012)
10. Horn, R.A.: Matrix Analysis. Cambridge University Press, Cambridge (2012)

Emerging Applications

Robust Frequency Estimation Under Additive Mixture Noise

Yuan Chen[1], Dingfan Zhang[2], and Longting Huang[3(✉)]

[1] University of Science and Technology Beijing, Beijing, China
[2] Beihang University, Beijing, China
zdf1999@buaa.edu.cn
[3] Wuhan University of Technology, Wuhan, China
huanglt08@whut.edu.cn

Abstract. In this paper, we address the frequency estimation problem of a single sinusoid embedded in the heavy-tailed noise, where the additive Cauchy-Gaussian mixture (ACG) model is considered. Here the ACG noise model is the sum of Gaussian and Cauchy variables. With the use of Metropolis-Hastings algorithm, an accurate frequency estimator is developed in the presence of ACG noise. Simulation results demonstrate that the mean square error performance of the proposed algorithm can attain the Cramér-Rao lower bound.

Keywords: Frequency estimation · Additive Cauchy-Gaussian mixture noise · Metropolis-Hastings algorithm · Cramér-Rao lower bound

1 Introduction

Heavy-tailed noise is commonly encountered in many areas such as wireless communication and image processing [1]. Typical models of this noise type are α-stable distribution, Student's t-distribution and generalized Gaussian distribution (GGD) [2–5]. Nevertheless, in the real world applications, those noise models cannot represent all kinds of the impulsive noise types, especially for that cannot be expressed by a single known function. To remedy this shortcoming, mixture models have been proposed including the Gaussian mixture model (GMM) and the Cauchy Gaussian mixture (CGM) [6,7] model, whose probability density function (PDF) is the sum of the weighted PDFs of the component distributions, such as Gaussian or Cauchy distributions. They can describe most impulsive noise types in the real-world, except the case where the interference is caused by the channel and device. In astrophysical imaging processing [8], the observation noise is the sum of an symmetric α-stable (SαS) distribution caused by the radiation from galaxies and a Gaussian noise due to the satellite antenna. In a communication network [9], the multi-access interference can be modelled as SαS distribution while the environmental noise is a Gaussian process. Therefore,

X. Wu et al. (Eds.): QShine 2020, LNICST 381, pp. 115–125, 2021.
https://doi.org/10.1007/978-3-030-77569-8_9

a new description of the mixture impulsive noise is proposed, referring to as the sum of SαS and Gaussian random variables in time domain.

In this work, the frequency estimation in the presence of the mixture noise is considered, where the model is the sum of a variable following the SαS with $\alpha = 1$ and a Gaussian noise with unknown variance. Since when $\alpha = 1$, the SαS stable noise is in fact the Cauchy distribution [10], the mixture noise can be referred to as additive Cauchy-Gaussian (ACG) noise [11,12]. Since the closed-form PDF expression of ACG noise is not existed, traditional frequency estimators cannot provide the optimum estimation. To fix this issue, a Bayesian method, namely, Markov chain Monte Carlo (MCMC) method [13,14] is employed, which is applicable for the distribution where directly sampling from the posterior PDF is difficult. The main idea of MCMC is sampling from a simple distribution, referred to as the proposal distribution, in the way of constructing a Markov chain. For a convergent Markov chain, the state of the chain can be utilized to describe a sample of the target PDF. It is worth pointing out that the convergence here means the state of the chain becomes stationary. As a series of sampled algorithms, the commonly used MCMC algorithms are the Metropolis-Hastings (M-H) [15] and Gibbs sampling [16] algorithms. The M-H algorithm provides a general sampling framework requiring the computations of an acceptance criterion to judge whether the samples come from the correct posterior or not. On the other hand, Gibbs sampling is utilized when the full conditionals are available and are easy to sample, where the calculation of acceptance ratios is avoid In this paper, to illustrate the frequency estimation clearly, the single complex sinusoid signal is taken as an example. Since the target PDF of ACG noise is complicated and involves Voigt profile, the Gibbs sampling algorithm cannot be employed, and hence the M-H algorithm is chosen as the sampling algorithm.

The rest of this paper is organized as follows. In Sect. 2, the details of M-H algorithm is reviewed. Then the details of the proposed method is presented in Sect. 3, where the PDF of additive impulsive noise is also included. Computer simulations are provided in Sect. 4 to evaluate the accuracy of the proposed scheme. Finally, conclusions are drawn in Sect. 5.

2 Review of M-H Algorithm

Before reviewing the M-H algorithm, some basic ideas of the Markov chain is introduced [17,18]. Then we define a Markov chain $\{x_l\}$ using a sequence of dependent random variables [20]

$$x_1,\ x_2,\ \cdots,\ x_l,\ x_{l+1},\ \cdots \tag{1}$$

where the probability of x_{l+1} relies only on x_l with the conditional PDF being defined by $\mathscr{P}(x_{l+1}|x_l)$. Hence, the PDF of x_{l+1}, namely, π_{l+1}, is expressed as

$$\pi_{l+1} = \int \mathscr{P}(x_{l+1}|x_l)\pi_l dx_l. \tag{2}$$

Let $\pi^* = \lim_{l\to\infty} \pi_{l+1}$. Then we can say that a Markov chain is stationary in the case that

$$\pi^* = \mathscr{P}(\cdot|\cdot)\pi^*, \tag{3}$$

is satisfied. To ensure (3), a sufficient, but not necessary condition can be written as

$$\pi_l \mathscr{P}(x_{l+1}|x_l) = \pi_{l+1}\mathscr{P}(x_l|x_{l+1}). \tag{4}$$

Among traditional MCMC algorithms, instead of directly sampling from a target PDF $f(x)$, samples are usually obtained with the use of a Markov chain. In principle, if a proper $\mathscr{P}(\cdot|\cdot)$ is chosen, whose stationary distribution align with the target PDF $f(x)$, samples generated from the Markov chain eventually tends to be from $f(x)$ accordingly.

In the following, the steps of the MCMC method can shown in Table 1. It is worth to point out that since the initialization x_1 is arbitrarily chosen since it only influences the convergence rate of MCMC method. Furthermore, we introduce L_1 as burn-in period, after which the chain reaches convergence.

Table 1. Steps of MCMC method

(i)	Initialize x_1
(ii)	Iteratively draw x_{l+1} from the conditional PDF $\mathscr{P}(x_{l+1}|x_l)$ until $l = L$
(iii)	Throw out the first L_1 burn-in period samples

Differ with the typical MCMC methods, the main idea of the M-H algorithm [21] is drawing samples from a proposal distribution with a rejection criterion, instead of sampling from $\mathscr{P}(x_{l+1}|x_l)$ directly. In M-H method, a candidate, denoted by x^* is generated from a proposal distribution $q(x^*|x_l)$. Then the acceptance probability

$$\mathscr{A}(x_l, x^*) = \min\left\{1, \frac{q(x_l|x^*)f(x^*)}{q(x^*|x_l)f(x_l)}\right\}, \tag{5}$$

will determine whether the candidate is accepted or not. It is noted that the proposal distributions are usually chosen as uniform, Gaussian or Student's t processes, which are easier to be sampled. The details of the M-H algorithm can be seen in Table 2.

In the M-H algorithm, in order to ensure a stationary distribution, the transition kernel can be defined as

$$\mathscr{P}(x_{l+1}|x_l) = q(x_{l+1}|x_l)\mathscr{A}(x_l, x_{l+1}) + \delta(x_{l+1} - x_l)\mathscr{B}(x_l), \tag{6}$$

where $\mathscr{B}(x_l) = \int q(x^*|x_l)\left(1 - \mathscr{A}(x_l, x^*)\right) dx^*$. According to [19], the balance condition (4) is easily to be proven being hold.

Table 2. Steps of M-H algorithm

(i)	Initialize x_1
(ii)	Sample $u \sim \mathcal{U}(0,1)$
(iii)	Sample $x^* \sim q(x^*\|x^{(l)})$, $l = 1, 2, \cdots$
(iv)	Calculate $\mathscr{A}(x_l, x^*)$ according to (5)
(v)	If $u < \mathscr{A}(x_l, x^*)$
	$x_{l+1} = x^*$
	else
	$x_{l+1} = x_l$
(vi)	Repeat Steps (ii)–(iv) until $l = L$
(vii)	Discard burn-in samples

3 Proposed Method

Without loss of generality, the observed data $\mathbf{y} = [y_1\ y_2\ \cdots\ y_N]^T$ can be modeled as:

$$y_n = s_n + q_n, \tag{7}$$

where $q_n = c_n + g_n$ denotes the ACG noise which is mixed by the independent and identically distributed (IID) Cauchy noise c_n with unknown median γ and the IID zero-mean Gaussian noise g_n with known variance σ^2, and

$$s_n = A\cos(\omega n + \phi) = a_1 \cos(\omega n) + a_2 \sin(\omega n), \tag{8}$$

is the noise-free signal, with $a_1 = A\cos(\phi), a_2 = -A\sin(\phi)$ and A, ω, ϕ being amplitude, frequency and phase, respectively. Our task is to find ω from observations $\{y_n\}_{n=0}^{N-1}$.

We express the PDFs of Cauchy and Gaussian distributions as:

$$f_C(c_n|\gamma) = \frac{\gamma}{\pi(c_n^2 + \gamma^2)}, \tag{9}$$

$$f_G(g_n|\sigma^2) = \frac{1}{\sqrt{2\pi}\sigma} \exp\left(-\frac{g_n^2}{2\sigma^2}\right). \tag{10}$$

Then the PDF of the mixture noise $\mathbf{q}$, denoted as the Voigt profile [22], can be computed according to (9) and (10), which is

$$\begin{aligned} f_Q(\mathbf{q}|\gamma, \sigma^2) &= \prod_{n=1}^{N} \int_{-\infty}^{\infty} \frac{\gamma}{\pi((q_n - \tau)^2 + \gamma^2)} \frac{1}{\sqrt{2\pi}\sigma} e^{-\frac{\tau^2}{2\sigma^2}} d\tau \\ &= \prod_{n=1}^{N} \frac{\mathrm{Re}\{w_n\}}{\sigma\sqrt{2\pi}}, \end{aligned} \tag{11}$$

where $\mathrm{Re}\{x\}$ is the real part of $x \in \mathbb{C}$ and

$$w_n = \exp\left(-\left(\frac{q_n + i\gamma}{\sigma\sqrt{2}}\right)^2\right)\left(1 + \frac{2i}{\sqrt{\pi}}\int_0^{\frac{q_n+i\gamma}{\sigma\sqrt{2}}} \exp\left(t^2\right) dt\right). \quad (12)$$

Let $\boldsymbol{\theta} = [a_1,\ a_2,\ \omega,\ \gamma]^T$ being unknown parameter vector. In general, the priors of γ and σ^2 are assumed to be following conjugate inverse-gamma distribution with shape and scaling parameters being close to zero [23]. Assume that the priors for a_1, a_2, ω, γ and q_n are statistically independent, they can be expressed as

$$f(\mathbf{y}|a_1, a_2, \omega, \gamma, \sigma^2) = f_Q(\mathbf{y} - \mathbf{s}|\gamma, \sigma^2), \quad (13)$$

$$f(a_1, a_2) = \frac{1}{(2\pi\delta^2)^N} \exp\left(-\frac{a_1^2 + a_2^2}{2\delta^2} N\right), \quad (14)$$

$$f(\omega) = \frac{1}{\pi^N},\ \omega \in [0, \pi], \quad (15)$$

$$f(\gamma) = \frac{\beta_1^{\alpha_1}}{\Gamma(\alpha_1)} \exp\left(-\frac{\beta_1}{\gamma}\right), \quad (16)$$

$$f(\sigma^2) = \frac{\beta_2^{\alpha_2}}{\Gamma(\alpha_2)} \exp\left(-\frac{\beta_2}{\sigma^2}\right) \quad (17)$$

where $\beta_1 = \beta_2 = 0.01$ and $\alpha_1 = \alpha_2 = 10^{-10}$.

According to Bayes' theorem [13], we have

$$\begin{aligned} & f(a_1, a_2, \omega, \gamma, \sigma^2|\mathbf{y}) \\ &= f(\mathbf{y}|a_1, a_2, \omega, \gamma, \sigma^2) f(a_1, a_2) f(\omega) f(\gamma) \\ &= C \prod_{n=1}^{N} \int_0^{\infty} \gamma^{\alpha-1} \exp\left(-\gamma(t+\beta) - \frac{\sigma^2}{2}t^2\right) \exp\left(-\frac{a_1^2 + a_2^2}{2\delta^2}\right) \\ &\qquad \times \cos\left((y_n - a_1\cos(\omega n) - a_2\sin(\omega n))t\right) dt. \end{aligned} \quad (18)$$

where $C = \frac{\beta_1^{\alpha_1}\beta_2^{\alpha_2}}{\sqrt{2\pi}\sigma\Gamma(\alpha_1)\Gamma(\alpha_2)}$.

Although the PDF expression is given, the maximum likelihood estimator cannot be employed due to the multimodality of the likelihood function and the high computational complexity of the grid search. Furthermore, other typical robust estimators, such as the ℓ_p-norm minimizer [24], cannot provide optimum estimation for the mixture noise. Moreover, even when the conditional PDFs of each unknown parameters are known, the Gibbs sampling algorithm cannot be applied because of the complicated expression of the posterior $f(\mathbf{y}|\boldsymbol{\theta})$.

Therefore, to estimate parameters accurately, the M-H algorithm is utilized. Here we choose the multivariate Gaussian distribution as the proposal distribution, whose expression is

$$q(\mathbf{x}|\mu) = \frac{1}{2\pi\sqrt{|\boldsymbol{\Sigma}|}} \exp\left(-\frac{1}{2}(\mathbf{x}-\boldsymbol{\mu})^T\boldsymbol{\Sigma}^{-1}(\mathbf{x}-\boldsymbol{\mu})\right), \tag{19}$$

where $\mathbf{x} = [x_1\ x_2\ x_3\ x_4 x_5]^T$ with x_1, x_2, x_3, x_4, x_5 corresponding to $a_1, a_2, \omega, \gamma, \sigma^2$, respectively, $\boldsymbol{\mu} = [\mu_1\ \mu_2\ \mu_3\ \mu_4\ \mu_5]^T$ denotes the proposal mean vector and $\boldsymbol{\Sigma}$ is the 5×5 proposal covariance matrix with $|\boldsymbol{\Sigma}|$ denoting the determinant of $\boldsymbol{\Sigma}$. The $\boldsymbol{\Sigma}$ is a diagonal matrix whose main diagonal entries, namely, proposal variances. It is noted that the larger variance will cause a faster convergence but possible oscillation around the correct value. While the smaller variance values lead to slower convergence but small fluctuation.

In our problem, the mean of proposed distribution is set to all zeros. While for the proposed covariance, to choose a proper values, we utilize the idea of batch-mode. That is to say, for the k-th estimated values, denoted by $\boldsymbol{\theta}^{(k)}$, the m-th elements of the corresponding proposed covariance matrix, namely, $\boldsymbol{\Sigma}^{(k)}(m,m)$ is defined as

$$\boldsymbol{\Sigma}^{(k)}(m,m) = \sum_{l=0}^{L-1}\left(\boldsymbol{\theta}^{(k-l)} - \boldsymbol{\theta}^{(k-l-1)}\right)^2, m = 1, \cdots, 5. \tag{20}$$

To start the algorithm, the initial estimate of $\boldsymbol{\theta}$ and the burn-in period P should be determined. As it is discussed before, $\boldsymbol{\theta}^{(1)}$ can be chosen arbitrarily because the initialization of the M-H method only affects the convergence rate. Then the first P samples from the M-H algorithm are threw away, in order to avoid the initial bias. Here P refers to as the burn-in period. In the kth iteration $(k = P+1, 2, \ldots, K+P)$, $\boldsymbol{\theta}^{(k)}$ is calculated from the $\boldsymbol{\theta}^{(k-1)}$ by following the steps in Table 3.

Table 3. The proposed algorithm

1. Initialize $\boldsymbol{\theta}$ as all ones;
2. generate P samples using M-H algorithm in Table 2 with the fixed $\boldsymbol{\Sigma} = \mathbf{I}_5 \times 5$, where $\mathbf{I}_5 \times 5$ is an identity matrix;
3. For $k = P+1, \cdots K+P$
   - 3.1 compute $\boldsymbol{\Sigma}^{(k)}$ using (20);
   - 3.2 obtain $\boldsymbol{\theta}^{(k)}$ using M-H algorithm in Table 2 and $\boldsymbol{\Sigma}^{(k)}$.

Finally, the estimates $\hat{a}_1$, $\hat{a}_2$ and $\hat{\omega}$ are obtained from the mean of the samples $\boldsymbol{\theta}^{(k)}(1)$, $\boldsymbol{\theta}^{(k)}(2)$ and $\boldsymbol{\theta}^{(k)}(1)$ $(k = P+1, \cdots, K+P)$, respectively. Utilize the definition of a_1 and a_2, we can obtain the estimates of amplitudes and phase, denoted by $\hat{A}$ and $\hat{\phi}$,

$$\hat{A} = \sqrt{\hat{a}_1^2 + \hat{a}_2^2}, \tag{21}$$

$$\hat{\phi} = \mathrm{atan}\left(\frac{\hat{a}_2}{\hat{a}_1}\right), \tag{22}$$

where atan$(\cdot)$ is arctangent function.

4 Cramér-Rao Lower Bound (CRLB)

Let $\boldsymbol{\psi} = [A\ \omega\ \phi\ \gamma\ \sigma^2]^T$. The Cramér-Rao lower bound (CRLB) of unknown parameters can be obtained from the diagonal elements of the inverse of the Fisher information matrix $\mathbf{I}$ and the (m, k) entry $(m, k = 1, \cdots, 5)$ of $\mathbf{I}$ has the form of

$$\mathbf{I}(m,k) = -E\left\{\frac{\partial \log f(\mathbf{y}|\boldsymbol{\psi})}{\partial \boldsymbol{\psi}}\left(\frac{\partial \log f(\mathbf{y}|\boldsymbol{\psi})}{\partial \boldsymbol{\psi}}\right)^T\right\}, \tag{23}$$

where

$$\frac{\partial \log f(\mathbf{y}|\boldsymbol{\psi})}{\partial \boldsymbol{\psi}} = \begin{bmatrix} \frac{1}{\sigma^2}\frac{\cos(\omega n+\phi)\mathrm{Re}\{(y_n - A\cos(\omega n+\phi)+i\gamma)w_n\}}{\mathrm{Re}\{w_n\}} \\ \frac{1}{\sigma^2}\frac{An\sin(\omega n+\phi)\mathrm{Re}\{(y_n - A\cos(\omega n+\phi)+i\gamma)w_n\}}{\mathrm{Re}\{w_n\}} \\ \frac{1}{\sigma^2}\frac{A\sin(\omega n+\phi)\mathrm{Re}\{(y_n - A\cos(\omega n+\phi)+i\gamma)w_n\}}{\mathrm{Re}\{w_n\}} \\ -\frac{\frac{1}{\sigma^2}\mathrm{Re}\{i(y_n - An - B + i\gamma)w_n\} + \frac{2}{\sqrt{2\pi\sigma^2}}}{\mathrm{Re}\{w_n\}} \\ \frac{\frac{1}{\sigma^2}\mathrm{Re}\{(y_n - An - B + i\gamma)^2 w_n\} + \frac{\gamma}{\sqrt{2\pi\sigma^2}\sigma^2}}{\mathrm{Re}\{w_n\}} - \frac{1}{2\sigma^2} \end{bmatrix}. \tag{24}$$

Due to the complicated integration in (23) and (23), the closed-form of CRLB is difficult to be obtained. As a result, we calculate (23) using a numerical method which is an approximation:

$$\hat{\mathbf{I}}(k,l) \approx \frac{1}{M}\sum_{m=1}^{M}\sum_{n=1}^{N}\frac{\partial \log f(y_n^m|\boldsymbol{\psi})}{\partial \boldsymbol{\psi}}\left(\frac{\partial \log f(y_n^m|\boldsymbol{\psi})}{\partial \boldsymbol{\psi}}\right)^T, \tag{25}$$

where M is the number of independent Monte Carlo runs and y_n^m denotes the observed signal at the mth trial. Apparently, a sufficiently large value of M will make (25) approaching (23).

5 Simulation Results

To assess the performance of the proposed method, computer simulations have been conducted. The mean square error (MSE), referred to as $E\{(\hat{\omega} - \omega)^2\}$, is utilized as the performance measure. Then the signal s_n is generated according to (8) with $A = 13.84$, $\omega = 1.79$ and $\phi = -0.33$. In the M-H algorithm, the initial estimate is set to as $[1\ 1\ 1\ 1\ 1]^T$, while the number of iterations is $K = 8000$. It is demonstrated that burn-in period P can be chosen as 2000 in this setting. Here comparison with the ℓ_1-norm estimator is provide due to its robust and suboptimal for the Cauchy noise, while the CRLB are also included as a benchmark. It is noted that the ℓ_1-norm minimizer is solved by the least absolute deviation [25]. All results are based on 100 independent runs with a data length of $N = 100$.

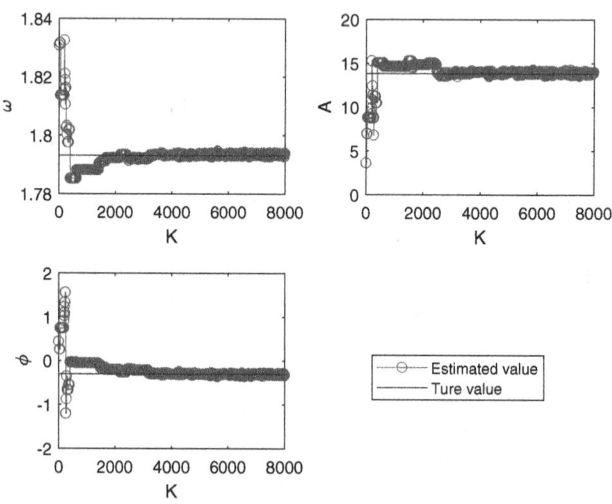

Fig. 1. Estimates of unknown parameters versus iteration number k

First of all, the convergence rate of the unknown parameters is investigated. Meanwhile, the burn-in period P can be determined, accordingly. In this test, the density parameters are set to $\gamma = 0.05$ and $\sigma^2 = 0.5$. Figures 1 and 2 indicates the estimates of all unknown parameters in different iteration number k, which are ω, A, ϕ, γ and σ^2. It can be seen in these figures that after the first 2000 samples, the sampled data approaches the true values of unknown parameters. Therefore, the burn-in period P can be chosen as 2000 in this parameter setting.

In the following, the MSE performance of the proposed estimator is considered. In the proposed method, the Σ is identical to the previous test and γ is scaled to produce different noise conditions. According to the study in the previous test, we throw away first 2000 samples to ensure the stable of the method. It is shown in Fig. 3 that the MSEs of the proposed attain the CRLB in the case of $\gamma \in [-20, 10]$ dB. Furthermore, the proposed method is superior to the ℓ_1-norm estimator.

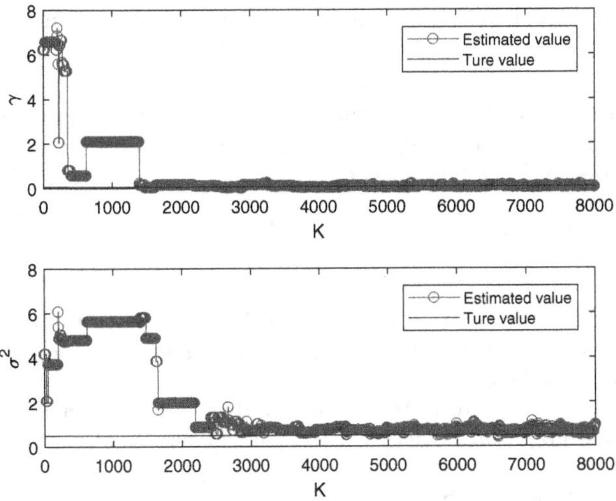

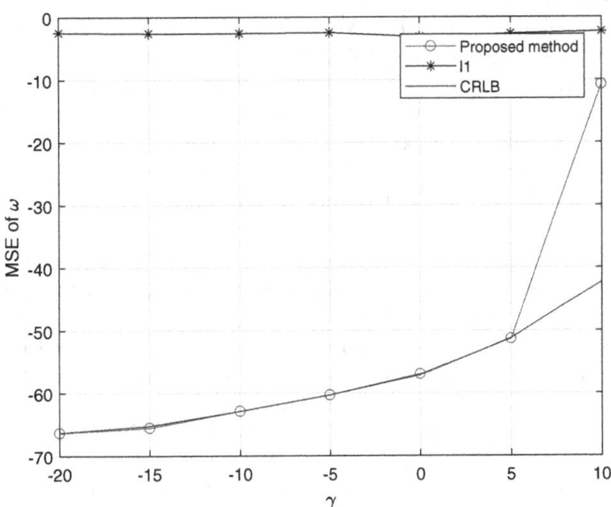

Fig. 2. Estimates of density parameters versus iteration number k

Fig. 3. Mean square error of ω versus γ

6 Conclusion

In this paper, the frequency estimation of a single sinusoid signal under additive Cauchy-Gaussian mixtures is studied. An accurate frequency estimator is developed, by employing a numerical Bayesian method, namely, Metropolis-Hastings algorithm. Simulation results show that the proposed method can provide an unbiased estimates with a long-term samples. Furthermore, the MSE of the proposed estimator can attain the CRLB after throw away the burn-in period samples. Our method can be extended to the family of signals with more complicated models.

Funding. The work was financially supported by National Natural Science Foundation of China (Grant No. 61701021) and Fundamental Research Funds for the Central Universities (Grant No. FRF-TP-19-006A3).

References

1. Zoubir, A.M., Koivunen, V., Chakhchoukh, Y., Muma, M.: Robust estimation in signal processing: a tutorial-style treatment of fundamental concepts. IEEE Signal Process. Mag. **29**(4), 61–80 (2012)
2. Zhang, T., Wiesel, A., Greco, M.S.: Multivariate generalized Gaussian distribution: convexity and graphical models. IEEE Trans. Signal Process. **61**(16), 4141–4148 (2013)
3. Nikias, C.L., Shao, M.: Signal Processing with Alpha-Stable Distribution and Applications. Wiley, New York (1995)
4. Aas, K., Haff, I.H.: The generalized hyperbolic skew student's t-distribution. J. Financ. Economet. **4**(2), 275–309 (2006)
5. Shynk, J.J.: Probability, Random Variables, and Random Processes: Theory and Signal Processing Applications. Wiley, Hoboken (2013)
6. Reynolds, D.A.: Gaussian mixture models. In: Encyclopedia of Biometrics, pp. 659–663 (2009)
7. Swami, A.: Non-Gaussian mixture models for detection and estimation in heavy tailed noise. In: Proceedings of IEEE International Conference on Acoustics, Speech, and Signal Processing, Istanbul, Turkey, June 2000, vol. 6, pp. 3802–3805 (2000)
8. Herranz, D., Kuruoglu, E.E., Toffolatti, L.: An α-stable approach to the study of the P(D) distribution of unresolved point sources in CMB sky maps. Astron. Astrophys. **424**(3), 1081–1096 (2004)
9. Ilow, J., Hatzinakos, D., Venetsanopoulos, A.N.: Performance of FH SS radio networks with interference modeled as a mixture of Gaussian and alpha-stable noise. IEEE Trans. Commun. **46**(4), 509–520 (1998)
10. Kahrari, F., Rezaei, M., Yousefzadeh, F.: On the multivariate skew-normal-Cauchy distribution. Stat. Probab. Lett. **117**, 80–88 (2016)
11. Chen, Y., Kuruoglu, E.E., So, H.C.: Estimation under additive Cauchy-Gaussian noise using Markov chain Monte Carlo. In: Proceedings of IEEE Workshop on Statistical Signal Processing (SSP 2014), Gold Coast, Australia, June–July 2014, pp. 356–359 (2014)
12. Chen, Y., Kuruoglu, E.E., So, H.C., Huang, L.-T., Wang, W.-Q.: Density parameter estimation for additive Cauchy-Gaussian mixture. In: Proceedings of IEEE Workshop on Statistical Signal Processing (SSP 2014), Gold Coast, Australia, June–July 2014, pp. 205–208 (2014)
13. Bretthorst, G.L.: Bayesian Spectrum Analysis and Parameter Estimation. Springer, New York (1989). https://doi.org/10.1007/978-1-4684-9399-3
14. Bishop, C.: Pattern Recognition and Machine Learning. Springer, Boston (2006). https://doi.org/10.1007/978-1-4615-7566-5
15. Andrieu, C., Freitas, N.D., Doucet, A., Jordan, M.I.: An introduction to MCMC for machine learning. Mach. Learn. **50**, 5–43 (2003)

16. Casella, G., George, E.: Explaining the Gibbs sampler. Am. Stat. **46**(3), 167–174 (1992)
17. Spall, J.C.: Estimation via Markov chain Monte Carlo. IEEE Control. Syst. **23**(2), 34–45 (2003)
18. Grinstead, C.M., Snell, J.L.: Introduction to Probability. American Mathematical Society, Providence (2012)
19. Cemgil, A.T.: Academic Press Library in Signal Processing, vol. 1. Signal Processing Theory and Machine Learning. Elsevier Science & Technology (2013)
20. Robert, C., Casella, G.: Introducing Monte Carlo Methods with R. Springer, New York (2009). https://doi.org/10.1007/978-1-4419-1576-4
21. Chib, S., Greenberg, E.: Understanding the Metropolis-Hastings algorithm. Am. Stat. **49**(4), 327–335 (1995)
22. Olver, F.W.J., Lozier, D.M., Boisvert, R.F.: NIST Handbook of Mathematical Functions, pp. 167–168. Cambridge University Press, Cambridge (2010)
23. Kohn, R., Smith, M., Chan, D.: Nonparametric regression using linear combinations of basis functions. Stat. Comput. **11**(4), 313–322 (2001)
24. Li, T.H.: A nonlinear method for robust spectral analysis. IEEE Trans. Signal Process. **58**(5), 2466–2474 (2010)
25. Li, Y., Arce, G.: A maximum likelihood approach to least absolute deviation regression. EURASIP J. Appl. Signal Process. **12**, 1762–1769 (2004)

Data Augmentation for Cardiac Magnetic Resonance Image Using Evolutionary GAN

Ying Fu[1,2](✉), Minxue Gong[1], Guang Yang[1], and Jiliu Zhou[1,2]

[1] School of Computer Science, Chengdu University of Information and Technology, Chengdu 610225, China
fuying@cuit.edu.cn

[2] Image and Spatial Information 2011 Collaborative Innovation Center of Sichuan Province, Chengdu 610225, China

Abstract. Generative adversarial networks (GAN) could synthesize semantically meaningful data from standard signal distribution, which make it have considerable potential to alleviate data scarcity. In this paper, based on Evolutionary GAN, cardiac magnetic resonance images enhancement method is proposed to solve over-fitting problem caused by training convolution network with small dataset. The most optimal generator which consider the quality and diversity of generated images simultaneously from many generator mutations is chosen. Meanwhile, to expand the whole training set distribution, we combine the linear interpolation of eigenvectors to synthesize new training samples and synthesize related linear interpolation labels, which can make the discrete sample space become continuous to improve the smoothness between domains. In this paper, the effectiveness of this method is verified by classification experiments, and the influence of the proportion of synthesized samples on the classification results of cardiac magnetic resonance images is explored.

Keywords: Evolutionary GAN · Cardiac magnetic resonance · Data augmentation · Linear interpolation

1 Introduction

Cardiac magnetic resonance imaging (MRI) is known as the gold standard for assessing cardiac function. Conventional cardiac MRI scanning technology has been relatively mature and has played a vital role in disease diagnosis. At present, many cardiac magnetic resonance image-assisted diagnosis tasks based on deep learning [1] have achieved good results, but cardiac magnetic resonance images not only require expensive medical equipment to obtain, but also require experienced radiologists to carry out a large number of manual data annotation, which is undoubtedly extremely time-consuming and labor-consuming. In addition, the privacy of patients in the field of medical images has always been very sensitive, so it costs a lot to obtain a large number of data sets that are balanced between positive and negative samples.

X. Wu et al. (Eds.): QShine 2020, LNICST 381, pp. 126–141, 2021.
https://doi.org/10.1007/978-3-030-77569-8_10

A great challenge in the field of medical imaging based on deep learning is how to deal with small-scale data sets and limited number of labeled data. Especially when using complex deep learning model, the data set is not sufficient or the data set sample is unbalanced, which will make the deep convolution neural network with huge parameters appear over fitting [2]. In the field of computer vision, scholars have proposed many effective methods for over fitting, such as batch regularization [3], dropout [4], early stopping method [5], weight sharing [6], weight attenuation [7], etc. The above method is to adjust the network structure. In addition, data enhancement [8] is an effective method to operate on the data itself, which alleviates the phenomenon of over fitting in image analysis and classification to a certain extent. The classical data enhancement techniques mainly include affine transformation methods such as translation, rotation, scaling, flipping and shearing [9, 10], and the original samples and new samples are mixed as training sets and input into convolutional neural network. Adjusting the color space of samples is also a data enhancement method. Wang et al. [11] used the method of changing the brightness value to expand the sample size. Although these methods have improved, only the operation on the original samples does not produce new features. The diversity of the original samples has not been substantially improved [12], and the promotion effect is weak when processing small-scale data.

Generative Adversarial Network (GAN) [13] is a generative model proposed by Ian Goodfellow and others. It consists of a generator G and a discriminator D. The generator G uses noise z sampled from uniform distribution or normal distribution as input to synthesize image $G(z)$. The discriminator D attempts to judge the synthetic image $G(z)$ as false as much as possible, and judges the real image x as true, and adjusts the parameters of each model through successive confrontation training. Finally, the generator obtains the distribution model of real samples and obtains the generation performance close to the real image. The specific structure of GAN is shown in Fig. 1.

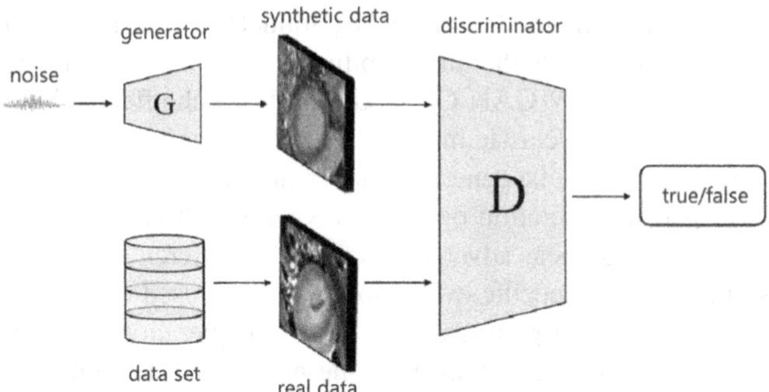

Fig. 1. The structure of GAN

The entire training process of GAN is to find the balance between the generating network and the discriminating network, which makes the discriminator unable to judge whether the samples generated by the generator are real or generated, so that the generating network can achieve the optimal performance. This process can be expressed as formula (1):

$$min_G max_D E_{xP_{data}}\left[logD(x)\right] + E_{zP_z}\left[log(1 - D(G(z)))\right] \tag{1}$$

The generative adversarial network generates new samples by fitting the original sample distribution. The new samples are generated from the distribution learned by the generative model, which makes it have new features that are different from the original samples. This feature makes it possible to use the samples generated by the generating network as new training samples to achieve data expansion. Although GAN has achieved good results in many computer vision fields, it has many problems in practical applications. On the one hand, GAN is very difficult to train. Once the data distribution and the distribution fitted by the generating network do not substantially overlap at the beginning of training, the gradient of the generating network can easily point to a random direction, resulting in the problem of gradient disappearance [14]. On the other hand, in order to make the discriminator give high scores, the generator will try to generate a relatively safe but lack of diversity of single samples, which will lead to the problem of pattern collapse [15].

In order to alleviate the gradient disappearance and model collapse, a large number of GAN variant models have been proposed. The more representative ones are: DCGAN [16] which combines convolutional neural network with GAN and Conditional GAN [17] which adds precondition control generator to the input data. There are also LSGAN [18] and WGAN [19], which have made great improvements to the loss function. Among them, WGAN uses Wasserstein distance to measure the distribution distance, which makes GAN more stable in training to a large extent. However, Ishaan et al. found that WGAN uses a forced phase method to make the parameters of the network mostly focus on $-0.01, 0.01$, which will waste the fitting ability of the convolutional neural network. Therefore, they proposed the WGAN-GP model [20], which effectively alleviated this problem, so it became a more classic model. The Evolutionary GAN [21] proposed by Zhang et al. is a variant model of a generative adversarial network based on evolutionary algorithms. It will perform mutation operations when the discriminator stops training to generate multiple generators as adversarial targets. In different environments (that is, the current discriminator), a specific evaluation method is used to evaluate the quality and diversity of the generated pictures. This series of operations can reserve one or more generators with strong performance for the next round of training. This method of overcoming the limitations of single adversarial target has been proven to be able to keep the best offspring all the time, effectively alleviate the problem of mode collapse and improve the quality of the generator.

Recently, many scholars use GAN to enhance training data samples. The article [22] uses GAN to enhance the data of human faces and handwritten fonts. Ibrahim et al. [23] used the improvement of PGGAN to expand the data set of skin injury and improved the classification accuracy. Maayan et al. [24] used DCGAN and ACGAN to expand the data of liver medical images, and proved that the classification effect of DCGAN in this data set is improved more. Compared with affine transformation, GAN can be used to generate images with new features by learning the real distribution.

Considering that Evolutionary GAN can improve the diversity and quality of generated samples, this paper uses Evolutionary GAN to enhance cardiac magnetic resonance image data. The main contributions of this paper are as follows:

1) A cardiac magnetic resonance medical image data enhancement method based on Evolutionary GAN is proposed, which generates high-quality and diverse samples to expand the training set, and finally improves the various indicators of the classification results;
2) Combining the linear interpolation of feature vectors in Evolutionary GAN to synthesize new training samples and generate related linear interpolation labels, which not only expands the distribution of the entire training set, but also makes the discrete sample space continuous and improves the smoothness between fields, so that the model can be better trained.
3) We use various indicators of downstream classification tasks to optimize the model and experimental details.

2 Evolutionary GAN

The training process of Evolutionary GAN can be divided into three stages: the first stage is mutation, that is, the parent generator is mutated into multiple offspring generators; the second stage is evaluation, that is, the adaptive score of each offspring generator of the current discriminator is calculated through the fitness function; the third stage is selection, that is, the offspring generator with the highest adaptive score is selected by sorting. The basic structure of the Evolutionary GAN is shown in Fig. 2:

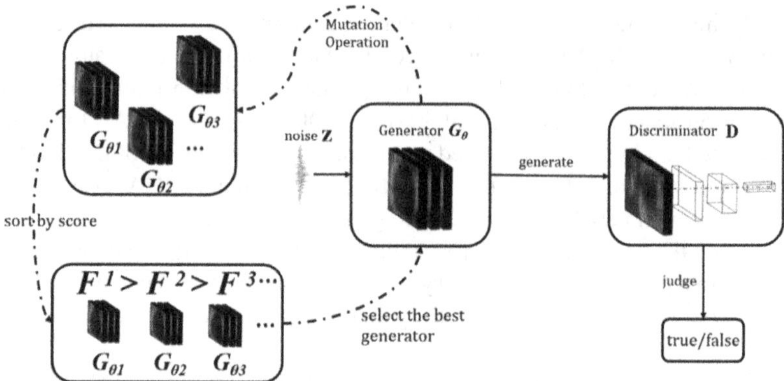

Fig. 2. The structure of Evolutionary GAN

2.1 Mutation

Evolutionary GAN uses different mutation methods to obtain offspring generators based on parent generators. These mutation operators are actually different training targets. The purpose is to reduce the distance between the generated distribution and the real data distribution through different angles. It should be noted that the best discriminator D* in formula (2) should be trained before each mutation operation.

$$D(x) = \frac{P_{data}(x)}{P_{data}(x) + P_g(x)} \tag{2}$$

Zhang et al. proposed three mutation methods:

1) Maximum and minimum value mutation: the mutation has little change to the original objective function, which can provide effective gradient and alleviate the phenomenon of gradient disappearance. It can be written as formula (3):

$$M_G^{minimax} = \frac{1}{2} E_{zP_z}[log(1 - D(G(z)))] \tag{3}$$

2) Heuristic mutation: heuristic mutation aims to maximize the log probability of the discriminator's error. When the discriminator judges the generated sample as false, the heuristic mutation will not be saturated, and can still provide effective gradient so that the generator can be continuously trained. It can be written as formula (4):

$$M_G^{heuristic} = \frac{1}{2} E_{zP_z} \tag{4}$$

3) Least squares mutation: inspired by ISGAN, least squares mutation can also avoid vanishing gradient. At the same time, compared with heuristic mutation, the least square mutation does not generate false samples at a very high cost, but it does not use very low cost to avoid punishment, which can avoid model collapse to a certain extent. It can be written as formula (5):

$$M_G^{least-square} = E_{zP_z}\left[D(G(z) - 1)^2\right] \tag{5}$$

2.2 Fitness Function

Evolutionary GAN uses the fitness function to evaluate the generator's performance and quantifies it to the corresponding adaptability score, which can be written as formula (6):

$$F = F_q + \gamma F_d \tag{6}$$

F_q is used to measure the quality of the generated samples, that is, whether the offspring generator can fool the discriminator, which can be written as formula (7):

$$F_q = E_z[D(G(z))] \tag{7}$$

F_q measures the diversity of the generated samples. It measures the gradient generated when the parameters of the discriminator are updated again according to the offspring generator. If the samples generated by the offspring generator are relatively concentrated (lack of diversity), it is easier to cause large gradient fluctuations when updating the discriminator parameters, which can be written as formula (8):

$$F_d = -log\left\| \nabla_D - E_x\left[logD(x)\right] - E_z\left[log(1 - D(G(z)))\right]\right\| \tag{8}$$

$\gamma(\geq 0)$ is a hyperparameter used to adjust the quality of samples generated and the weight of diversity, which can be adjusted freely in the experiment.

3 Method

In this paper, we design a data enhancement model of cardiac magnetic resonance medical image based on Evolutionary GAN, which can generate high-quality and diverse samples to expand the training set. The linear interpolation of related labels is generated by combining the linear interpolation of feature vector, which expands the distribution of training set and makes the discrete sample space continuous, so that the model can be trained better. The specific network structure is shown in Fig. 3:

3.1 DAE GAN

Using GAN for data enhancement requires high quality and diversity of samples. Evolutionary GAN can be stably trained and can generate high-quality and diverse samples, so it is very suitable for data enhancement. By adjusting the parameters in the fitness function, you can choose to focus on diversity or quality according to your needs, which can make the data enhancement process more operative. This article improves the Evolutionary GAN and names the improved model *Data Augmentation Evolutionary GAN* (DAE GAN).

There is no difference between the input and output of Evolutionary GAN and Vanilla GAN, except that after the discriminator parameters are fixed, multiple offspring generators are mutated based on the parent generator for training. After the evaluation of the fitness function, the optimal one or more generators are selected as the parent generator in the next discriminator environment.

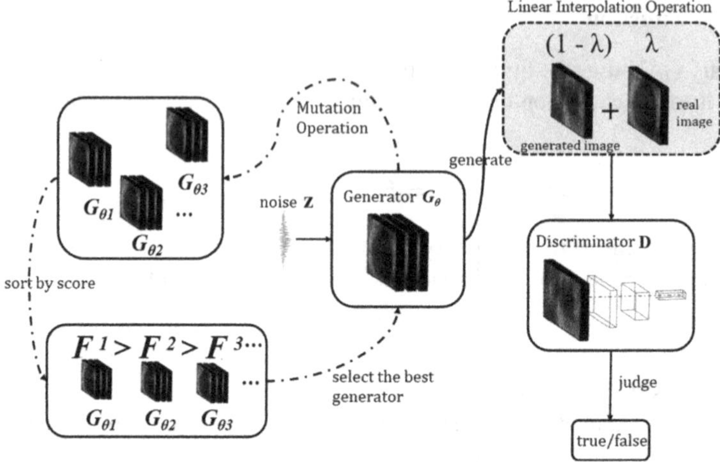

Fig. 3. The proposed network

Although Evolutionary GAN greatly improves the diversity of generated samples, a certain number of training samples are required if you want to fully train the GAN model. In the case of too few training samples, the generator and discriminator are prone to reach the equilibrium point prematurely, which will also cause the phenomenon of model collapse in the generated data. In order to alleviate this problem, this paper uses the traditional affine transformation data enhancement method before training GAN, and expands the data through horizontal flip, vertical inversion, translation, rotation and other operations. Due to the security of medical images, the original data is not added with noise, crop and other operations, and the texture and edge features of the original data are retained as far as possible. But traditional data enhancement only makes small changes to the original data, and does not generate new features, and the samples are also discrete. Thus, this article introduces the linear interpolation.

Zhang et al. proposed a data-independent data enhancement method in the article [25]. This method constructs virtual training samples from original samples, combines linear interpolation of feature vectors to synthesize new training samples and generates related linear interpolations Labels to expand the distribution of the entire training set. The specific formula is as formula (9):

$$\begin{cases} \tilde{x} = \lambda x_i + \left(1 - \lambda x_j\right) \\ \tilde{y} = \lambda y_i + \left(1 - \lambda y_j\right) \end{cases} \tag{9}$$

x_i, x_j is the original input vector, y_i, y_j is the label code, $(x_i, x_j), (y_i, y_j)$ are two samples randomly sampled from the original sample, $\lambda \in Beta[\alpha, \alpha]$ is the weight vector, and $\alpha \in (0, +\infty)$ is the hyperparameter that controls the interpolation strength between the feature and the target vector. The linear interpolation method enables the model to behave linearly when processing the area between the original sample and the sample, so as to reduce the inadaptability of predicting test samples other than the training sample, and enhance the generalization ability. At the same time, the discrete sample space can be continuous and the smoothness between fields can be improved.

When the generator parameters are fixed, the original input of the Evolutionary GAN discriminator is two samples: one is the generated sample, the discriminator tries to minimize the distance between the predicted label of the sample and "0"; the other is the real sample, the discriminator is minimized as much as possible the distance between the predicted label of this sample and "1". The discriminator loss function of the original Evolutionary GAN is as formula (10):

$$L_D = L_{Real} + L_{Fake} \tag{10}$$

The discriminator loss function is expanded into formula (11) as follow:

$$\underset{x,z}{E}\, L(D(x), 1) + \underset{x,z}{E}\, L(D(G(z)), 0) \tag{11}$$

This paper uses linear interpolation operation in Evolutionary GAN to modify the discriminator input from the original two pictures to one picture, and the discriminator task is changed to minimize the distance between the predicted label of the fusion sample and "λ". The loss function of discriminator is modified as formula (12):

$$\underset{x,z,\lambda}{E}\, L(D(\lambda x + (1-\lambda)G(z)), \lambda) \tag{12}$$

3.2 Algorithm

Usually GAN will use the noise z that obeys the multivariate uniform distribution or multivariate normal distribution as the input of the model. Matan et al. [26] believe that multiple Gaussian distributions can better adapt to the inherent multi-modality of the real training data distribution, so a multi-modal distribution is used as input in GAN and it is proved that this method can improve the quality and variety of generated images. The algorithm combined with Gaussian mixture model in this paper is as follows:

Input: **N, K, D, BS, M** and β_1, β_2. The total number of iterations **N**, the number of Gaussian distributions **K**, latent spatial dimension **D**, batch size **BS**, the number of mutation operations **M**, hyper-parameters β_1, β_2.

Output: the updated weight of discriminator and generator parameters.

```
Step1. Initialize the discriminator parameter ω_0 and generator parameter θ_0
Step2. Gaussian distribution operation
cycle 1 start:  for k = 1 : K
   Step2.1. Sample the initial mean of the Gaussian distribution K
   Step2.2. Initialize the covariance matrix of Gaussian distribution K
cycle 1 end
cycle 2 start:  for n = 1 : N
   Step3. Train discriminator
   cycle 3 start:  for j = 1 : BS
      Step3.1. Sample a real image
      Step3.2. Sample Gaussian index
      Step3.3. Sample noise from the k-th Gaussian distribution
      Step3.4. Input the noise into the generator to synthesize a sample
      Step3.5. Perform linear interpolation on real samples and synthetic samples,
   interpolate new samples and labels
      Step3.6. Calculate the loss of discriminator
   cycle 3 end
   Step3.7. Calculate the average loss of discriminator
   Step3.8. Update the discriminator parameter
   Step4. Train generator
   cycle 4 start:  for m = 1 : M
      cycle 5 start:  for j = 1 : BS
         Step4.1. Sample Gaussian index
         Step4.2. Sample noise from the k-th Gaussian distribution
         Step4.3. Input the noise into the generator to synthesize a sample
         Step4.5. Perform a mutation operation on the parent generator and use the
      offspring generator to generate samples
         Step4.6. Calculate the loss of the offspring generator
      cycle 5 end
      Step4.7. Calculate the average loss of the offspring generator
      Step4.8. Update the parameter of the offspring generator
      Step4.9. Calculate the adaptive score of the offspring generator
   cycle 4 end
   Step4.10. Sort the offspring generators in descending order according to the
adaptive score
   Step4.11. Leave the offspring generator with the highest adaptive score in the
current environment and use it as the parent generator for the next iteration
cycle 2 end
```

4 Results Analysis

4.1 Data Set and Preprocessing

The cardiovascular magnetic resonance data in this experiment comes from a partner hospital. All samples are 2D short-axis primary T1 mapped images. The spatial distance of these cardiac magnetic resonance images ranges from $1.172 \times 1.172 \times 1.0\ \text{mm}^3$ to $1.406 \times 1.406 \times 1.0\ \text{mm}^3$, and the original pixel size was $256 \times 218 \times 1$. The benign and malignant labeling and segmentation areas of the image are manually labelled and drawn by senior experts. The original image data is in the ".mha" format. The original image data was in ". MHA" format. After preprocessing, such as resampling, selection of regions of interest, normalization and final selection of interest, a total of 298 images were obtained, including 221 cardiomyopathy images and 77 non-diseased images. The image size after preprocessing was $80 \times 80 \times 1$. The pretreated cardiac magnetic resonance image is shown in Fig. 4.

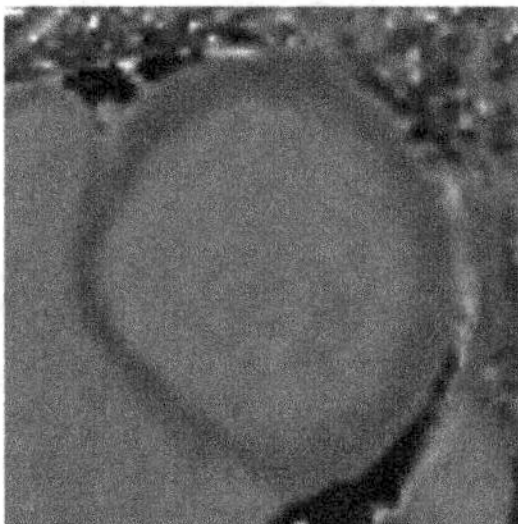

Fig. 4. Cardiac magnetic resonance image

In order to ensure the consistency of training data, all samples are normalized in this experiment. Before training GAN, this experiment performed affine transformation data enhancement on the training set, including: horizontal flip, vertical flip, 0°–20° random amplification and rotation, 90°, 180°, 270° rotation, 0–2% random amplification and translation of vertical and horizontal axes, small and specific amplitude amplification and rotation, and amplification translation, so as to make the data not lose the original image information. After the training set is enhanced once, it is divided into two types of operations: one is to put it into the classifier for training directly, and then use the test set to get the classification results; the other is to put it into different GAN for training, and finally generate new samples to train the classifier again.

4.2 Training DAE GAN

The original evolutionary GAN uses the structure of DCGAN. In this paper, we consider that the residual structure [27] can alleviate the gradient vanishing problem and accelerate the convergence speed of the model, so as to train the high-performance generator more quickly in the same training time. The residual structure as shown in Fig. 5 is used in the generator and discriminator in this article.

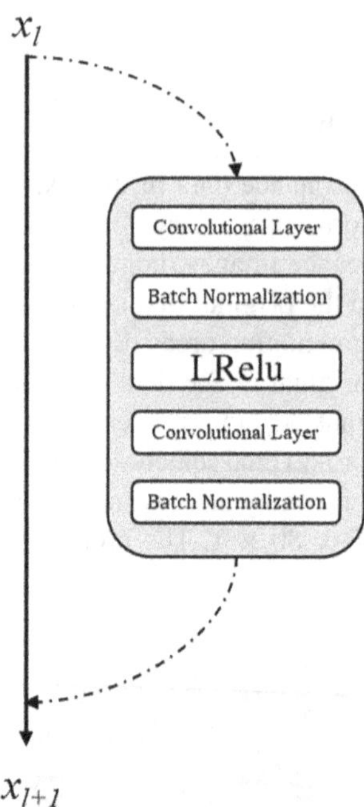

Fig. 5. Residual block structure

Combined with the self-attention module [28], the detailed structure of the generator and discriminator and the output size of each layer are shown in Table 1.

DAE GAN experimental environment: Ubuntu 16.04.1 TLS, Tensorflow 1.14.0, two Nvidia Tesla M40 GPU with 12 GB video memory used to train the generative models of diseased and non-diseased samples respectively. The maximum storage capacity of the model is set to 4, taking into account the space occupation and preventing accidental interruption.

4.3 The Generation Results of DAE GAN

This experiment uses 5-fold cross validation to dynamically divide the heart magnetic resonance image into a training set and a test set at a ratio of 0.8:0.2. Training DAE Gan only uses the training set. Due to the uncertainty of deep convolution model in the training process, each model was trained several times ($\geq$5), and the specific effect of data enhancement method was verified by average classification results.

After normalization and affine transformation, the training set of the cardiac magnetic resonance image data is expanded. We train DAE GAN model following the steps of Algorithm 1. In order to intuitively show the training process of the generative model, Fig. 6 shows the changing process of the samples generated in the training process of the model.

The comparison between the samples generated by the trained model generator and the real samples is shown in Fig. 7.

Table 1. The structure of DAE GAN

Generator	Kernel size
Noise z	—
Fully Connected Mapping	—
Residual Structure	3 × 3
Residual Structure	3 × 3
Self-attention Module	—
Residual Structure	3 × 3
Residual Structure	3 × 3
Convolutional Layer & tanh	3 × 3
Discriminator	Kernel size
Input Image	—
Convolutional Layer	3 × 3
Residual Structure	3 × 3
Self-attention Module	—
Residual Structure	3 × 3
Residual Structure	3 × 3
Residual Structure	3 × 3
Fully Connected Layer	—

4.4 Classification Experiment and Analysis of Experimental Results

Observation method has strong subjectivity. In this experiment, data enhancement is performed on small sample medical images, as a result, the observation method can only be used as a reference evaluation standard. In order to evaluate the effect of data enhancement, this article uses the ResNet50 model and the Xception model [29] as a classifier, the classification results are used to uniformly evaluate the effects of various data enhancement methods.

In addition to the conventional accuracy index, the sensitivity and specificity of two medical image classification indexes are also calculated. These indicators are briefly explained below.

The accuracy rate, that is, the probability that the diseased sample and the non-diseased sample are judged correctly. The calculation method is as formula (13):

$$Accuracy = \frac{TP + TN}{TP + TN + FP + FN} \tag{13}$$

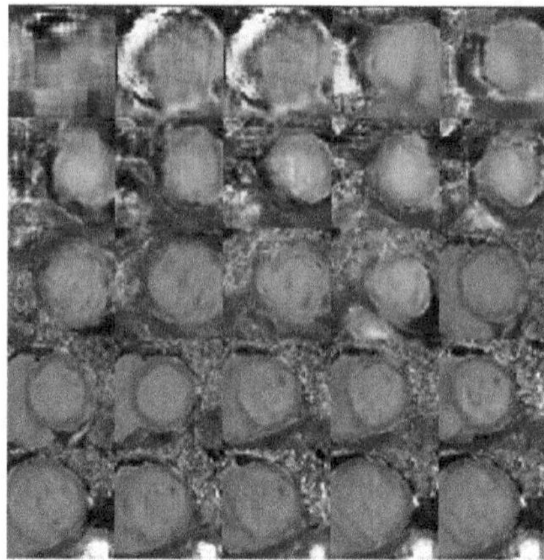

Fig. 6. The changing process of generated sample

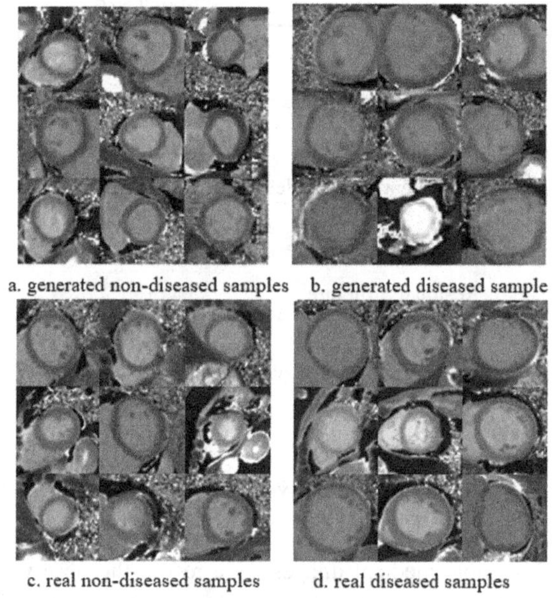

Fig. 7. Comparison of generated samples with real samples

Sensitivity, namely the probability that a diseased sample is judged to be diseased. The calculation method is as formula (14):

$$Sensitivity = \frac{TP}{TP + FN} \tag{14}$$

Specificity, namely the probability of judging a non-diseased sample as non-diseased. The calculation method is as formula (15):

$$Specificity = \frac{TP}{TP + FN} \tag{15}$$

TP represents True Positive, namely the classifier judges it to be a diseased sample, which is in fact also a diseased sample; **TN** represents True Negative, that is, the classifier judges it to be a non-diseased sample, which is in fact not a diseased sample. **FP** is short for False Positive, namely the classifier judges it to be a diseased sample, which is in fact a non-diseased sample; **FN** is short for False Negative, that is, the classifier judges that the sample is not diseased, but is actually a diseased sample.

In this article, the classification experiment uses the Keras framework under the Ubuntu 16.04.1 TLS system environment, the version number is 2.24; the training process uses a Tesla M40. The learning rate is set to 1e−4, and we use the RMSprop optimizer, setting early stopping method to prevent over-fitting, and the 5-fold cross-validation method is used to find the average classification result of the classifier. Table 2 details the average classification results of each enhancement methods in the ResNet50 and Xception classification models.

Table 2. The classification results of enhancement methods

Enhancement method	Accuracy	Sensitivity	Specificity
Classification Network 1: ResNet50			
No Enhancement	0.7767	0.9674	0.6964
Affine Transformation	0.8093	**0.9806**	0.7300
DCGAN	0.8140	0.9760	0.7436
ACGAN	0.7915	0.9728	0.7127
Evolutionary GAN	0.8288	0.9726	0.7591
Our Method	**0.8478**	0.9772	**0.7822**
Classification Network 2: Xception			
No Enhancement	0.7953	0.9765	0.6833
Affine Transformation	0.8279	0.9700	0.7696
DCGAN	0.8326	0.9672	0.7239
ACGAN	0.8054	0.9731	0.7082
Evolutionary GAN	0.8514	0.9780	0.7821
Our Method	**0.8698**	**0.9798**	**0.8116**

Through the experiments, we found that compared with the classification results without any data enhancement method, in ResNet50 model, the classification accuracy increased from 0.7767 to 0.8478, the sensitivity from 0.9674 to 0.9772, the specificity from 0.6964 to 0.7822; in Xception model, the classification accuracy increased from

0.7953 to 0.8698, the sensitivity from 0.9765 to 0.9798, and the specificity from 0.6833 to 0.8116.

5 Conclusion

The DAE GAN model proposed in this paper can effectively expand the amount of cardiac magnetic resonance image data, and effectively alleviate the problem that the classification network cannot be fully trained due to the small amount of medical image data and the uneven data. Compared with any data enhancement method, the classification accuracy of DAE GAN in ResNet50 and Xception models has been improved by 7.11% and 7.45% respectively; compared with affine transformation data enhancement, the method proposed in this paper has been improved by 3.85% and 4.19% respectively, and the experimental results show that the method is effective in different classification models.

Funding Statement. This work was supported in part by the Sichuan Science and Technology Program under Grant 2019ZDZX0005 and the Chinese Scholarship Council under Grant 201908515022.

References

1. Hinton, G.E., Srivastava, N., Krizhevsky, A., et al.: Improving Neural Networks by Preventing Co-adaptation of Feature Detectors. arXiv preprint arXiv:1207.0580 (2012)
2. Geman, S., Bienenstock, E., Doursat, R.: Neural networks and the bias/variance dilemma. Neural Comput. **4**(1), 1–58 (1992). https://doi.org/10.1162/neco.1992.4.1.1
3. Ioffe, S., Szegedy, C.: Batch Normalization: Accelerating Deep Network Training by Reducing Internal Covariate Shift. arXiv:1502.03167
4. Srivastava, N., Hinton, G., Krizhevsky, A., et al.: Dropout: a simple way to prevent neural networks from overfitting. J. Mach. Learn. Res. **15**(1), 1929–1958 (2014)
5. Morgan, N., Bourlard, H.: Generalization and parameter estimation in feedforward nets: some experiments. In: [22], pp. 630–637 (1990)
6. Nowlan, S.J., Hinton, G.E.: Simplifying neural networks by soft weight-sharing. Neural Comput. **4**(4), 473–493 (1992)
7. Krogh, A., Hertz, J.A.: A simple weight decay can improve generalization. In: [16], pp. 950–957 (1992)
8. Krizhevsky, A., Sutskever, I., Hinton, G.E.: ImageNet classification with deep convolutional neural networks. In: Advances in Neural Information Processing Systems, pp. 1097–1105 (2012)
9. Roth, H.R., Lu, L., Liu, J., et al.: Improving computer-aided detection using convolutional neural networks and random view aggregation. IEEE Trans. Med. Imaging **35**(5), 1170–1181 (2015)
10. Setio, A.A.A., Ciompi, F., Litjens, G., et al.: Pulmonary nodule detection in CT images: false positive reduction using multi-view convolutional networks. IEEE Trans. Med. Imaging **35**(5), 1160–1169 (2016)
11. Wang, S.: Facial Affect Detection Using Convolutional Neural Networks. Stanford University (2016)

12. Medical Image Synthesis for Data Augmentation and Anonymization using Generative Adversarial Networks
13. Goodfellow, I.J., Pouget-Abadie, J., Mirza, M., et al.: Generative adversarial nets. In: Advances in Neural Information Processing Systems (NIPS) (2014)
14. Arjovsky, M., Bottou, L.: Towards principled methods for training generative adversarial networks. In: Proceedings of the International Conference on Learning Representations (ICLR) (2017)
15. Radford, A., Metz, L., Chintala, S.: Unsupervised representation learning with deep convolutional generative adversarial networks. In: Proceedings of the International Conference on Learning Representations (ICLR) (2016)
16. Radford, A., Metz, L., Chintala, S.: Unsupervised Representation Learning with Deep Convolutional Generative Adversarial Networks. arXiv preprint arXiv:1511.06434 (2015)
17. Mirza, M., Osindero, S.: Conditional Generative Adversarial Nets. arXiv:1411.1784
18. Mao, X., Li, Q., Xie, H., et al.: Least squares generative adversarial networks. In: Proceedings of the IEEE International Conference on Computer Vision, pp. 2794–2802 (2017)
19. Arjovsky, M., Chintala, S., Bottou, L.: Wasserstein GAN. arXiv preprint arXiv:1701.07875
20. Gulrajani, I., Ahmed, F., Arjovsky, M., et al.: Improved Training of Wasserstein GANs. In: Advances in Neural Information Processing Systems, pp. 5767–5777 (2017)
21. Wang, C., Xu, C., Yao, X., et al.: Evolutionary generative adversarial networks. IEEE Trans. Evol. Comput. **23**(6), 921–934 (2019)
22. Antoniou, A., Storkey, A., Edwards, H.: Data Augmentation Generative Adversarial Networks. arXiv preprint arXiv:1711.04340 (2017)
23. Alia, I.S., Mohameda, M.F., Mahdya, Y.B.: Data Augmentation for Skin Lesion Using Self-Attention Based Progressive Generative Adversarial Network. arXiv:1910.11960
24. Frid-Adar, M., Diamant, I., Klang, E., Amitai, M., Goldberger, J., Greenspan, H.: GAN-based synthetic medical image augmentation for increased CNN performance in liver lesion classification. Neurocomputing **321**, 321–331 (2018)
25. Zhang, H., Cisse, M., Dauphin, Y.N., et al.: mixup: Beyond Empirical Risk Minimization. arXiv preprint arXiv:1710.09412 (2017)
26. Ben-Yosef, M., Weinshall, D.: Gaussian Mixture Generative Adversarial Networks for Diverse Datasets, and the Unsupervised Clustering of Images. arXiv preprint arXiv:1808.10356 (2018)
27. He, K., Zhang, X., Ren, S., et al.: Deep residual learning for image recognition. In: Proceedings of the IEEE Conference on Computer Vision and Pattern Recognition, pp. 770–778 (2016)
28. Zhang, H., Goodfellow, I., Metaxas, D., Odena, A.: Self-Attention Generative Adversarial Networks. arXiv:1805.08318
29. Chollet, F.: Xception: deep learning with depthwise separable convolutions. In: Proceedings of the IEEE Conference on Computer Vision and Pattern Recognition, pp. 1251–1258 (2017)

Analysis of Spectrum Detection and Decision Using Machine Learning Algorithms in Cognitive Mobile Radio Networks

Pablo Palacios Játiva[1(✉)], Cesar Azurdia-Meza[1], Iván Sánchez[2], David Zabala-Blanco[3], and Milton Román Cañizares[4]

[1] Department of Electrical Engineering, University of Chile, Santiago, Chile
pablo.palacios@ug.uchile.cl, cazurdia@ing.uchile.cl

[2] Department of Telecommunications Engineering, Universidad de las Américas, Quito, Ecuador
ivan.sanchez.salazar@udla.edu.ec

[3] Department of Computing and Industries, Universidad Católica del Maule, Talca, Chile

[4] Departamento de Ingeniería de Comunicaciones, Universidad de Málaga, Málaga, Spain
0610744939@uma.es

Abstract. In this work, the performance of four Machine Learning Algorithms (MLAs) applied to Cognitive Mobile Radio Networks (CMRNs) are analyzed. These algorithms are Coalition Game Theory (CGT), Naive Bayesian Classifier (NBC), Support Vector Machine (SVM), and Decision Trees (DT). The numerical results of the performance analysis of these algorithms are presented based on two metrics. These metrics are commonly used in CMRNs which are Probability of Detection (P_d) and Probability of False Alarm (P_{fa}) against Signal-to-Noise Ratio (SNR). Furthermore, outcomes regarding the Classification Quality (CQ) and the simulation time are exposed. Theoretical and numerical results show that the SVM outperforms the rest of the algorithms in each of the metrics. The reasons behind this come from the SVM features, namely high precision, fast learning, and simplicity in the realization stage.

Keywords: Cognitive mobile radio networks (CMRNs) · Coalition game theory (CGT) · Support vector machine (SVM) · Decision tree (DT) · Machine learning algorithms (MLAs) · Naive bayesian classifier (NBC)

1 Introduction

Communications based on Cognitive Radio (CR) have been studied in recent years because they use the electromagnetic spectrum efficiently [1]. This use

X. Wu et al. (Eds.): QShine 2020, LNICST 381, pp. 142–153, 2021.
https://doi.org/10.1007/978-3-030-77569-8_11

efficiency occurs when performing frequency band jumps between wireless protocols and technologies. Furthermore, by complying with this paradigm, CR is considered an enabling technology for 5G communications. Another feature of CR is that its physical layer is a radio that changes its transmission characteristics depending on the communications environment. This adaptation occurs by detecting spectral holes and efficiently using the available frequencies. These benefits present CR in the short term as the best performing solution to achieve high data rates in wireless communications and enable large-scale user mobility. However, CR's biggest challenge is identifying primary users (PUs) who are using a wide range of spectrum, at a certain time, and in a specific geographic location. On the other hand, CR implementations must meet the following criteria: there is no interference between secondary (SU) (unlicensed) users and PUs [2].

For a communications system to be considered CR-based, it must fulfill a cognitive process, which requires four steps: spectrum detection, spectrum decision, spectrum sharing, and spectrum mobility [3]. In the last decade, investigations regarding the CR topic have been oriented in the field of spectrum detection and decision. Consequently, several techniques have been proposed, such as Energy Detection (ED), Cycle-Stationary Detection (CD), Singular Value Decomposition (SVD) [4], and Eigen-Value Decomposition (EVD). Nevertheless, in order to ensure that the CR devices to be truly conscious of the frequency changes that occurs in the mobility stage, just to improve the efficiency, is imperative that equipped with learning and reasoning functionalities.

In the search for mechanisms to mitigate problems in the spectrum detection and decision stages in CR systems, Machine Learning Algorithms (MLAs) have received a lot of attention from the scientific community. In the context of future networks (CR, femto/small cells, and heterogeneous networks), in [5], the authors present a problem formulation and methodology of several MLAs in terms of effectiveness in the testing stage. This analysis is done because MLAs present a new paradigm of proactive, self-aware, self-adaptive, and predictive networks. The authors conclude that the benefits of the MLAs will be verified in the next generation networks. In [6], the authors present a multiple antenna CR system, in which Support Vector Machine (SVM) algorithms are used to solve the spectrum detection problem. This work shows that the SVM algorithms applied to spectrum detection, specifically of spectral holes detection, are robust in terms of temporal and spatio-temporal detection. In [7], another application proposes an MLA-based solution for CR at the end-user device level. The authors conclude that in terms of the terminal service experience and user behavior, the complexity of the central CR network can be reduced.

The main contribution of this manuscript is the performance evaluation of the Coalition Game Theory (CGT), Naive Bayesian Classifier (NBC), Support Vector Machine (SVM), and Decision Trees (DT) methods applied and adapted to a single Cognitive Mobile Radio Networks (CMRNs) by using the Network Simulator 3 (NS-3.23) modules [8]. This software validates several Machine Learning (ML) Approaches in a functional mobile network. The performance of these algorithms are analyzed in terms of the Probability of Detection (P_d), Probability of False Alarm (P_{fa}), Classification Quality (CQ) and simulation time, by employing numerical simulations and by obtaining Cumulative Probability Distributions (CDFs).

The organization of this manuscript is presented as follows: the CMRN, PUs and SUs models are explained in Sect. 2. In Sect. 3, the proposed MLAs are described with their respective mathematical formulation. Then, in Sect. 4, The performance of the CR with the MLAs applied and the numerical results of the simulations are discussed. Finally, the conclusions are presented in Sect. 5.

2 Cognitive Radio System Model

To evaluate the spectrum detection and spectrum decision, we use the standard MLA, which is composed by numerous input, hidden layers with different number of neurons, and various output applied to CMRNs, implemented to coexist with a primary wireless network composed of two state-of-the-art technologies, which are Wireless Fidelity (WiFi) and Long Term Evolution (LTE). We supposed an area covered by CMRN like the propose scenario, composed of m source-destination PUs pairs. The primary transmitters set is as follow $P_p = (P_{1p}, P_{2p}..., P_{mp})$, while the corresponding receivers set is $P_r = (P_{1r}, P_{2r}...., P_{mr})$. We assume the coexistence of l secondary transmitter in the set $S_s = (S_{1s}, S_{2s}..., S_{ls})$, and their corresponding receivers in the set $S_r = (S_{1r}, S_{2r}..., S_{lr})$. This scenario is presented in Fig. 1.

2.1 Secondary User Model

In a PU network, a single SU is considered to access the licensed bands without interfering with the communication of the PUs. We define t as the time slot and i as the frequency bin, where $t = 1, 2, ..., n$ and $i = 1, 2, ..., k$ respectively. We also define n as the number of time slots and k as the number of frequency bins. By using the SVD detection method [4], the spectrum sensing problem can be formulated as follows

$$x_i(t) = \begin{cases} n_i(t) & H_0 \\ h_i(t) * s_i(t) + n_i(t) & H_1 \end{cases}, \tag{1}$$

here, $x_i(t)$ is the signal received by the SU at the tth time slot in the ith frequency bin, $s_i(t)$ is the signal transmitted by the PU, $n_i(t)$ is the Additive White Gaussian Noise (AWGN), and $h_i(t)$ is the channel gain. H_0 and H_1 are the hypothesis test that indicates whether the SU is using the corresponding channel or not.

We used the SVD detection method as the spectrum detection technique because of its easy to design and efficiency in terms of P_d over other detection methods. The spectrum detection process carried out by the SVD method and its operating characteristics are explained in more detail in [4]. The spectrum status $SS_i(t)$ is given as follows

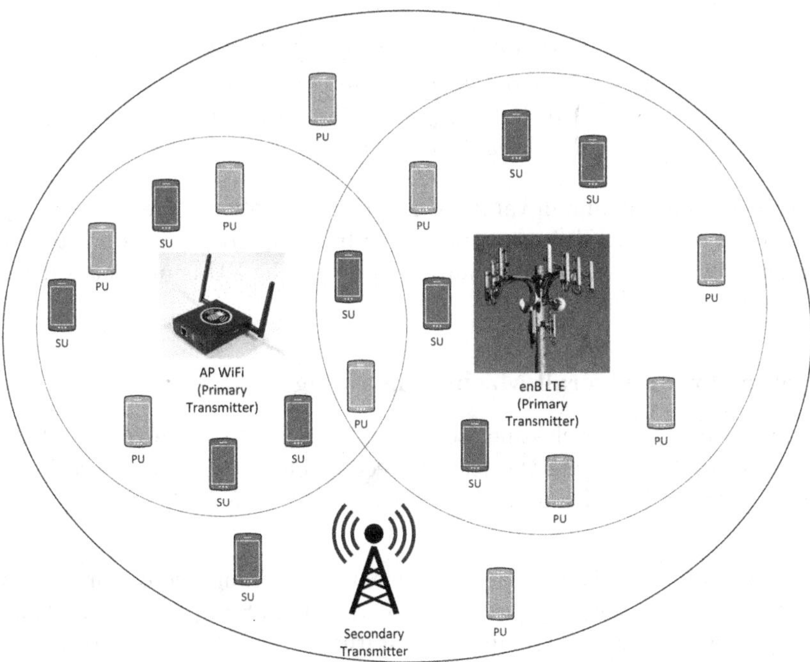

Fig. 1. Scheme of the proposed CMRN.

$$SS_i(t) = \begin{cases} 1 \; x_i(t) > \lambda \\ 0 \; x_i(t) < \lambda \end{cases}. \tag{2}$$

The occupancy O_c^i for the ith time slot acquires the form of

$$O_c^i = (\sum_{t=1}^{k} SS_i(t))/k, \tag{3}$$

which means the average state of the spectrum between the used frequencies bins. When a PU uses some frequency bin that is being sensed by some SU, the $SS_i(t)$ is increased. Consequently, the O_c^i will take large values, by highlighting that the probability of using that frequency bin for the SU is low. Conversely, if $SS_i(t)$ decreases, O_c^i also decreases and it is more likely that this frequency bin is idle and can be used.

2.2 Primary User Model

The PU status is represented as PU^i. This status, for the ith time slot, can be chosen according the rules that follow

$$PU^i = \begin{cases} 1 & (O_c^i > U_{oc}) \ or \ (L_{oc} \le O_c^i \le U_{oc} \\ & and \ ffb^i < \overline{ffb}) \\ 0 & (O_c^i < L_{oc}) \ or \ (L_{oc} \le O_c^i \le U_{oc} \\ & and \ ffb^i \ge \overline{ffb}) \end{cases}, \tag{4}$$

The maximum and minimum values of occupancy for the n time slots are represented by U_{oc} and L_{oc}. The consecutive free frequency bins in the ith time slot are denoted by ffb^i, whose maximum value when the PU is present is denoted as $\overline{ffb}$.

2.3 Steps for an Overall Machine Learning

First, MLA constructs a classifier to map SS_i to PU^i, where $SS_i = (SS_i(1), SS_i(2)..., SS_i(k))$ represents the feature vector. There are three steps for constructing a classifier, which are:

Training: For the training stage, we denote a training vector for the spectral state, $SS_{i(train)} = (SS_{i(train)}(1), SS_{i(train)}(2)....., SS_{i(train)}(k))^T$. Then, we denote a variable for the training PU status, PU^i_{train}. For the total cases, $i = 1, 2, ..., n_1$, where n_1 is the number of training time slots fed into the classifier.

Testing: For the training stage, we denote a testing vector for the spectral state, $SS_{i(test)} = (SS_{i(test)}(1), SS_{i(test)}(2)....., SS_{i(test)}(k))^T$, Then, we denote a variable fot the testing PU status, PU^i_{test}. For the total cases, $i = n_1 + 1, n_1 + 2, ..., n_2$, n_2 refers to the length of testing sequence. In this work, the matrix of size $n * k$ is divided into 10% training data matrix of size $n_1 \times k$ and 90% testing data matrix of size $n_2 \times k$ [9].

Classification Quality: For the classification quality stage, we denote the PU status for the ith time slot as PU^i_{eval}. In this stage it is categorized the testing vector $SS_{i(test)}$ as an occupied class ($PU^i_{eval} = 1$) or unoccupied class ($PU^i_{eval} = 0$). The PU status is correctly determined, when $PU^i_{eval} = PU^i_{test}$, by producing $CQ^i = 1$. This scenario will be represented as $P_d = 1$ or 0 depending on the value of PU^i_{eval}. The no-detection occurs when $PU^i_{eval} = 0$ and $PU^i_{test} = 1$, whereas false alarm occurs when $PU^i_{eval} = 1$ and $PU^i_{test} = 0$, by giving $CQ^i = 0$. This situation will be represented as P_{fa}.

3 Proposed Machine Learning Algorithms

In this work, four MLA are analyzed to predict the PU status via the occupancy data. The motivation to use these algorithms is to find the most efficient MLA for predicting future status.

3.1 Coalition Game Theory

We formulate the cooperative problem as a coalition game $G = (N; u)$, where $N = SS_i$ and u represents the payoff function that transform a user contribution in a coalition into its profit. We formulate this scheme via two steps, as follows [2].

The Cooperation Phase: First, local detection is required, which is done using the SVD method. In a Rayleigh fading environment, the P_d and P_{fa} of the i-th SU are respectively represented as $P_{d,i,j}$ and $P_{f,i,j}$, which are given by [2]

$$P_{d,i,j} = [PY_{i,j} > \lambda | H_1] = e^{-\frac{\lambda}{2}} \sum_{n=0}^{w-2} \frac{1}{n!}\left(\frac{\lambda}{2}\right)^n + \left(\frac{1+\gamma_{i,j}}{\gamma_{i,j}}\right)^{w-1} \left[e^{-\frac{\lambda}{2(1+\gamma_{i,j})}} - e^{-\frac{\lambda}{2}} \sum_{n=0}^{w-2} \frac{1}{n!}\left(\frac{\lambda * \gamma_{i,j}}{2(1+\gamma_{i,j})}\right)^n\right], \tag{5}$$

and

$$P_{fa,i,j} = [PY_{i,j} > \lambda | H_0] = \frac{\Gamma\left(w, \frac{\lambda}{2}\right)}{\Gamma(w)}, \tag{6}$$

where $Y_{i,j}$ is the normalized output of the i-th SU sensing the status of the j-th PU, λ is the detection threshold for the j-th PU, w is the time-bandwidth product, and $\gamma_{i,j}$ denotes the average SNR of the received signal from the PU to the SU. Furthermore, $\Gamma(.,.)$ and $\Gamma(.)$ are the incomplete and complete Gamma functions respectively.

In addition, the missing probability P_m for the i-th SU is considered as follows

$$P_{m,i} = 1 - P_{d,i,j}. \tag{7}$$

By reducing the $P_{m,i}$ directly maps to increasing the P_d and, consequently, interference on the PU decreases.

Within each coalition Ω, a single SU, named as the coalition head, k, collects the sensing bits from the coalition SUs, and acts as a fusion center to decide on the presence or not of the PUs in the channel.

The missing and false alarm probabilities are as follows

$$Q_{m,\Omega} = \prod_{i \in \Omega} [P_{m,i} * (1 - P_{e,i,k}) + (1 - P_{m,i}) * P_{e,i,k}], \tag{8}$$

and

$$Q_{f,\Omega} = 1 - \prod_{i \in \Omega} [(1 - P_{fa}) * (1 - P_{e,i,k}) + P_{fa} * P_{e,i,k}]. \tag{9}$$

A suitable function acquires the form of

$$u(\Omega) = Q_{d,\Omega} - C(Q_{f,\Omega}) = (1 - Q_{m,\Omega}) - C(Q_{f,\Omega}), \tag{10}$$

where $Q_{d,\Omega}$ denotes the probability detection of the coalition Ω and $C(Q_{f,\Omega})$ represent a cost function of the P_{fa} within the coalition Ω. The latter can be written as

$$C(Q_{f,\Omega}) = \begin{cases} -\alpha^2 log(1 - \left(\left(\frac{Q_{f,\Omega}}{\epsilon}\right)^2\right) & Q_{f,\Omega} < \epsilon \\ +\infty & Q_{f,\Omega} \geq \alpha \end{cases}, \tag{11}$$

where ϵ denotes a false alarm constraint per coalition, namely per SU.

The SU Transmission Phase. For the SU transmission phase we assume a time division multiplexing [10]. Then, the transmission is divided according the SU contribution in Ω. The time allocated for SU is given by $(1 - \alpha_P) * t_i^{\Omega}$. Its reward is made proportional to the energy spent by the SU.

3.2 Naive Bayesian Classifier

This algorithm is named as the "independent feature model" because it does not consider the features interdependence. In this model, the total samples are contained in the feature vector for the ith time, Furthermore, these samples are independent of each other, because of every feature represents a specific frequency bin. However, the variable of the PU status PU^i, results a function of the frequency bin. The probability of SS_i by using the Bayes theorem is defined as [11]

$$\mathrm{p(PU^i,\, SS_i)} = \mathrm{p(PU^i)} * \mathrm{p(SS_i|PU^i)}. \tag{12}$$

When $PU^i = 0$, SS_i is classified as an idle class; otherwise SS_i is an occupied class. The goal is to obtain the class with the largest posterior probability in the classification phase. The classification rule is represent as follows

$$classify(S\hat{S}_i) = \underset{SS_i}{\arg\max} \left\{\mathrm{p(PU^i,\, SS_i)}\right\}, \tag{13}$$

where $S\hat{S}_i = \{S\hat{S}_i\,(1)\,,\ S\hat{S}_i\,(2)\ldots\{S\hat{S}_i(k)\}$.

3.3 Support Vector Machine

This algorithm results in a discriminative classifier with high accuracy. In addition, SVM tends to be resistant to over-fitting and. Generally, two types of classifiers in SVM are presented in the literature: linear and non-linear SVM. In this work, for simplicity but without losing the generality, linear SVM is employed.

The training feature and response vectors are represented as $D = \left(\text{PU}^i,\ \text{SS}_i\right)$, where $PU^i \in \{0,\ 1\}$. By definition, the two classes of SVM are separated defining a hyperplane H, which is represented as $x * SS_i = \rho$, where x represents the normal vector and ρ represents the constant separating occupied and idle classes ($PU^i \in \{0,\ 1\}$), which in turn is defined as [12]

$$PU^i = \begin{cases} 1 \rightarrow x * SS^i > \rho(\text{Occupied class}) \\ 0 \rightarrow x * SS^i < \rho(\text{Idle class}) \end{cases} \tag{14}$$

3.4 Decision Trees

In this work, decision trees are represented with a classificatory approach, where the leaves of the tree define the class labels. A benefit of DTs is that they can handle interactions and feature dependency. Regarding the decision made by this algorithm, it is made at each node internally, which allows the data division into two own subsets. The data is represented ad follows

$$(\text{SS}_i, \text{PU}^i) = \{(SS_i\,(1)\,,\ SS_i\,(2)\ldots\ SS_i(k)), \text{PU}^i\}, \tag{15}$$

where PU^i is the dependent variable, which is assigned by calculating the entropy of the feature as follows [13]

$$\text{Entropy(t)} = -\sum_{idi=0}^{Z} p(idi|t) * log_2(p(idi|t), \tag{16}$$

where $p(idi|t)$ is the fraction of records belonging to class idi for a certain node t, and Z represents the total classes.

4 Results

4.1 Simulation Parameters

Each of the MLAs were implemented in a simulated CR environment. To generate the simulations, we use the NS-3 software, because it provides executable models of signal propagation and user mobility. We have chosen for the propagation model, the range propagation loss model, due to its unique end-user and transmitter dependency. For the mobility model we chose the random waypoint model. To create a more realistic environment, we created two types of SU: SUs with cognitive capacity to work only on LTE or WiFi and SUs with cognitive capacity for both technologies (dual SU). The most important technical parameters used in the simulation scenarios are shown in Table 1.

Table 1. Technical parameters of the simulation

Parameter	Value
AP coverage	50 m
Channel model	Slow Rayleigh fading
CR LTE/WiFi (SU)	5
Dual CR (SU)	10
eNB cells	3
eNB coverage	350 m
LTE frequency	729 MHz [4]
LTE bandwidth	20 MHz [4]
Mobility model	Random way-point
Noise model	AWGN
Propagation model	Range propagation loss
PU LTE/WiFi	5
Receiver power	0.06 mW
Samples	Variable
Transmitter power	0.037 mW
WiFi bandwidth	20 MHz [4]
WiFi frequency	2400 MHz [4]

4.2 Numerical Simulation Results

The MLA curves were obtained through the implementation and simulation of the algorithms in NS-3. The P_d vs SNR and P_{fa} vs SNR are presented as a Cumulative Distribution Function (CDF) for the MLAs, as shown in Fig. 2 and Fig. 3. As can be seen in both figures, the algorithm that presents the worst performance is the NBC, due to its features of not having complete information and making decisions based on statistics. CGT and DT algorithms have a similar behavior, because the first takes a cooperative detection between nodes, and the second divides the decisions into subsets, which are similar processes for the system. Finally, the algorithm with the best performance is the SVM, due to its high precision and accuracy when recognizing the use patterns of frequency bins for detection. For low levels of SNR, specifically −10 dB, SVM has a P_d of 50% and a P_{fa} of 5%, while NBC has a P_d of 20% and a P_{fa} of 7%.

Figure 4 shows the plot of the Classification Quality as a function of Number of Samples. As N_s increases, they have more chance to sense and sensing accuracy improves. However, by increasing the N_s, the algorithms have a greater amount of data to process and become slower.

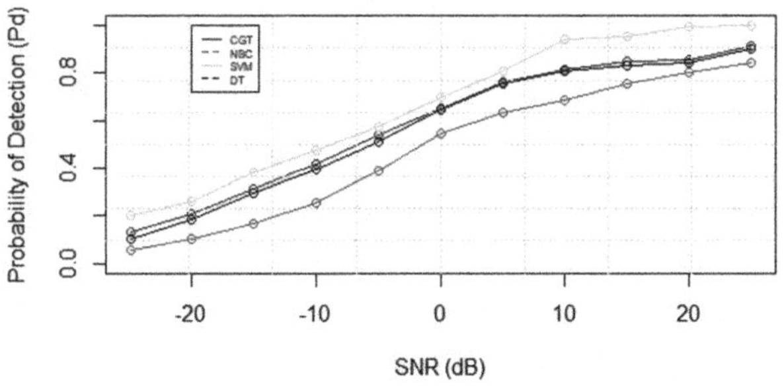

Fig. 2. Probability of detection CDF.

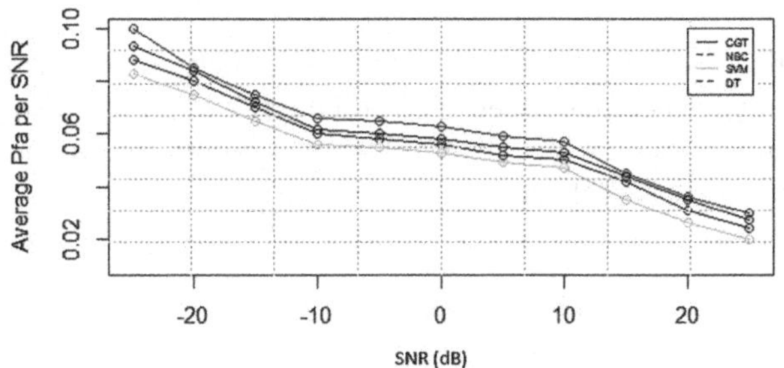

Fig. 3. Probability of false alarm CDF.

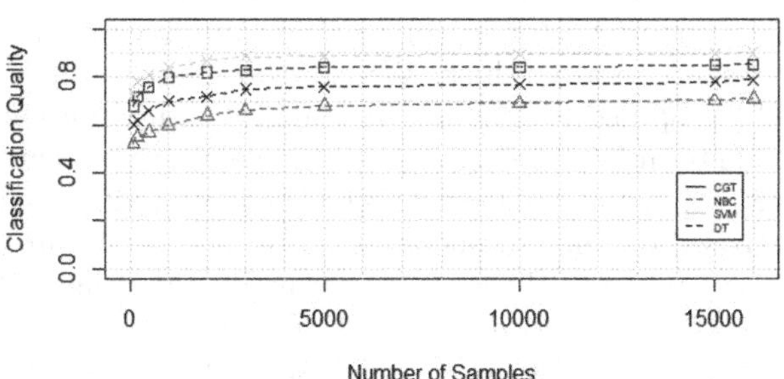

Fig. 4. Classification quality curve.

An important factor that must be considered by the simulator to develop the experiments in a controlled way is the simulation time, since it is not the same magnitude as the real time. To determine these times, several simulations were run with different N_s, maintaining the basic technical parameters indicated in Table 1 in each of them. We defined the number of simulations for the experiment using the Monte Carlo method, with 21 iterations for all variations of SUs. This process was done to have reliable and valid statistics of the generated data [14]. We observe the linear and increasing behavior of Simulation time vs Number Of Samples in Fig. 5. The SVM algorithm presents slightly better

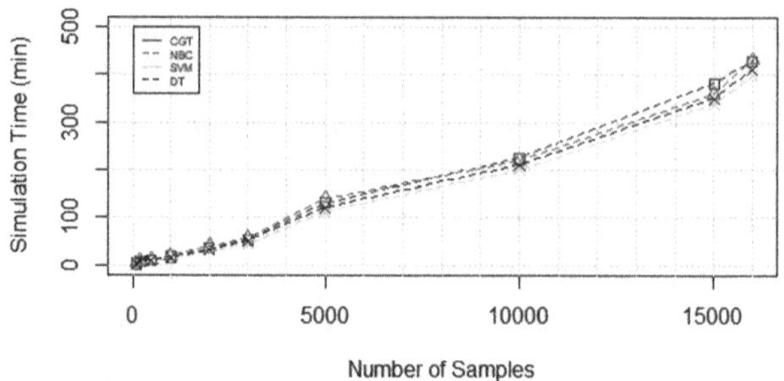

Fig. 5. Simulation time as a function of number of samples.

performance than the other MLAs, due to its simplicity of learning. Specifically the biggest difference is found when N_s is 15000, the SVM has a simulation time of 340 min approximately, while the CGT that presents the worst performance, has a simulation time of 390 min.

5 Conclusions

In this work, some MLA implemented in the detection and decision stage of a CMRN have been analyzed and compared. We modeled the detection and decision stage of the "Cognitive Cycle" as a Coalition Game (CG), Naive Bayesian Classifier (NVC), Support Vector Machine (SVM) and Decision Tree (DT) problem with their respective algorithms. We characterized the network structure resulting from the proposed techniques, its stability and performance was studied and observed in terms of P_d, P_{fa}, CQ and simulation time. Simulation results showed that SVM algorithm, compared with CG, NVC and DT, outperform the CMRN system, based on the parameters studied, specifically in 40 min less in simulation time, compared to the worst performance algorithm (CGT), keeping the N_s fixed at 15000. It also increases the P_d by 30% and decreases the P_{fa} by 2%, compared to the NBC algorithm, which showed the worst performance in these parameters, keeping the SNR fixed at −10 dB.

Acknowledgment. This work was supported by ANID PFCHA/Beca de Doctorado Nacional/2019 21190489, SENESCYT "Convocatoria abierta 2014-primera fase Acta CIBAE-023-2014", UDLA Telecommunications Engineering Degree, Project FONDECYT No. 11160517, and Grupo de Investigación en Inteligencia Artificial y Tecnologías de la Información (IA&TI).

References

1. Sasipriya, S., Vigneshram, R.: An overview of cognitive radio in 5G wireless communications. In: 2016 IEEE International Conference on Computational Intelligence and Computing Research (ICCIC), pp. 1–5, December 2016

2. Palacios, P., Saavedra, C.: Coalition game theory in cognitive mobile radio networks. In: Botto-Tobar, M., Pizarro, G., Zúñiga-Prieto, M., D'Armas, M., Zúñiga Sánchez, M. (eds.) CITT 2018. CCIS, vol. 895, pp. 3–15. Springer, Cham (2019). https://doi.org/10.1007/978-3-030-05532-5_1
3. Palacios Játiva, P., Saavedra, C., Freire, J.J., Román Cañizares, M., Zabala-Blanco, D.: Comparative analysis of cooperative routing protocols in cognitive radio networks. In: Botto-Tobar, M., Zambrano Vizuete, M., Torres-Carrión, P., Montes León, S., Pizarro Vásquez, G., Durakovic, B. (eds.) ICAT 2019. CCIS, vol. 1195, pp. 43–56. Springer, Cham (2020). https://doi.org/10.1007/978-3-030-42531-9_4
4. Palacios, P., Castro, A., Azurdia-Meza, C., Estevez, C.: SVD detection analysis in cognitive mobile radio networks. In: 2017 Ninth International Conference on Ubiquitous and Future Networks (ICUFN), July 2017, pp. 222–224 (2017)
5. Kibria, M.G., Nguyen, K., Villardi, G.P., Zhao, O., Ishizu, K., Kojima, F.: Big data analytics, machine learning, and artificial intelligence in next-generation wireless networks. IEEE Access **6**, 32 328–32 338 (2018)
6. Awe, O.P., Deligiannis, A., Lambotharan, S.: Spatio-temporal spectrum sensing in cognitive radio networks using beamformer-aided SVM algorithms. IEEE Access **6**, 25 377–25 388 (2018)
7. Perez, J.S., Jayaweera, S.K., Lane, S.: Machine learning aided cognitive RAT selection for 5G heterogeneous networks. In: 2017 IEEE International Black Sea Conference on Communications and Networking (BlackSeaCom), June 2017, pp. 1–5 (2017)
8. Patel, R., Kamboj, P.: Investigation of network simulation tools and comparison study: NS3 vs NS2. Transactions **14**, 15 (2015)
9. Shavlik, J.W., Dietterich, T., Dietterich, T.G.: Readings in Machine Learning. Morgan Kaufmann (1990)
10. Lee, J., Noh, H., Lim, J.: TDMA-based cooperative MAC protocol for multi-hop relaying networks. IEEE Commun. Lett. **18**(3), 435–438 (2014)
11. Shen, B., Su, X., Greiner, R., Musilek, P., Cheng, C.: Discriminative parameter learning of general Bayesian network classifiers. In: Proceedings. 15th IEEE International Conference on Tools with Artificial Intelligence, November 2003, pp. 296–305 (2003)
12. A. Ben-Hur and J. Weston, "A user's guide to support vector machines", in Data mining techniques for the life sciences. Springer, 2010, pp. 223–239
13. Dahan, H., Cohen, S., Rokach, L., Maimon, O.: Proactive Data Mining with Decision Trees. Springer, Heidelberg (2014)
14. Alfonso, U.M., Carla, M.V.: Modelado y simulación de eventos discretos. Editorial UNED (2013)

AutoMTS: Fully Autonomous Processing of Multivariate Time Series Data from Heterogeneous Sensor Networks

Ricardo Sousa[1,2], Conceição Amado[1], and Rui Henriques[2(✉)]

[1] CEMAT and Instituto Superior Técnico, Universidade de Lisboa, Lisbon, Portugal
{ricardo.filipe.sousa,conceicao.amado}@tecnico.ulisboa.pt

[2] INESC-ID and Instituto Superior Técnico, Universidade de Lisboa, Lisbon, Portugal
rmch@tecnico.ulisboa.pt

Abstract. Heterogeneous sensor networks, including water distribution systems and traffic monitoring systems, produce abundant time series data with an arbitrarily-high multivariate order for monitoring network dynamics and detecting events of interest. Nevertheless, errors and failures in the calibration, data storage or acquisition can occur on some of the sensors installed in those systems, producing missing and/or anomalous values. This work proposes a computational system, referred as AutoMTS, for the fully autonomous cleaning of multivariate time series data using strict quality criteria assessed against ground truth extracted from the targeted series data. The proposed methodology is parameter-free as it relies on robust principles for the assessment, hyperparameterization and selection of methods. AutoMTS coherently supports an extensive set state-of-the-art methods for (multivariate) time series imputation and outlier detection-and-treatment, considering both point and segment/serial occurrences. A comprehensive evaluation of AutoMTS is accomplished using heterogeneous sensors from two water distribution systems with varying sampling rates, water consumption patterns, and inconsistencies. Results confirm the relevance of the proposed AutoMTS system. AutoMTS is provided as an open-source tool available at https://github.com/RicardoFLNSousa/AutoMTS/tree/master.

Keywords: Parameter-free learning · Multivariate time series · Missing values imputation · Outlier detection · Heterogeneous sensor networks

1 Introduction

The placement of heterogeneous sensors within complex systems – whether physiological, mechanical, digital, geophysical, environmental or urban – offers the possibility to acquire comprehensive views of their behavior along time. Sensorized systems produce abundant time series data, used for monitoring purposes

X. Wu et al. (Eds.): QShine 2020, LNICST 381, pp. 154–178, 2021.
https://doi.org/10.1007/978-3-030-77569-8_12

or the detection of events of interest. However, the placed sensors are susceptible to failures and errors associated with sensor calibration and data acquisition-transmission-storage [1], producing time series data with missing and anomalous values. In this context, time series data are generally subjected to initial processing stages for leveraging their quality for the subsequent mining stages.

Processing time series data produced by networks of heterogeneous sensors is, nevertheless, a laborious process due to four major reasons. First, the selection and parameterization of the processing methods is highly dependent on the regularities of the target series data and challenged by the wide diversity of approaches currently available. Second, the profile of errors can be diversified, each leading to different processing choices. In this context, the type and amount of anomalies and missing values can largely affect decisions. Third, different types of sensors – such as water flow, pressure and water quality sensors in water distribution systems – may benefit from dissimilar processing methods. In fact, sensors of the same type but with singular calibrations, sampling rates, or positioning within the monitored system can as well benefit from different choices. Fourth and finally, different systems equipped with identical sensors do not necessarily benefit from the same processing options. Consider water distribution network (WDN) systems, water consumption patterns can highly vary between WDNs or along time, impacting decisions. Also, different WDNs may be susceptible to unique externalities, affecting the profile of observed errors.

In addition, time series data processing generally yields suboptimal results. First, cross-variable relationships in multivariate time series data are commonly disregarded. For instance, flow and pressure sensors in WDNs are generally correlated, and thus co-located or nearby sensors can guide the treatment of low-quality series data. Second and understandably, optimal decisions are challenged by the wide diversity of available processing approaches, multiplicity of sensors, and profile of errors observed per sensor.

This work proposes a methodology for the fully autonomous cleaning of multivariate time series that is able to address the introduced challenges. The proposed methodology, referred as AutoMTS (**Auto**nomous **M**ultivariate **T**ime **S**eries data processing), offers three major contributions. First, AutoMTS provides strict guarantees of optimality as it places robust processing decisions against ground truth extracted from the targeted series data. To this end, series data are automatically explored in order to detect conserved segments and identify the profile of observed errors, which are then planted in the conserved segments for the sound comparison of available processing choices.

Second, AutoMTS provides a comprehensive coverage of available processing options, currently providing over twenty state-of-the-art methods for missing imputation, outlier detection and gross-error removal from time series data. Particular attention was placed to guarantee the presence of state-of-the-art methods able to consider cross-variable dependencies in the presence of multivariate time series data. Also, we further guarantee the presence of methods able to deal with both point and segment/serial missing and outlier values.

Third, AutoMTS is parameter-free as it relies on robust principles to assess, hyperparameterize and select state-of-the-art processing methods.

To assess the significance of the proposed contributions, AutoMTS is extensively evaluated in two water distribution network systems with heterogeneous sensors, producing observations at varying sampling rates, and subjected to unique water consumption patterns and error profiles.

The gathered results confirm the relevance of the proposed AutoMTS methodology, highlighting that processing choices are highly specific to each sensor and thus guarantees of optimality can only be provided under comprehensive and robust assessments. Also, results further offer a thorough comparison of state-of-the-art imputation and outlier detection methods, assessing their ability to handle diverse error profiles in real-world series data with varying regularities.

AutoMTS is provided as both a graphical and programmatic tool satisfying strict usability criteria.

The manuscript is structured as follows. Section 2 provides essential background and surveys recent contributions on time series data processing. Section 3 described the AutoMTS approach. Section 4 comprehensively assesses the adequacy of AutoMTS using two real-world heterogeneous networks as study cases. Finally, concluding remarks and major implications are synthesized.

2 Background and Related Work

This section offers a structured view on how to process inconsistencies in (multivariate) time series, providing essential *background*, surveying *recent contributions*, and describing the preprocessing *methods* implemented in AutoMTS.

Time Series Data Processing. Signals produced by sensors are generally represented as *time series*, an ordered set of observations $\mathbf{x}_{1..T} = (\mathbf{x}_1, ..., \mathbf{x}_T)$, each $\mathbf{x}_t$ being recorded at a specific time point t. Time series can be *univariate*, $\mathbf{x}_t \in \mathbb{R}$, or *multivariate*, $\mathbf{x}_t \in \mathbb{R}^m$, where $m > 1$ is the order (number of variables).

Errors associated with the calibration, measurement, storage, logger communication and synchronization of sensors are associated with inconsistencies on the produced time series. As a result different types of errors can be observed, including: 1) anomalous values, 2) missing values; 3) duplicate values; 4) atypical values or gross errors (impossibilities in a given domain); and 5) incorrectly timestamped observations (arbitrarily-high sampling delays).

Low-quality data can be rectified. The task of *preprocessing time series* is the process of leveraging quality data to facilitate the subsequent extraction of useful information from the time series. In this context, cleaning the identified inconsistencies is an important step, and the one targeted in this work.

Time series can be decomposed into *trend*, *seasonal*, *cyclical*, and *irregular components* using additive or multiplicative models [2]. Processing can take place on the original series or separately on each component. Classical approaches for time series analysis generally rely on statistical principles, including *autoregression*, *differencing* and *exponential smoothing* operations to either detect deviations from expectations as well as to impute missing values [3].

Time series typically have an internal structure with domain-specific meanings. In this context, normalization, resampling, piecewise aggregate approximation, symbolic aggregate approximation, and transformations (including Fourier, Wavelet and other forms of window-based feature extraction) can support the analysis of the internal structure of time series. However, finding suitable representations is highly dependent on the subsequent mining ends and therefore is not considered part of the processing pipeline proposed in our work.

Missing Value Imputation. Missing observations, commonly referred as missing values, can be characterized by the underlying stochastic processes that describe their occurrence: i) missing completely at random (MCAR) where there is no distribution characterizing their occurrence, generally caused by punctual problems on data transmission-storage-acquisition; ii) missing at random (MAR) where missings are independent of the value of the observation but dependent on the other non-missing observations (e.g. sensor malfunction under high temperatures); and iii) not missing at random (NMAR) where missings essentially depend on the value of the observation (e.g. sensors failing measuring high pressure). Complementary, missing values can be described by their *type* – whether point, sequential or mixed similarly to outliers – and *amount* from a given period.

There are three typical choices to deal with missing values: i) force removal, leading to gaps on the time series to be handled along the subsequent time series processing steps; ii) replace them with a dedicated value or symbol; and iii) estimate their values using imputation principles. Missing removal can be listwise (indiscriminate missing deletion) or pairwise (controlled deletion in accordance with the amount) [4]. Missing imputation can either produce hot-deck estimates from similar/nearby observations or from matched segments of the time series; or cold-deck estimates from external time series datasets [4].

Last observation carried forward (LOCF) and next observation carried backward (NOCB) are simplistic methods based on the closest available observation. Linear interpolation linearly combines last and next observations. Usually, the seasonal component is removed at the beginning and included after linear interpolation is done. Moving average (MA) can include further observations to estimate the missing value, $\hat{\mathbf{x}}_t = \frac{1}{m}\sum_{j=-k}^{k}\mathbf{x}_{t+j}$ where $[t-k, t+k]$ is a centered window of *2k+1* length (also termed order). When the sequential values are all missing observations, the window size can dynamically expand until two non-missing values occur. In this context, linear interpolation is a moving average or order 2. Average (median) imputation corresponds to a moving average (median) with unbounded order, imputing the average (median) of all non-missing occurrences. The expectation maximization algorithm (EM) has been also suggested for estimating missing observations within multivariate time series data, although in its original form disregards time dependencies. Amelia combines the EM method with bootstrapping to impute missing values in time series data using principles from multiple imputation. Classical approaches for time series modeling, including SARIMA and Holt-Winters [3], are also viable imputation candidates when time series have well-established regularities.

k-nearest neighbors (kNN) can be applied to impute both point and sequential missings from (multivariate) time series. To this end, time series are subjected to segmentation, and the value estimates inferred from the closest neighbor subsequences. Particular attention should be paid to its parameterization, as kNN performance highly depends on the selected distance (e.g. ability to tolerate shift and scale misalignments on the time and amplitude axes) and number of neighbors. In the presence of multivariate time series data, MissForests [5] uses principles from random forest approaches to deal with mixed-variables (relevant when dealing with heterogeneous sensors) in accordance with the frequency of missing values (chained principle). Despite its role, it neglects time dependencies between observation. The time-extended version of multivariate imputation by chained equations (MICE) [6] is able to addresses such drawback while still accounting for cross-variable dependencies.

Osman et al. [4] proposed an ensemble approach that selects between classical imputation techniques (such as moving average) and modern alternatives in accordance with the type (MAR or MCAR) and amount of missings. In addition to some of the surveyed methods, modern imputation techniques further include reconstruction methods based on principal component analysis [7] and machine learning techniques such as Gaussian process regression, tensor-based methods [8], and neural networks, specially auto-associative neural networks [9].

Moritz et al. [10] extensively compares multiple-imputation approaches by deleting observations from time series with varying trend and seasonal characteristics. Multiple-imputation approaches rely on multiple estimates to reduce biases. For instance, Aggregated values [11] is an estimator from mean estimates collected at multiple temporal granularities (overall, yearly, monthly and daily mean). Seasonal Kalman filters and model-based approaches have been also applied within multiple-imputation settings [10,12].

Imputation methods have been also proposed in the context of specific domains. In water-energy-gas distribution systems, the well-recognized Quevedo method [13] estimates missings from observations collected at similar periods from previous days, weeks, months and years. Barrela et al. [14] further proposed a estimator that combines both forecast and backcast missing observations values generated by TBATS and ARIMA models, accommodating multiple seasonality.

Time Series Outlier Detection. *Outliers* are observations significantly deviating from expectations as to arouse suspicion of being generated by a different mechanism [15]. Outliers can occur in point or serial forms. *Point outliers* (also referred as punctual or singular outliers) can be detected against the whole series (*global outliers*) or against observations that occur on nearby time points or share the same context (local/*contextual outliers*). *Sequential outliers* (also referred as segment or serial outliers) are anomalous subsequences of contiguous observations. Outliers can be further characterized in accordance with their causation and impact [16]: additive outliers affect the time series for a single time period; level shift outliers have preserved/continuous effects; temporary change outliers show an exponential decaying over time; and innovational outliers affect the nearest subsequent observations. *Outlier analysis* generally comprises anomaly *scoring*, *detection* and *treatment* steps. Treatment either denotes the removal (planting missing values) or re-estimation of outlier values. Approaches for outlier analysis are generally categorized according to *distribution*-based, *depth*-based, *distance*-based, *density*-based and *clustering*-based *approaches* [17].

Outlier analysis can be applied on the raw time series or over its irregular component once decomposed. Simple methods for point outlier detection rely on *deviation criteria* or *inter-quartile ranges* assessed on the irregular component. Generally, this class of methods fits empirical or statistical distributions and fix thresholds on what it is expected to occur. Despite their simplicity, time dependencies are disregarded. *Local outlier factor* (LOF) [18] approach minimizes this drawback by computing anomaly scores based on the local density of an observation with respect to its neighbours where the neighborhood criteria can include temporal and cross-variable distances. *Isolation forests* [19] recursively generate partitions from multivariate series data by randomly selecting a feature and a split value for the feature. Presumably the anomalies need fewer partitions to be isolated compared to "normal" points, thus yielding smaller trees. *Parametric models* from maximum likelihood estimates are also available [20].

Gupta et al. [21] provide a comprehensive survey of contributions on outlier detection over temporal data structures, including (geolocalized) time series data. The approaches to detect *point outliers* are grouped into five major categories: predictive, profile-based models, information-theoretic, classification and clustering approaches. In the context of predictive models, a score is assigned to each observation as a deviation from the estimated value. Estimates can be computed using imputation techniques for univariate and multivariate time series data previously covered. Profile-based approaches trace a normal profile for the time series using classical time series models [3] and more recent advances, including recurrent neural networks that act as auto-encoders [22]. Anomaly scores are then inferred by testing deviations against the approximated profile. The principle behind the less common information theoretic approaches is that the removal of outlier results in higher abstraction ability (time series representations with lower error bound) [23].

Approaches for sequential outlier detection traditionally compare subsequences segmented under multi-scale sliding windows to identify dissimilar subsequences. Keogh et al. [24] outlines principles to surpass the computational complexity of computing pairwise time series distances between all subsequences, including heuristics to reorder candidate subsequences, locality sensitive hashing, Haar wavelets, and joint use of symbolic aggregations with augmented tries. These are used for an improved ordering of subsequences. An additional challenge is the fact that sequential outliers may have an arbitrary length. Chen et al. [25] proposed a new class of approaches that satisfy this premise: a pattern (subsequence of two consecutive points) is defined and outliers are composed of infrequent patterns on either the original time series or compressed time series recovered after wavelet transform.

Time series *clustering* algorithms are as well used to detect sequential outliers. Generally, these approaches segment the inputted series to identify anomalous segments, paying particular attention to distance metrics between time series (including metrics to tolerate misalignments) and barycenter criteria whenever applicable. Understandably, traditional clustering algorithms can be also applied to detect outliers from (multivariate) time series by assuming independence between observations. HOT SAX [26] also offers the possibility to detect sequential outliers, referred as time series discords, from symbolic representations of the time series. HOT SAX, originally prepared to detect global sequential outliers, was later on extended towards local sequential outliers [27].

Other Inconsistencies. In the presence of domain knowledge, *atypical values* or gross errors in time series can be detected by fixing upper and/or lower bounds on the acceptable values. *Duplicate values* are harder to detect as they may not necessarily result in anomalous values. Duplicates can have different causes: 1) accumulation of values from previous observations (generally preceded by missing occurrences), and 2) multiplicity of measurements within a single time step. Density-based outlier approaches are generally considered for the former case, while rule-based analysis of timestamps against sampling expectations are pursued for the latter case. Finally, *irregular sampling rates* observed within or between sensors or between sensors often result from faulty sensor synchronization. Diverse transforms and dedicated time series analysis algorithms have been proposed to deal with irregular measurements [28,29].

Parameter-Free and Autonomous Processing. The literature on autonomous selection of either parametric or non-parametric methods for time series processing is scarce, generally providing series-dependent contributions and focusing on a single processing task. Rayana et al. [30] and Zimek et al. [31] proposed ensemble principles to infer anomaly scores from multiple estimates, validated in specific data domains. Similarly, ensemble principles for imputing missing observations in time series have been proposed [32,33]. Böhm et al. [34] introduced CoCo, a parameter-free method for detecting outliers in data with unknown underlying distributions. Despite the relevance of these contributions, to our knowledge there are not yet methodologies for autonomously assessing, parameterizing and selecting methods able to treat time series unsupervisedly.

3 Solution: Autonomous Time Series Data Processing

Despite the relevance of the surveyed contributions, existing time series preprocessing methods are generally oriented towards specific data regularities and types of errors. Thorough comparisons are thus necessary to place proper decisions, a generally laborious and difficult process due to the difficulty of performing objective assessments in the absence of ground truth. In this context, we propose a novel approach for the fully Autonomous processing of Multivariate Time Series data, referred as AutoMTS. AutoMTS receives as input a pointer to a database or file with the raw time series data, and produces as output the processed data without inconsistencies in accordance with strict quality criteria. Annotations, including bounds associated with the estimated anomaly scores, and performance statistics can be optionally outputted.

The AutoMTS is a parameter-free methodology, a composition of steps that guarantee the robust assessment, hyperparameterization and selection of state-of-the-art processing methods in accordance with the regularities and inconsistencies observed in the inputted series data. The major idea behind AutoMTS is to generate precise ground truth for the sound and quality-driven evaluation of available processing options. To this end, AutoMTS relies on two major principles: i) detection of conserved segments within the inputted series data, and ii) modeling the type and amount of observed errors. Under these principles, the assessment can be conducted by purposefully planting inconsistencies along the conserved segments and, depending on their length, on synthetically generated series using the approximated component-wise regularities. In this way, available processing options can be objectively assessed.

AutoMTS provides a good coverage of available processing options, providing over twenty state-of-the-art methods for missing imputation, outlier detection and gross-error removal from time series data. With the aim of handling errors of varying profile, AutoMTS incorporates processing methods able to deal with both point and serial missing and outlier values. In addition, AutoMTS is able explore the aided processing guidance provided by correlated variables within multivariate time series data. To this end, state-of-the-art processing methods able to capture cross-variable dependencies are further supported in AutoMTS.

3.1 Methodology

AutoMTS is a sequential approach for preprocessing time series produced from heterogeneous networks. The four major steps are depicted in Fig. 1. Given a (multivariate) time series, the *first step* is to treat non-cumulative duplicates through a rule-based inspection of sampling irregularities (see Sect. 2). After the time series is cleansed of duplicates, the *second step* is the detection of atypical values against background knowledge. For instance, in the context of water flow and pressure sensors, lower bounds are generally zero and upper bounds fixed in accordance with pipe specifications. Atypical values are then translated into missing values to be dealt later in the process. On the *third step*, we detect outlier observations. This is a core step in our pipeline as the wide-diversity of state-of-the-art methods for outlier detection needs to be robustly assessed using the methodology proposed in Sect. 3.2. The selected method, already hyperparameterized, is then applied to detect outliers in the target (multivariate) time series. The detected outliers, along with their anomaly scores, will be given to the user and he may opt to either discard the outliers (default option) or mark some of the outputted outliers to be retained in the time series. The *fourth step* is to impute values on the missing observations, including originally missing occurrences as well as the removed outliers and atypical values. Similarly with the third step, this is another core step within the AutoMTS process. The assessment methodology for hyperparameterizing and selecting imputation methods is introduced in Sect. 3.3. Once missing occurrences are imputed, the treated time series is returned by AutoMTS.

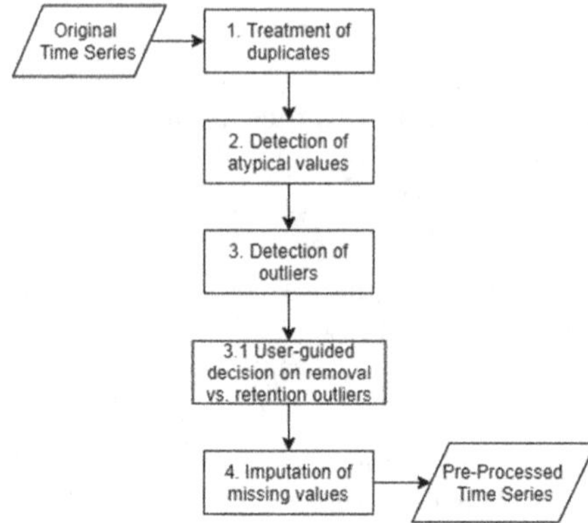

Fig. 1. Time series preprocessing methodology.

3.2 Autonomous Outlier Detection *(Step 3)*

The third step purposefully plants artificial outliers in the conserved segments of the inputted time series in accordance with the signal regularities observed along those segments. The regularities reveal information related with the point-wise and segment-wise distribution of values to guide the planting of point and segment outliers. The robust planting of artificial outliers is essential to gather ground truth for the objective assessment of the methods, necessary to their hyperparameterization and comparison.
For generating the ground truth, five major steps are undertaken:

1. the time series is decomposed into trend, seasonal, cyclical and noise/irregular components;
2. the distribution of values observed along the irregular component is dynamically fitted into a well-known probability distribution using both the Kolmogorov-Smirnov and χ^2 statistical tests;
3. the tails of the approximated distributions are used to plant point outlier values randomly distributed along the irregular component;
4. sequential outliers are further planted by guaranteeing a residual joint probability of the observed values along the artificial subsequence;
5. the irregular component with the planted point and sequential outliers is added to the original trend, seasonal and cyclical components.

The statistical properties of this five-step process guarantee the presence of non-trivial outliers resembling the characteristics of real-world anomalies. AutoMTS runs by default 30 process simulations to collect performance estimates.

Some of the outlier detection methods available in the AutoMTS are standard deviation, inter quartile range, isolation forests, LOF, DBScan and HOT SAX.

Let *TP* (true positives) be the correctly detected outliers, *TN* (true negatives) be observations correctly identified as non-outliers, *FP* (false positives) be the incorrectly detected outliers, and *FN* (false negatives) be the non-detected outliers wrongly. To evaluate the behavior of outlier detection methods, we suggest as essential performance views the analysis of recall,

$$\text{recall} = \frac{TP}{TP + FN},$$

to understand the percentage of correctly identified outliers, as well as precision,

$$\text{precision} = \frac{TP}{TP + FP},$$

to understand whether the retrieved outliers were identified at the cost of retrieving non-outlier observations (false positives). To objectively guide the hyperparameterization and selection steps, these complementary views can be combined within scores, such as the F1-score,

$$\text{F1-score} = 2 \times \frac{\text{precision} \times \text{recall}}{\text{precision} + \text{recall}},$$

which is not free of criticisms [35] due to the inherent characteristics of the harmonic mean. Complementary integrative scores able to reconcile recall and precision views at alternative anomaly score thresholds, including the area under the ROC curve (AUC), can be alternatively selected [35].

3.3 Autonomous Missing Imputation *(Step 4)*

The fourth step wittingly generates missing observations within conserved segments of the inputted time series in accordance with the profile of missing data observed along the non-conserved segments. The profile of missing observations essentially discloses information on their temporal distribution, nature (point versus sequential), length, amount and periodicity (well-defined versus random). Similarly to the generation of artificial outliers, the removal of observations is essential to gather ground truth for objective assessments required for the hyperparameterization and selection of imputation methods.

For generating the ground truth, three major steps are undertaken. First, AutoMTS verifies whether the largest conserved segment satisfies a minimum length assumption (four times the seasonal factor as default). If the largest conserved segment does not satisfies the assumption, the segment is replaced by an artificial time series. To generate the artificial time series, the original time series should be decomposed in order to approximate its core components. The irregular component is regenerated in accordance with the underlying distribution and added to the remaining components to produce a synthetic time series without missing occurrences. Second, the approximated percentage amount and temporal distribution of punctual missings in the original time series is used to remove observations from the conserved segment or synthesized time series. Third, and finally, sequential missings are planted in accordance with the distribution of their extension and recurrence on the original time series.

The statistical properties of this three-step process guarantee the presence of missing observations resembling real-world characteristics. By default, 30 process simulations are considered to collect performance estimates.

Some the univariate imputation methods available in the AutoMTS are: random sample, interpolation, LOCF, NOCB and moving average. Some of the supported multivariate methods are: random forests, EM, kNN, Mice and Amelia.

To evaluate the performance of imputation methods, residue-based scores are considered, including the mean absolute error (MAE),

$$\text{MAE} = \Sigma_{i=1}^{n} |\hat{\mathbf{x}}_{t_i} - \mathbf{x}_{t_i}|,$$

where $\mathbf{x}$ and $\hat{\mathbf{x}}$ are the observed and imputed time series respectively, and n is the number of missings; the root mean squared error (RMSE),

$$\text{RMSE} = \sqrt{\sum_{i=1}^{n} \frac{(\hat{\mathbf{x}}_{t_i} - \mathbf{x}_{t_i})^2}{n}},$$

the symmetric mean absolute percentage error (SMAPE); and the percentage of missing values imputed since not all imputation methods may not encounter necessary conditions for imputing certain missing observations.

3.4 Computational Complexity

Considering the presence of k_1 preprocessing methods, each with $O(T_i)$ complexity, then the complexity of executing them is $\sum_i^{k_1} O(T_i) = O(k_1 T_{max})$. Assuming that the conducted Bayesian optimization per method converges in a bounded number of k_2 iterations for each method, then $O(k_1 k_2 T_{max})$. Finally, considering the presence of k_3 testing settings in accordance with the detected error profiles in the original series (e.g. k_3=2 for missing and outlier segments with well-defined rate and length distributions), then AutoMTS has $O(k_1 k_2 k_3 T_{max})$ complexity. k_1 and k_3 are constants. Given a window of bounded size w, the majority of preprocessing methods are linear on the window size, yielding $O(k_1 k_2 k_3 w)$.

3.5 Final Remarks on the Behavior of AutoMTS

The state-of-the-art methods supported along the third and fourth steps of the AutoMTS pipeline are tested one by one. A good portion of these methods require the input of parameter values. In this context, hyperparameterization is conducted using the planted inconsistencies in order to identify the best parameters. To this end, we rely on Bayesian optimization [36] due to its inherent ability to traverse only the most promising areas of the search space, thus promoting efficiency. The hyperparameterization should be driven by one of the performance views previously introduced. By default, F1-score is selected for the hyperparameterization of outlier detection methods, while RMSE is the default criteria to guide the hyperparameterization of missing imputation methods.

Once parameterized, methods are then evaluated using the same performance views. If the length of the largest conserved segment (or synthesized time series) permits, the segment is further segmented into two subsequences, one for hyperparameterization and other for the final method evaluation. In this way, we prevent the overfitting of the selected parameters.

4 Results

Results are organized in three major steps. First, we describe the networks of heterogeneous sensors that will be used as study cases, exploring some of the produced time series. Second, we provide a thorough comparison of state-of-the-art methods to detect outliers and impute missings, showing that their adequacy is highly dependent on the time series regularities and error profiles. Finally, we assess AutoMTS, quantifying its performance gains.

Study Cases: Beja and Barreiro Water Distribution Systems

A Water Distribution Network (WDN) is a system composed of pumps, pipelines, tanks and other elements for delivering water in adequate quantities, pressure and quality for the everyday needs. WDNs can be equipped with an arbitrarily-high number of heterogeneous sensors, including water flow and pressure sensors.

The results of this article were obtained in collaboration with two major water utilities: Barreiro city Council and Beja city Council, which provided time series representative of their telemetry systems.

Barreiro WDN is composed by 14 sensors of water flow and pressure that provide aggregated measurements on an hourly basis along 2018. The time series has 8473 observations, an amount inferior to the total yearly hours given the presence of weekly periods without measurements – real sequential missings – and the presence of a scarce number of punctual missings. Beja WDN offers water flow and pressure measurements along a two-year period (5/2017 to 4/2019) with an approximate 5-minute sampling rate. Each time series has over 200.000 observations, a irregular sampling rate and the presence of missing values along segments of lower extension than those observed in the Barreiro WDN.

Figure 2a depicts the water flow series from sensors located near the principal tanks in the Barreiro and Beja WDNs, while Fig. 2b depicts the time series

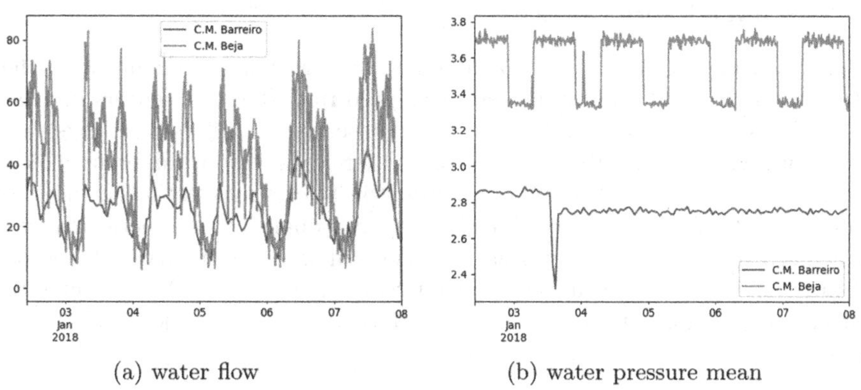

Fig. 2. Sensor measurements over 5 illustrative days for both Barreiro and Beja WDNs.

produced by the approximately co-located water pressure sensors. As one can observe, the pressure and flow series from show highly dissimilar structure. In addition, sensors of the type show considerably different regularities for different water distribution systems. These observations motivate the need to perform processing decisions separately for each sensor from the monitored systems.

Experimental Setting
To assess the impact of placing appropriate choices along the processing stages in accordance with the characteristics and inconsistencies observed along time series, we consider the water flow and pressure time series from Barreiro and Beja WDNs and applied the proposed AutoMTS methodology to generate ground truth. To facilitate the interpretability of results, we further varied the profile of the planted inconsistencies for some of the conducted analyzes. The major parameters controlling the experimental setting are:

- available methods for point outlier detection (e.g. isolation forests) and sequential outlier detection (e.g. SAX), and the corresponding parameters;
- planted outlier profiles, including: i) frequency of outliers (1% to 10%); ii) type of outliers (point versus sequential); and iii) length of sequential outliers;
- available methods for missing imputation from univariate series (e.g. moving average) or multivariate series (e.g. MICE), and corresponding parameters;
- planted missing profiles, including: i) frequency of missing values (from 1% to 20%); ii) type of missings (point versus sequential); and iii) length of sequential missing observations.

The presented results provide the average performance collected from 30 simulations. A stochastic process to generate inconsistencies in accordance with the introduced parameters is used to produce each simulation. Random seeds are considered to guarantee fair comparisons between methods.

4.1 AutoMTS Performance

Table 1 provides a comprehensive analysis of the performance of multiple outlier detection methods on time series data produced from different sensors installed within the Barreiro and Beja WDNs. We can observe that different settings – different sensors, water distribution systems, outlier types – propel different choices. Considering F1-score and recall, while isolation forests appears to be the most promising option for water pressure sensors, inter-quartile range performance is particularly good on water flow sensors. The recall of the most surveyed methods significantly differs between WDNs. Understandably, as AutoMTS selects the best choice available, it shows optimal performance across major performance views.

Figures 3a and 3b offer a complementary graphical description of previous results, further showing how the performance of different outlier detection methods vary with the amount of planted outliers. Illustrating, HOT SAX is not competitive when considering a low amount of outliers (offers a good recall yet low precision due its focus on outlier segments), yet performance improves with a medium-to-high amount of outliers. The analysis of these figures further highlights that there are significant changes in performance associated with changes on the amount of outlier values. These variations can affect processing decisions (e.g. isolation forests versus inter-quartile range in pressure sensors), further supporting the relevance of the proposed AutoMTS methodology.

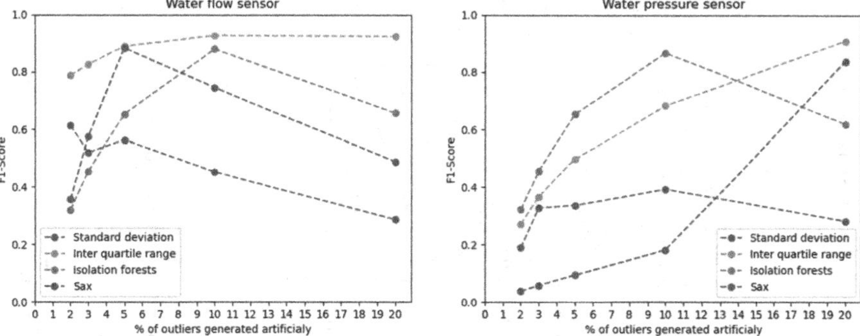

(a) water flow sensor at Barreiro WDN (b) water pressure mean at Barreiro WDN

Fig. 3. Performance of outlier detection methods with varying percentage of planted point outliers in time series produced from heterogeneous sensors.

Similarly to Table 1, Table 2 gathers results on the performance of missing imputation methods on time series data produced from different sensors placed within the Barreiro and Beja WDNs. Decisions are similarly dependent on the target sensor, network and missing profile (type and amount). For instance, while interpolation shows generally good performance on water flow sensors is not competitive on water pressure sensors. The characteristics of the Beja WDN, where measurements are collected under a smaller sampling rate, presents compelling evidence towards the use moving average imputation technique. Finally, we can observe a decreased performance of single-value estimators such as LOCF and NOCB for imputing missing segments and an increased performance of multi-point estimators such as moving average estimators for this sequential type of missings. These remarks underline the role of AutoMTS.

Table 1. Performance of outlier detection methods for water pressure and flow sensors from Barreiro and Beja WDNs with planted point and sequential outliers on up to 10% of observations.

		Barreiro WDN				Beja WDN			
		F1-score	Accuracy	Precision	Recall	F1-score	Accuracy	Precision	Recall
pressure: point	Standard deviation	0.393 ± 0.06	0.921 ± 0.01	0.259 ± 0.05	0.832 ± 0.05	0.414 ± 0.02	0.926 ± 0.00	0.261 ± 0.02	$\mathbf{1.0 \pm 0.00}$
	Inter quartile range	0.684 ± 0.01	0.908 ± 0.00	1.0 ± 0.00	0.52 ± 0.01	0.686 ± 0.02	0.952 ± 0.00	0.522 ± 0.03	$\mathbf{1.0 \pm 0.00}$
	Isolation forests	$\mathbf{0.867 \pm 0.03}$	0.973 ± 0.01	$\mathbf{0.873 \pm 0.03}$	$\mathbf{0.861 \pm 0.03}$	$\mathbf{0.75 \pm 0.03}$	0.95 ± 0.01	$\mathbf{0.751 \pm 0.03}$	0.75 ± 0.03
	Local outlier factor	0.179 ± 0.07	0.835 ± 0.01	0.18 ± 0.07	0.178 ± 0.07	0.0 ± 0.00	0.801 ± 0.00	0.0 ± 0.00	0.0 ± 0.00
	Dbscan	0.0 ± 0.00	0.899 ± 0.00	0.0 ± 0.00	0.0 ± 0.00	0.0 ± 0.00	0.9 ± 0.00	0.0 ± 0.00	0.0 ± 0.00
	SAX	0.182 ± 0.00	0.106 ± 0.00	1.0 ± 0.00	0.1 ± 0.00	0.61 ± 0.11	0.945 ± 0.01	0.448 ± 0.11	$\mathbf{1.0 \pm 0.00}$
	AutoMTS	0.867 ± 0.03	0.973 ± 0.01	0.873 ± 0.03	0.861 ± 0.03	0.75 ± 0.03	0.95 ± 0.01	0.751 ± 0.03	0.75 ± 0.03
flow: point	Standard deviation	0.452 ± 0.04	0.93 ± 0.00	0.293 ± 0.03	$\mathbf{1.0 \pm 0.00}$	0.477 ± 0.02	0.932 ± 0.00	0.313 ± 0.02	$\mathbf{1.0 \pm 0.00}$
	Inter quartile range	$\mathbf{0.926 \pm 0.02}$	0.986 ± 0.00	0.864 ± 0.04	$\mathbf{1.0 \pm 0.00}$	$\mathbf{0.964 \pm 0.01}$	0.993 ± 0.00	$\mathbf{0.981 \pm 0.01}$	0.949 ± 0.00
	Isolation forests	0.879 ± 0.05	0.976 ± 0.01	$\mathbf{0.886 \pm 0.05}$	0.873 ± 0.05	0.855 ± 0.02	0.971 ± 0.00	0.856 ± 0.02	0.854 ± 0.02
	Local outlier factor	0.15 ± 0.05	0.829 ± 0.01	0.151 ± 0.05	0.149 ± 0.05	0.084 ± 0.03	0.817 ± 0.01	0.084 ± 0.03	0.083 ± 0.03
	Dbscan	0.837 ± 0.06	0.966 ± 0.01	0.881 ± 0.09	0.801 ± 0.03	0.774 ± 0.01	0.944 ± 0.00	0.954 ± 0.02	0.651 ± 0.01
	SAX	0.745 ± 0.10	0.96 ± 0.01	0.603 ± 0.12	$\mathbf{1.0 \pm 0.00}$	0.598 ± 0.11	0.944 ± 0.01	0.435 ± 0.10	$\mathbf{1.0 \pm 0.00}$
	AutoMTS	0.926 ± 0.02	0.986 ± 0.00	0.864 ± 0.04	1.0 ± 0.00	0.964 ± 0.01	0.993 ± 0.00	0.981 ± 0.01	0.949 ± 0.00
pressure: segment	Standard deviation	0.373 ± 0.05	0.918 ± 0.00	0.243 ± 0.04	0.807 ± 0.04	0.415 ± 0.03	0.926 ± 0.00	0.262 ± 0.02	$\mathbf{1.0 \pm 0.00}$
	Inter quartile range	0.665 ± 0.00	0.898 ± 0.00	$\mathbf{1.0 \pm 0.00}$	0.498 ± 0.00	0.692 ± 0.02	0.953 ± 0.00	0.529 ± 0.03	$\mathbf{1.0 \pm 0.00}$
	Isolation forests	$\mathbf{0.854 \pm 0.03}$	0.97 ± 0.01	0.853 ± 0.03	$\mathbf{0.855 \pm 0.03}$	$\mathbf{0.75 \pm 0.04}$	0.95 ± 0.01	$\mathbf{0.751 \pm 0.04}$	0.749 ± 0.04
	Local outlier factor	0.151 ± 0.07	0.828 ± 0.01	0.15 ± 0.07	0.151 ± 0.07	0.0 ± 0.00	0.801 ± 0.00	0.0 ± 0.00	0.0 ± 0.00
	Dbscan	0.0 ± 0.00	0.897 ± 0.00	0.0 ± 0.00	0.0 ± 0.00	0.0 ± 0.00	0.9 ± 0.00	0.0 ± 0.00	0.0 ± 0.00
	SAX	0.185 ± 0.00	0.108 ± 0.00	$\mathbf{1.0 \pm 0.00}$	0.102 ± 0.00	0.617 ± 0.11	0.946 ± 0.01	0.455 ± 0.11	$\mathbf{1.0 \pm 0.00}$
	AutoMTS	0.854 ± 0.03	0.97 ± 0.01	0.853 ± 0.03	0.855 ± 0.03	0.75 ± 0.04	0.95 ± 0.01	0.751 ± 0.04	0.749 ± 0.04
flow: segment	Standard deviation	0.45 ± 0.05	0.928 ± 0.00	0.292 ± 0.04	$\mathbf{1.0 \pm 0.00}$	0.486 ± 0.02	0.932 ± 0.00	0.322 ± 0.01	$\mathbf{1.0 \pm 0.00}$
	Inter quartile range	$\mathbf{0.929 \pm 0.03}$	0.987 ± 0.01	0.869 ± 0.05	$\mathbf{1.0 \pm 0.00}$	$\mathbf{0.956 \pm 0.01}$	0.991 ± 0.00	$\mathbf{0.982 \pm 0.03}$	0.934 ± 0.03
	Isolation forests	0.889 ± 0.05	0.978 ± 0.01	$\mathbf{0.889 \pm 0.05}$	0.889 ± 0.05	0.855 ± 0.02	0.971 ± 0.00	0.856 ± 0.02	0.854 ± 0.02
	Local outlier factor	0.157 ± 0.05	0.829 ± 0.01	0.157 ± 0.05	0.157 ± 0.05	0.091 ± 0.04	0.819 ± 0.01	0.091 ± 0.04	0.091 ± 0.04
	Dbscan	0.816 ± 0.10	0.962 ± 0.02	0.864 ± 0.14	0.779 ± 0.07	0.773 ± 0.03	0.944 ± 0.01	0.953 ± 0.05	0.651 ± 0.02
	SAX	0.795 ± 0.08	0.966 ± 0.01	0.667 ± 0.11	$\mathbf{1.0 \pm 0.00}$	0.61 ± 0.11	0.945 ± 0.01	0.448 ± 0.11	$\mathbf{1.0 \pm 0.00}$
	AutoMTS	0.929 ± 0.03	0.987 ± 0.01	0.869 ± 0.05	1.0 ± 0.00	0.956 ± 0.01	0.991 ± 0.00	0.982 ± 0.03	0.934 ± 0.03

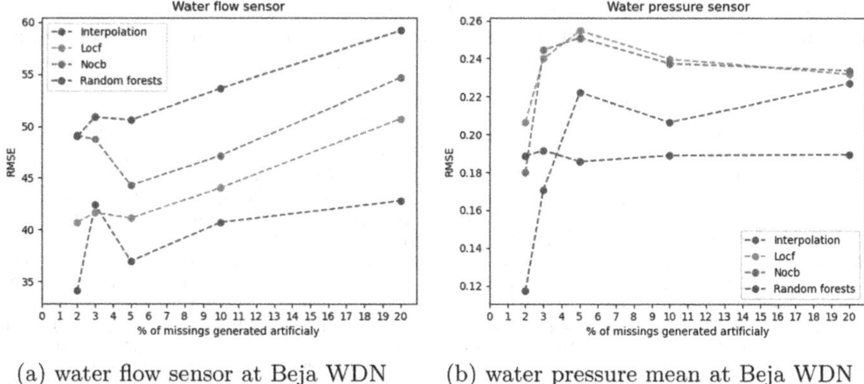

(a) water flow sensor at Beja WDN (b) water pressure mean at Beja WDN

Fig. 4. Performance of missing imputation methods with varying percentage of point missings planted in time series from heterogeneous sensors.

Figures 4a and 4b extend some of the presented settings, offering a complementary graphical description sensitive to the amount of planted missing values. Generally, the higher the amount of missing observations, the higher the imputation difficulty. These figures highlight the presence of significant performance differences related with the amount of missing observations, further suggesting the relevance of understanding the missing profiles when placing preprocessing decisions. For instance, while random forests is generally a non-competitive method for a small amount of missings, it is the suggested option to impute high amounts of missing observations in water pressure series. This last remark further pinpoints the relevance of considering cross-variable dependencies.

Tables 3 and 4 in appendix provide complementary results on the behavior of both outlier detection and missing imputation methods to handle point and sequential inconsistencies.

4.2 AutoMTS Tool

Figure 5 provides a snapshot of the AutoMTS tool. On the left panel it is possible to upload the file which contains the time series dataset. Different file formats are supported, including .xlsx and .csv, as well as different data representations. An illustrative representation of the input data is a table with timestamped rows containing the measurements and as many columns as the number of sensors (time series). If sensors have temporally misaligned measurements, each row can alternatively describe a single event, identifying the timestamp, sensor and collected measurement. To guarantee that ground truth is assessed over the provided series data, each sensor needs to have at least one period of four weeks without missing observations. Otherwise, synthetic series are generated for the parameterization and selection of methods. Once the uploaded dataset passes the initial validation process, it is possible to filter the dataset by selecting the time series (sensors) that we want to process. This can be done using *sensor*

Table 2. Performance of missing imputation methods for water pressure and flow sensors from Barreiro and Beja WDNs with planted point and sequential missing values on 2% of observations.

		Barreiro WDN				Beja WDN			
		RMSE	MAE	SMAPE	%	RMSE	MAE	SMAPE	%
pressure: point	Mean	0.051 ± 0.07	0.025 ± 0.02	0.941 ± 0.89	1.00	0.166 ± 0.01	0.154 ± 0.01	4.325 ± 0.21	1.00
	Median	0.051 ± 0.07	0.024 ± 0.02	0.927 ± 0.89	1.00	0.19 ± 0.02	0.124 ± 0.02	3.495 ± 0.52	1.00
	Random sample	0.058 ± 0.07	0.034 ± 0.04	1.268 ± 1.43	1.00	0.23 ± 0.05	0.169 ± 0.06	4.759 ± 1.64	1.00
	Interpolation	0.052 ± 0.08	0.023 ± 0.02	0.897 ± 1.04	1.00	$\mathbf{0.048 \pm 0.01}$	$\mathbf{0.03 \pm 0.01}$	0.842 ± 0.17	1.00
	Locf	$\mathbf{0.039 \pm 0.07}$	$\mathbf{0.019 \pm 0.02}$	0.747 ± 0.93	1.00	0.067 ± 0.02	0.036 ± 0.01	1.017 ± 0.22	1.00
	Nocb	0.073 ± 0.11	0.03 ± 0.03	1.247 ± 1.53	1.00	0.057 ± 0.02	0.033 ± 0.01	0.923 ± 0.22	1.00
	Moving average	0.047 ± 0.07	0.021 ± 0.02	0.822 ± 0.88	1.00	0.08 ± 0.02	0.042 ± 0.01	1.168 ± 0.27	1.00
	Random forests	0.074 ± 0.07	0.037 ± 0.02	1.425 ± 1.00	1.00	0.173 ± 0.01	0.132 ± 0.01	3.709 ± 0.39	1.00
	EM	0.092 ± 0.06	0.065 ± 0.02	2.422 ± 0.87	1.00	0.231 ± 0.02	0.19 ± 0.02	5.343 ± 0.50	1.00
	Knn	0.059 ± 0.05	0.032 ± 0.02	1.216 ± 0.82	1.00	0.164 ± 0.01	0.132 ± 0.01	3.707 ± 0.41	1.00
	Mice	0.083 ± 0.06	0.046 ± 0.02	1.722 ± 0.95	1.00	0.22 ± 0.02	0.152 ± 0.02	4.304 ± 0.50	1.00
	Amelia	0.095 ± 0.06	0.069 ± 0.02	2.547 ± 0.95	0.98	0.229 ± 0.01	0.187 ± 0.01	5.247 ± 0.37	0.97
	AutoMTS	0.039 ± 0.07	0.019 ± 0.02	0.747 ± 0.93	1.00	0.048 ± 0.01	0.03 ± 0.01	0.842 ± 0.17	1.00
flow: point	Mean	8.89 ± 1.25	7.461 ± 1.31	34.015 ± 6.62	1.00	47.609 ± 5.38	36.619 ± 3.97	48.716 ± 4.41	1.00
	Median	9.061 ± 1.27	7.47 ± 1.35	33.91 ± 7.02	1.00	49.51 ± 6.59	34.352 ± 4.72	45.846 ± 4.99	1.00
	Random sample	11.894 ± 4.25	9.956 ± 4.17	41.997 ± 12.05	1.00	58.912 ± 19.86	46.49 ± 20.79	60.353 ± 29.31	1.00
	Interpolation	$\mathbf{2.801 \pm 0.86}$	$\mathbf{2.0 \pm 0.66}$	10.056 ± 4.14	1.00	$\mathbf{16.871 \pm 2.52}$	$\mathbf{11.96 \pm 1.36}$	9.655 ± 2.54	1.00
	Locf	4.221 ± 1.22	3.264 ± 0.90	15.714 ± 4.51	1.00	20.677 ± 3.41	14.189 ± 1.94	23.204 ± 3.36	1.00
	Nocb	4.52 ± 1.17	3.484 ± 0.91	16.934 ± 4.57	0.99	20.526 ± 2.83	14.425 ± 1.73	23.604 ± 3.12	1.00
	Moving average	6.68 ± 2.04	5.314 ± 1.63	25.299 ± 6.89	1.00	20.534 ± 3.08	14.683 ± 1.64	23.764 ± 3.29	1.00
	Random forests	10.115 ± 1.86	8.315 ± 1.73	37.123 ± 8.07	1.00	46.03 ± 5.63	34.514 ± 3.76	46.435 ± 3.94	1.00
	EM	12.125 ± 2.88	9.982 ± 2.50	47.819 ± 11.22	1.00	64.264 ± 6.33	49.785 ± 4.78	77.618 ± 6.87	1.00
	Knn	9.771 ± 1.60	8.051 ± 1.46	36.148 ± 7.26	1.00	48.345 ± 5.78	36.213 ± 3.97	48.019 ± 4.41	1.00
	Mice	12.69 ± 2.57	10.433 ± 2.42	48.719 ± 11.95	1.00	72.407 ± 6.63	54.85 ± 5.43	67.096 ± 5.80	1.00
	Amelia	12.548 ± 2.03	10.346 ± 1.76	46.246 ± 8.02	0.98	67.114 ± 5.31	53.254 ± 5.21	75.529 ± 6.72	0.97
	AutoMTS	2.801 ± 0.86	2.0 ± 0.66	10.056 ± 4.14	1.00	16.871 ± 2.52	11.96 ± 1.36	19.655 ± 2.54	1.00
pressure: sequential	Mean	0.025 ± 0.06	0.014 ± 0.02	0.535 ± 0.80	1.00	0.166 ± 0.03	0.156 ± 0.02	4.399 ± 0.66	1.00
	Median	$\mathbf{0.024 \pm 0.06}$	$\mathbf{0.013 \pm 0.02}$	0.518 ± 0.80	1.00	0.185 ± 0.07	0.129 ± 0.06	3.653 ± 1.69	1.00
	Random sample	0.035 ± 0.06	0.024 ± 0.03	0.908 ± 1.20	1.00	0.227 ± 0.07	0.174 ± 0.08	4.924 ± 2.23	1.00
	Interpolation	0.03 ± 0.06	0.021 ± 0.04	0.834 ± 1.82	1.00	0.117 ± 0.03	0.09 ± 0.03	2.538 ± 0.92	1.00
	Locf	0.039 ± 0.10	0.029 ± 0.07	1.176 ± 3.10	1.00	0.207 ± 0.08	0.15 ± 0.08	4.228 ± 2.31	1.00
	Nocb	0.029 ± 0.06	0.018 ± 0.02	0.67 ± 0.87	1.00	0.18 ± 0.07	0.123 ± 0.08	3.489 ± 2.17	1.00
	Moving average	0.03 ± 0.07	0.022 ± 0.05	0.875 ± 2.12	0.67	$\mathbf{0.061 \pm 0.06}$	$\mathbf{0.044 \pm 0.05}$	1.224 ± 1.42	0.13
	Random forests	0.074 ± 0.09	0.034 ± 0.03	1.35 ± 1.30	1.00	0.189 ± 0.03	0.149 ± 0.03	4.187 ± 0.77	1.00
	EM	0.086 ± 0.05	0.064 ± 0.02	2.371 ± 0.75	1.00	0.227 ± 0.03	0.188 ± 0.03	5.286 ± 0.71	1.00
	Knn	0.049 ± 0.06	0.026 ± 0.02	1.013 ± 0.83	1.00	0.177 ± 0.03	0.145 ± 0.03	4.073 ± 0.79	1.00
	Mice	0.047 ± 0.06	0.027 ± 0.02	0.994 ± 0.81	1.00	0.229 ± 0.03	0.164 ± 0.03	4.633 ± 0.84	1.00
	Amelia	0.082 ± 0.05	0.063 ± 0.02	2.302 ± 0.88	1.00	0.236 ± 0.03	0.194 ± 0.03	5.441 ± 0.74	1.00
	AutoMTS	0.024 ± 0.06	0.013 ± 0.02	0.518 ± 0.80	1.00	0.117 ± 0.03	0.09 ± 0.03	2.538 ± 0.92	1.00
flow: sequential	Mean	8.922 ± 3.13	7.691 ± 3.31	33.079 ± 12.44	1.00	49.262 ± 15.83	38.472 ± 11.40	44.656 ± 7.20	1.00
	Median	8.956 ± 3.00	7.574 ± 3.26	32.503 ± 12.70	1.00	52.412 ± 20.38	38.268 ± 14.57	44.13 ± 10.56	1.00
	Random sample	10.477 ± 4.33	8.774 ± 3.75	37.703 ± 16.27	1.00	61.061 ± 22.51	49.609 ± 20.98	59.152 ± 29.97	1.00
	Interpolation	$\mathbf{5.889 \pm 2.26}$	$\mathbf{4.855 \pm 2.15}$	21.426 ± 8.18	1.00	34.086 ± 11.53	25.531 ± 7.72	31.003 ± 8.26	1.00
	Locf	9.482 ± 3.27	7.23 ± 2.73	32.741 ± 11.95	1.00	40.719 ± 13.46	31.17 ± 9.71	37.961 ± 13.23	1.00
	Nocb	8.971 ± 3.37	7.244 ± 3.08	31.694 ± 12.54	1.00	49.136 ± 19.13	37.041 ± 13.07	43.945 ± 12.47	1.00
	Moving average	9.006 ± 3.95	7.262 ± 3.38	33.103 ± 13.17	0.67	$\mathbf{24.745 \pm 12.15}$	$\mathbf{19.515 \pm 9.65}$	23.337 ± 12.06	0.13
	Random forests	10.211 ± 2.87	8.412 ± 3.00	35.549 ± 11.01	1.00	49.06 ± 14.73	37.128 ± 10.54	43.422 ± 6.87	1.00
	EM	12.111 ± 3.38	10.346 ± 3.01	47.673 ± 11.92	1.00	69.439 ± 16.02	54.84 ± 12.28	78.177 ± 9.67	1.00
	Knn	10.035 ± 2.98	8.375 ± 3.01	35.499 ± 11.47	1.00	51.228 ± 14.47	39.216 ± 10.56	46.073 ± 6.67	1.00
	Mice	12.611 ± 4.13	10.549 ± 3.47	47.185 ± 14.90	1.00	72.33 ± 13.95	56.269 ± 11.41	64.379 ± 8.45	1.00
	Amelia	12.232 ± 3.02	10.263 ± 2.76	45.0 ± 11.82	1.00	68.419 ± 13.26	55.239 ± 10.25	72.444 ± 7.02	1.00
	AutoMTS	5.889 ± 2.26	4.855 ± 2.15	21.426 ± 8.18	1.00	34.086 ± 11.53	25.531 ± 7.72	31.003 ± 8.26	1.00

type and *sensor name* fields. It is possible to further filter the observations by time period on the *period* field, the days of the week on the *calendar* field (e.g. weekdays, holidays, saturdays), as well as the desirable time granularity for the target time series.

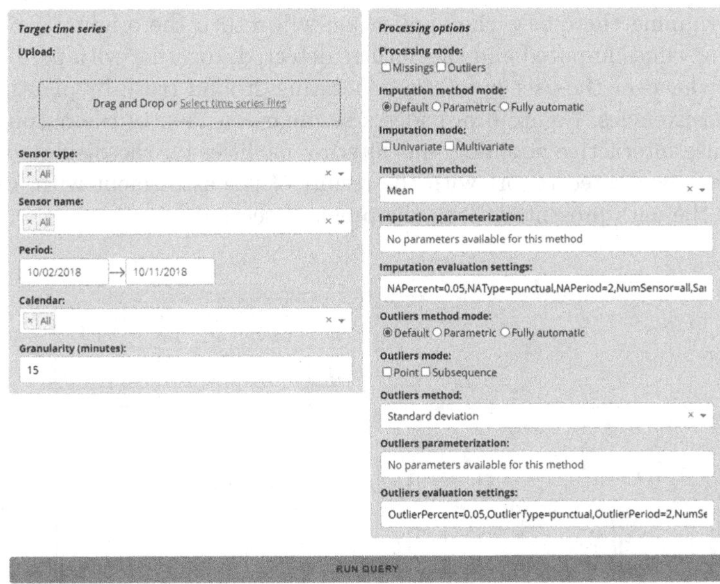

Fig. 5. AutoMTS tool: graphical user interface.

On the right panel it is possible to select the steps along the AutoMTS pipeline to be accomplished, in particular whether we want to conduct missing imputation and/or outlier detection. For both options, it is possible to select one of three distinct modes: i) the *default* mode which provides a simple rule-based decision on what is the most appropriate method given the general characteristics of the inputted series data; ii) the *parametric* mode which allows the user to select a desirable method method and its parameters; and, at last, iii) the *fully automatic* mode which runs AutoMTS (Sect. 3) to autonomously identify the best method for each one of the sensors selected in the left panel.

The user can optionally specify the profile of the artificially planted missing values and outlier values to be considered along the evaluation stage of AutoMTS (as well as to provide statistics whenever the user opts to select default and parametric modes). Here the user can select the type, percentage and duration of artificial missings and outliers. It is also possible to select the number of sensors on where we want to plant the artificial inconsistencies. Finally, the user can also specify whether the inconsistencies must occur at the same time for the inputted set of sensors or planted for each sensor individually, thus mimicking different real-world problems in heterogeneous networks.

After running the query, the application will return the original series with the missing values imputed and the outliers detected, together with performance statistics whenever the user opted for generating ground truth by planting artificial inconsistencies. Figure 6 provides a summarized view of the outputs. The user can use interactive zooming and filtering facilities on the displayed series, and access a generated report with the results of the assessment with a similar format as the ones presented along the previous section.

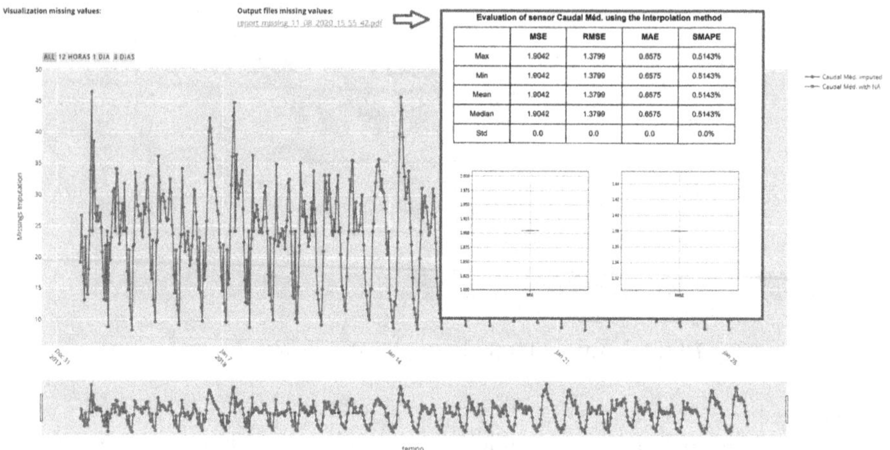

	MSE	RMSE	MAE	SMAPE
Max	1.9042	1.3799	0.6575	0.5143%
Min	1.9042	1.3799	0.6575	0.5143%
Mean	1.9042	1.3799	0.6575	0.5143%
Median	1.9042	1.3799	0.6575	0.5143%
Std	0.0	0.0	0.0	0.0%

Fig. 6. AutoMTS tool: output overview.

5 Conclusion

This work proposed a methodology, AutoMTS, for the fully-autonomous and quality-driven processing of time series data produced by networks of heterogeneous sensors. AutoMTS is parameter-free and offers strict guarantees of optimality as it places robust principles to assess, hyperparameterize and select state-of-the-art processing methods. To this end, ground truth is produced from conserved series segments in accordance with the eligible error profiles. AutoMTS further provides a comprehensive coverage of state-of-the-art methods for missing imputation, outlier detection and gross-error removal from time series data. AutoMTS implements processing methods able to explore the aided guidance from cross-variable dependencies in the presence of multivariate time series data. In addition, we guarantee the presence of methods able to deal with varying types and amount of missing and outlier values, including both point and serial occurrences of varying duration and recurrence.

The experimental assessment of AutoMTS over two real-world study cases – water distribution network systems with different sampling rates, water consumption patterns and error profiles – confirm the significance of the above contributions. The gathered results confirm the relevance of the proposed AutoMTS methodology, highlighting that processing choices are highly specific to each sensor and thus guarantees of optimality can only be provided under comprehensive and robust assessments. Also, results further offer a thorough comparison of state-of-the-art imputation and outlier detection methods, evidencing inherent strengths and limitations to handle diverse error profiles in real-world series data with varying regularities.

This work opens up possibilities for the processing of networks of sensors, particularly those networks that are large in size, heterogeneous in nature, or whose regularities are subjected to significant changes along time. AutoMTS surpasses the need for laborious processing decisions in these contexts, autonomously leveraging time series data quality for subsequent analytics.

As future work, we aim to extend the proposed methodology to guarantee the online processing of time series data streams for real-time monitoring tasks.

Acknowledgements. The authors thank the support of *Câmara Municipal do Barreiro*, *Câmara Municipal de Beja*, *Infraquinta* and *Câmara Municipal de Lisboa*. This work is further supported by national funds through *Fundação para a Ciência e Tecnologia* under projects WISDOM (DSAIPA/DS/0089/2018), ILU (DSAIPA/DS/0111/ 2018), FARO (PTDC/EGE-ECO/30535/2017), and the INESC-ID (UIDB/50021/2020) and CeMAT/IST-ID (UID/Multi/04621/2019) pluriannuals.

A Supplementary Material

Table 3. Performance of outlier detection methods for water pressure and flow sensors from Barreiro and Beja WDNs with planted point and sequential outliers on 2% of observations.

		Barreiro WDN				Beja WDN			
		F1-score	Accuracy	Precision	Recall	F1-score	Accuracy	Precision	Recall
pressure: point	Standard deviation	0.189 ± 0.11	0.976 ± 0.00	0.149 ± 0.09	**0.262 ± 0.13**	0.132 ± 0.06	0.982 ± 0.00	0.072 ± 0.03	**1.0 ± 0.00**
	Inter quartile range	0.272 ± 0.00	0.896 ± 0.00	**1.0 ± 0.00**	0.157 ± 0.00	0.058 ± 0.04	0.981 ± 0.00	0.031 ± 0.02	0.767 ± 0.42
	Isolation forests	**0.322 ± 0.00**	0.918 ± 0.00	**1.0 ± 0.00**	0.192 ± 0.00	0.337 ± 0.01	0.922 ± 0.00	**1.0 ± 0.00**	0.202 ± 0.00
	Local outlier factor	**0.322 ± 0.00**	0.918 ± 0.00	**1.0 ± 0.00**	0.192 ± 0.00	0.0 ± 0.00	0.881 ± 0.00	0.0 ± 0.00	0.0 ± 0.00
	Dbscan	0.0 ± 0.00	0.979 ± 0.00	0.0 ± 0.00	0.0 ± 0.00	0.0 ± 0.00	0.98 ± 0.00	0.0 ± 0.00	0.0 ± 0.00
	SAX	0.038 ± 0.00	0.023 ± 0.00	**1.0 ± 0.00**	0.019 ± 0.00	**0.684 ± 0.29**	0.941 ± 0.11	**1.0 ± 0.00**	0.582 ± 0.29
	AutoMTS	0.322 ± 0.00	0.918 ± 0.00	1.0 ± 0.00	0.192 ± 0.00	0.684 ± 0.29	0.941 ± 0.11	1.0 ± 0.00	0.582 ± 0.29
flow: point	Standard deviation	0.615 ± 0.11	0.989 ± 0.00	0.454 ± 0.12	**1.0 ± 0.00**	**0.779 ± 0.03**	0.991 ± 0.00	0.772 ± 0.04	**0.788 ± 0.02**
	Inter quartile range	**0.787 ± 0.10**	0.993 ± 0.00	0.659 ± 0.13	**1.0 ± 0.00**	0.445 ± 0.01	0.95 ± 0.00	**1.0 ± 0.00**	0.287 ± 0.01
	Isolation forests	0.321 ± 0.00	0.918 ± 0.00	**1.0 ± 0.00**	0.191 ± 0.00	0.332 ± 0.00	0.92 ± 0.00	**1.0 ± 0.00**	0.199 ± 0.00
	Local outlier factor	0.309 ± 0.02	0.917 ± 0.00	0.964 ± 0.07	0.184 ± 0.01	0.089 ± 0.02	0.891 ± 0.00	0.269 ± 0.07	0.054 ± 0.01
	Dbscan	0.602 ± 0.06	0.976 ± 0.00	0.959 ± 0.12	0.439 ± 0.04	0.415 ± 0.02	0.949 ± 0.00	0.912 ± 0.06	0.269 ± 0.01
	SAX	0.357 ± 0.12	0.916 ± 0.05	**1.0 ± 0.00**	0.224 ± 0.09	0.504 ± 0.08	0.959 ± 0.01	**1.0 ± 0.00**	0.341 ± 0.07
	AutoMTS	0.787 ± 0.10	0.993 ± 0.00	0.659 ± 0.13	1.0 ± 0.00	0.779 ± 0.03	0.991 ± 0.00	0.772 ± 0.04	0.788 ± 0.02
pressure: segment	Standard deviation	0.141 ± 0.12	0.977 ± 0.00	0.114 ± 0.10	**0.189 ± 0.15**	0.136 ± 0.06	**0.981 ± 0.00**	0.074 ± 0.03	**0.967 ± 0.18**
	Inter quartile range	0.253 ± 0.00	0.895 ± 0.00	**1.0 ± 0.00**	0.145 ± 0.00	0.054 ± 0.05	0.98 ± 0.00	0.028 ± 0.03	0.733 ± 0.44
	Isolation forests	**0.301 ± 0.00**	0.917 ± 0.00	**1.0 ± 0.00**	0.177 ± 0.00	**0.341 ± 0.01**	0.922 ± 0.00	**1.0 ± 0.00**	0.205 ± 0.00
	Local outlier factor	0.3 ± 0.00	0.917 ± 0.00	**1.0 ± 0.00**	0.177 ± 0.00	0.0 ± 0.00	0.881 ± 0.00	0.0 ± 0.00	0.0 ± 0.00
	Dbscan	0.0 ± 0.00	0.981 ± 0.00	0.0 ± 0.00	0.0 ± 0.00	0.0 ± 0.00	0.98 ± 0.00	0.0 ± 0.00	0.0 ± 0.00
	SAX	0.035 ± 0.00	0.021 ± 0.00	**1.0 ± 0.00**	0.018 ± 0.00	0.64 ± 0.31	0.925 ± 0.13	0.976 ± 0.13	0.561 ± 0.31
	AutoMTS	0.301 ± 0.00	0.917 ± 0.00	1.0 ± 0.00	0.177 ± 0.00	0.341 ± 0.01	0.922 ± 0.00	1.0 ± 0.00	0.205 ± 0.00
flow: segment	Standard deviation	0.653 ± 0.11	0.991 ± 0.00	0.494 ± 0.12	**1.0 ± 0.00**	**0.787 ± 0.03**	0.992 ± 0.00	0.778 ± 0.05	**0.799 ± 0.02**
	Inter quartile range	**0.827 ± 0.07**	0.995 ± 0.00	0.711 ± 0.10	**1.0 ± 0.00**	0.441 ± 0.03	0.948 ± 0.01	**1.0 ± 0.00**	0.283 ± 0.02
	Isolation forests	0.3 ± 0.00	0.917 ± 0.00	**1.0 ± 0.00**	0.177 ± 0.00	0.336 ± 0.00	0.92 ± 0.00	**1.0 ± 0.00**	0.202 ± 0.00
	Local outlier factor	0.294 ± 0.01	0.916 ± 0.00	0.981 ± 0.04	0.173 ± 0.01	0.093 ± 0.03	0.891 ± 0.00	0.277 ± 0.10	0.056 ± 0.02
	Dbscan	0.597 ± 0.06	0.976 ± 0.00	0.983 ± 0.09	0.429 ± 0.04	0.416 ± 0.03	0.948 ± 0.00	0.91 ± 0.07	0.27 ± 0.01
	SAX	0.366 ± 0.12	0.928 ± 0.04	**1.0 ± 0.00**	0.231 ± 0.09	0.508 ± 0.09	0.958 ± 0.01	**1.0 ± 0.00**	0.346 ± 0.08
	AutoMTS	0.827 ± 0.07	0.995 ± 0.00	0.711 ± 0.10	1.0 ± 0.00	0.787 ± 0.03	0.992 ± 0.00	0.778 ± 0.05	0.799 ± 0.02

Table 4. Performance of imputation methods for water pressure and flow sensors from Barreiro and Beja WDNs with planted point and sequential missing values on 10% of observations.

		Barreiro WDN				Beja WDN			
		RMSE	MAE	SMAPE	%	RMSE	MAE	SMAPE	%
pressure: point	Mean	0.064 ± 0.05	0.025 ± 0.01	0.933 ± 0.43	1.00	0.166 ± 0.00	0.154 ± 0.00	4.335 ± 0.08	1.00
	Median	0.065 ± 0.05	0.024 ± 0.01	0.921 ± 0.42	1.00	0.192 ± 0.01	0.125 ± 0.01	3.517 ± 0.18	1.00
	Random sample	0.07 ± 0.05	0.033 ± 0.03	1.247 ± 0.91	1.00	0.23 ± 0.04	0.168 ± 0.05	4.737 ± 1.45	1.00
	Interpolation	0.058 ± 0.05	0.019 ± 0.01	0.731 ± 0.37	0.99	$\mathbf{0.05 \pm 0.00}$	$\mathbf{0.031 \pm 0.00}$	0.857 ± 0.06	1.00
	Locf	0.062 ± 0.05	0.021 ± 0.01	0.796 ± 0.36	0.99	0.068 ± 0.01	0.036 ± 0.00	0.999 ± 0.10	1.00
	Nocb	0.062 ± 0.06	0.021 ± 0.01	0.808 ± 0.43	0.99	0.065 ± 0.01	0.035 ± 0.00	0.973 ± 0.09	1.00
	Moving average	$\mathbf{0.055 \pm 0.04}$	$\mathbf{0.018 \pm 0.01}$	0.712 ± 0.33	0.99	0.078 ± 0.01	0.04 ± 0.00	1.113 ± 0.10	1.00
	Random forests	0.089 ± 0.04	0.037 ± 0.01	1.414 ± 0.41	1.00	0.175 ± 0.01	0.133 ± 0.01	3.752 ± 0.18	1.00
	EM	0.098 ± 0.03	0.065 ± 0.01	2.404 ± 0.28	1.00	0.228 ± 0.01	0.188 ± 0.01	5.275 ± 0.23	1.00
	Knn	0.069 ± 0.04	0.032 ± 0.01	1.19 ± 0.38	1.00	0.166 ± 0.01	0.135 ± 0.01	3.806 ± 0.15	1.00
	Mice	0.084 ± 0.04	0.043 ± 0.01	1.574 ± 0.49	1.00	0.221 ± 0.01	0.154 ± 0.01	4.337 ± 0.24	1.00
	Amelia	0.104 ± 0.04	0.068 ± 0.01	2.508 ± 0.54	0.90	0.234 ± 0.01	0.19 ± 0.01	5.342 ± 0.21	0.90
	AutoMTS	0.055 ± 0.04	0.018 ± 0.01	0.712 ± 0.33	0.99	0.05 ± 0.00	0.031 ± 0.00	0.857 ± 0.06	1.00
flow: point	Mean	8.878 ± 0.58	7.331 ± 0.57	32.584 ± 2.57	1.00	47.573 ± 2.71	36.25 ± 1.68	48.104 ± 1.96	1.00
	Median	8.976 ± 0.57	7.312 ± 0.58	32.392 ± 2.70	1.00	49.678 ± 3.18	34.009 ± 2.14	45.249 ± 2.21	1.00
	Random sample	12.186 ± 4.31	10.31 ± 4.14	42.831 ± 12.28	1.00	59.523 ± 17.85	46.715 ± 19.59	60.721 ± 28.42	1.00
	Interpolation	$\mathbf{2.927 \pm 0.32}$	$\mathbf{2.055 \pm 0.25}$	9.821 ± 1.23	0.99	$\mathbf{17.246 \pm 1.34}$	$\mathbf{12.137 \pm 0.79}$	19.849 ± 1.08	1.00
	Locf	4.602 ± 0.47	3.49 ± 0.34	16.146 ± 1.65	0.99	21.231 ± 1.60	14.692 ± 0.85	23.877 ± 1.62	1.00
	Nocb	4.878 ± 0.55	3.649 ± 0.35	16.924 ± 1.43	0.99	21.133 ± 1.57	14.565 ± 1.01	24.047 ± 1.59	1.00
	Moving average	7.021 ± 0.79	5.48 ± 0.59	25.245 ± 2.28	0.99	21.595 ± 1.46	15.242 ± 0.83	24.077 ± 1.57	1.00
	Random forests	10.086 ± 0.83	8.085 ± 0.78	35.154 ± 3.40	1.00	47.097 ± 3.20	34.74 ± 2.29	46.244 ± 2.00	1.00
	EM	12.08 ± 0.85	9.806 ± 0.72	46.242 ± 3.74	1.00	66.136 ± 3.22	50.566 ± 2.36	79.039 ± 3.73	1.00
	Knn	9.639 ± 0.73	7.857 ± 0.66	34.557 ± 2.72	1.00	47.803 ± 2.11	35.747 ± 1.66	47.494 ± 2.00	1.00
	Mice	13.291 ± 1.05	10.893 ± 1.03	49.257 ± 5.83	1.00	72.356 ± 5.04	54.984 ± 4.83	66.829 ± 5.40	1.00
	Amelia	12.11 ± 0.94	9.78 ± 0.82	42.972 ± 4.07	0.90	65.755 ± 2.65	51.807 ± 2.21	71.872 ± 3.35	0.90
	AutoMTS	2.927 ± 0.32	2.055 ± 0.25	9.821 ± 1.23	0.99	17.246 ± 1.34	12.137 ± 0.79	19.849 ± 1.08	1.00
pressure: sequential	Mean	0.021 ± 0.02	0.013 ± 0.00	0.464 ± 0.13	1.00	$\mathbf{0.169 \pm 0.01}$	0.157 ± 0.00	4.416 ± 0.13	1.00
	Median	$\mathbf{0.02 \pm 0.02}$	$\mathbf{0.012 \pm 0.00}$	0.427 ± 0.13	1.00	0.197 ± 0.01	$\mathbf{0.13 \pm 0.01}$	3.684 ± 0.34	1.00
	Random sample	0.03 ± 0.03	0.022 ± 0.03	0.816 ± 0.95	1.00	0.23 ± 0.04	0.168 ± 0.05	4.729 ± 1.31	1.00
	Interpolation	0.03 ± 0.06	0.021 ± 0.04	0.817 ± 1.58	1.00	0.207 ± 0.01	0.164 ± 0.01	4.618 ± 0.39	1.00
	Locf	0.043 ± 0.11	0.032 ± 0.08	1.265 ± 3.34	1.00	0.24 ± 0.03	0.177 ± 0.04	4.998 ± 1.15	1.00
	Nocb	0.022 ± 0.02	0.014 ± 0.01	0.533 ± 0.20	1.00	0.237 ± 0.03	0.175 ± 0.03	4.955 ± 0.91	1.00
	Moving average	0.031 ± 0.07	0.022 ± 0.05	0.887 ± 2.12	0.12	0.061 ± 0.07	0.044 ± 0.05	1.225 ± 1.42	0.03
	Random forests	0.064 ± 0.04	0.027 ± 0.01	1.013 ± 0.33	1.00	0.189 ± 0.01	0.147 ± 0.01	4.155 ± 0.24	1.00
	EM	0.078 ± 0.01	0.061 ± 0.01	2.21 ± 0.19	1.00	0.23 ± 0.01	0.189 ± 0.01	5.326 ± 0.22	1.00
	Knn	0.045 ± 0.02	0.024 ± 0.00	0.871 ± 0.16	1.00	0.178 ± 0.01	0.146 ± 0.01	4.112 ± 0.19	1.00
	Mice	0.056 ± 0.02	0.032 ± 0.01	1.167 ± 0.31	1.00	0.229 ± 0.01	0.162 ± 0.01	4.587 ± 0.30	1.00
	Amelia	0.073 ± 0.02	0.056 ± 0.01	2.026 ± 0.41	0.89	0.238 ± 0.01	0.194 ± 0.01	5.433 ± 0.33	0.89
	AutoMTS	0.02 ± 0.02	0.012 ± 0.00	0.427 ± 0.13	1.00	0.169 ± 0.01	0.157 ± 0.00	4.416 ± 0.13	1.00
flow: sequential	Mean	8.839 ± 1.09	7.415 ± 1.25	32.556 ± 4.45	1.00	53.752 ± 13.36	40.987 ± 9.06	46.883 ± 3.85	1.00
	Median	8.884 ± 0.98	7.321 ± 1.20	32.073 ± 4.38	1.00	56.946 ± 18.43	41.172 ± 12.99	46.95 ± 8.74	1.00
	Random sample	11.589 ± 3.59	9.644 ± 3.29	41.054 ± 11.91	1.00	65.0 ± 21.24	51.602 ± 19.29	60.692 ± 29.12	1.00
	Interpolation	10.273 ± 1.21	8.493 ± 1.13	37.44 ± 4.24	1.00	40.72 ± 10.25	31.124 ± 7.20	37.789 ± 6.21	1.00
	Locf	12.074 ± 2.19	9.705 ± 1.90	43.785 ± 8.47	1.00	44.045 ± 11.33	33.286 ± 8.19	40.576 ± 11.90	1.00
	Nocb	11.092 ± 2.30	9.055 ± 1.90	39.8 ± 8.14	1.00	47.163 ± 15.20	37.099 ± 12.08	44.922 ± 12.72	1.00
	Moving average	$\mathbf{8.803 \pm 3.96}$	$\mathbf{7.065 \pm 3.39}$	31.923 ± 13.69	0.12	$\mathbf{24.396 \pm 12.08}$	$\mathbf{19.166 \pm 9.47}$	23.783 ± 12.51	0.03
	Random forests	9.913 ± 1.07	7.961 ± 1.05	34.648 ± 3.79	1.00	53.638 ± 13.67	39.78 ± 9.41	45.504 ± 4.33	1.00
	EM	12.193 ± 1.44	9.957 ± 1.25	46.583 ± 5.41	1.00	73.736 ± 13.64	57.226 ± 10.29	79.993 ± 5.61	1.00
	Knn	9.431 ± 1.13	7.696 ± 1.17	33.592 ± 4.34	1.00	54.104 ± 12.18	40.735 ± 8.19	47.224 ± 3.77	1.00
	Mice	12.148 ± 1.79	9.931 ± 1.55	44.663 ± 7.75	1.00	75.86 ± 6.55	58.838 ± 5.76	67.721 ± 4.48	1.00
	Amelia	12.125 ± 1.38	9.771 ± 1.29	43.056 ± 5.90	0.89	70.062 ± 8.17	55.106 ± 6.29	70.747 ± 3.56	0.89
	AutoMTS	8.839 ± 1.09	7.415 ± 1.25	32.556 ± 4.45	1.00	40.72 ± 10.25	31.124 ± 7.20	37.789 ± 6.21	1.00

References

1. Gill, P., Jain, N., Nagappan, N.: Understanding network failures in data centers: measurement, analysis, and implications. In: Proceedings of the ACM SIGCOMM 2011 Conference, SIGCOMM 2011, pp. 350–361. Association for Computing Machinery, New York (2011). ISBN 9781450307970. https://doi.org/10.1145/2018436.2018477
2. Jain, R.K.: A state space model-based method of seasonal adjustment. Monthly Lab. Rev. **124**, 37 (2001)
3. We, W.W.S.: Time series analysis. In: The Oxford Handbook of Quantitative Methods in Psychology, vol. 2 (2006)
4. Osman, M.S., Abu-Mahfouz, A.M., Page, P.R.: A survey on data imputation techniques: water distribution system as a use case. IEEE Access **6**, 63279–63291 (2018). ISSN 2169–3536. https://doi.org/10.1109/ACCESS.2018.2877269
5. Stekhoven, D.J., Bühlmann, P.: MissForest–non-parametric missing value imputation for mixed-type data. Bioinformatics **28**(1), 112–118 (2011). ISSN 1367–4803. https://doi.org/10.1093/bioinformatics/btr597
6. van Buuren, S., Groothuis-Oudshoorn, K.: MICE: multivariate imputation by chained equations in R. J. Stat. Softw. **45**(3), 1–67 (2011). ISSN 1548–7660. https://doi.org/10.18637/jss.v045.i03
7. Ilin, A., Raiko, T.: Practical approaches to principal component analysis in the presence of missing values. J. Mach. Learn. Res. **11**, 1957–2000 (2010). ISSN 1532–4435. http://dl.acm.org/citation.cfm?id=1756006.1859917
8. Garg, L., Dauwels, J., Earnest, A., Leong, K.P.: Tensor-based methods for handling missing data in quality-of-life questionnaires. IEEE JBHI **18**(5), 1571–1580 (2014). ISSN 2168–2208. https://doi.org/10.1109/JBHI.2013.2288803
9. Luo, Y., Cai, X., Zhang, Y., Xu, J., et al.: Multivariate time series imputation with generative adversarial networks. In: Advances in Neural Information Processing Systems, pp. 1596–1607 (2018)
10. Moritz, S., et al.: Comparison of different Methods for Univariate Time Series Imputation in R, October 2015
11. Zeileis, A., Grothendieck, G.: zoo: S3 infrastructure for regular and irregular time series. J. Stat. Softw. Articles **14**(6), 1–27 (2005). ISSN 1548–7660. https://doi.org/10.18637/jss.v014.i06. https://www.jstatsoft.org/v014/i06
12. Kowarik, A., Templ, M.: Imputation with the R package VIM. J. Stat. Softw. Articles **74**(7), 1–16 (2016). ISSN 1548–7660. https://doi.org/10.18637/jss.v074.i07. https://www.jstatsoft.org/v074/i07
13. Quevedo, J., et al.: Validation and reconstruction of flow meter data in the barcelona water distribution network. Control Eng. Practice **18**(6), 640–651 (2010). ISSN 0967–0661. https://doi.org/10.1016/j.conengprac.2010.03.003. http://www.sciencedirect.com/science/article/pii/S0967066110000791
14. Barrela, R., et al.: Data reconstruction of ow time series in water distribution systems - a new method that accommodates multiple seasonality. J. Hydroinformatics **19**(2), 238–250 (2016). ISSN 1464–7141
15. Hawkins, D.M.: Identification of outliers. Monographs on applied probability and statistics. Chapman and Hall, London [u.a.] (1980). ISBN 041221900X. http://gso.gbv.de/DB=2.1/CMD?ACT=SRCHA&SRT=YOP&IKT=1016&TRM=ppn+02435757X&sourceid=fbw_bibsonomy
16. Chan, W.-S.: Understanding the effect of time series outliers on sample autocorrelations. TEST **4**(1), 179–186 (1995)

17. Aggarwal, C.C.: Outlier analysis. In: Data Mining, pp. 237–263. Springer, Heidelberg (2015)
18. Breunig, M., Kriegel, H.-P., Ng, R., Sander, J.: LOF: Identifying density-based local outliers. **29**, 93–104 (2000). https://doi.org/10.1145/342009.335388
19. Liu, F.T., Ting, K.M., Zhou, Z.: Isolation forest. In: ICDM 2008: Proceedings of the 2008 Eighth IEEE International Conference on Data Mining, pp. 413–422. IEEE Computer Society (2008)
20. Chen, C., Liu, L.-M.: Joint estimation of model parameters and outlier effects in time series. J. Am. Stat. Assoc. **88**(421), 284–297 (1993). https://doi.org/10.1080/01621459.1993.10594321
21. Gupta, M., Gao, J., Aggarwal, C.C., Han, J.: Outlier detection for temporal data: a survey. IEEE Trans. Knowl. Data Eng. **26**(9), 2250–2267 (2014). ISSN 2326–3865. https://doi.org/10.1109/TKDE.2013.184
22. Guo, Y., et al.: Multidimensional time series anomaly detection: a GRU-based Gaussian mixture variational autoencoder approach. In: Asian Conference on Machine Learning, pp. 97–112 (2018)
23. Jagadish, H.V., Koudas, N., Muthukrishnan, S.: Mining deviants in a time series database. In: Proceedings of the 25th International Conference on Very Large Data Bases, VLDB 1999, pp. 102–113. Morgan Kaufmann Publishers Inc., San Francisco (1999). ISBN 1-55860-615-7. URL http://dl.acm.org/citation.cfm?id=645925.758373
24. Keogh, E., Lin, J., Lee, S.-H., Van Herle, H.: Finding the most unusual time series subsequence: algorithms and applications. Knowl. Inf. Syst. **11**(1), 1–27 (2007.) ISSN 0219–3116. https://doi.org/10.1007/s10115-006-0034-6
25. Chen, X.-Y., Zhan, Y.-Y.: Multi-scale anomaly detection algorithm based on infrequent pattern of time series. J. Comput. Appl. Math. **214**(1), 227–237 (2008). ISSN 0377–0427. https://doi.org/10.1016/j.cam.2007.02.027. http://www.sciencedirect.com/science/article/pii/S0377042707001100
26. Keogh, E., Lin, J., Fu, A.: Hot sax: efficiently finding the most unusual time series subsequence. In: Fifth IEEE International Conference on Data Mining (ICDM 2005), p. 8. IEEE (2005)
27. Toshniwal, D., Yadav, S.: Adaptive outlier detection in streaming time series. Proc. Int. Conf. Asia Agric. Animal ICAAA Hong Kong **13**, 186–192 (2011)
28. Fatehi, A., Huang, B.: Kalman filtering approach to multi-rate information fusion in the presence of irregular sampling rate and variable measurement delay. J. Process Control **53**, 15–25 (2017)
29. Jin, X.-B., Du, J.-J., Bao, J.: Target tracking of a linear time invariant system under irregular sampling. Int. J. Adv. Robot. Syst. **9**(5), 219 (2012)
30. Rayana, S., Akoglu, L.: Less is more: building selective anomaly ensembles. ACM Trans. Knowl. Discov. Data (TKDD) **10**(4), 1–33 (2016)
31. Zimek, A., Campello, R.J.G.B., Sander, J.: Ensembles for unsupervised outlier detection: challenges and research questions a position paper. ACM Sigkdd Explor. Newslett. **15**(1), 11–22 (2014)
32. Li, L., Zhang, J., Wang, Y., Ran, B.: Missing value imputation for traffic-related time series data based on a multi-view learning method. IEEE Trans. Intell. Transp. Syst. **20**(8), 2933–2943 (2018)
33. Oehmcke, S., Zielinski, O., Kramer, O.: kNN ensembles with penalized DTW for multivariate time series imputation. In: 2016 International Joint Conference on Neural Networks (IJCNN), pp. 2774–2781. IEEE (2016)

34. Böhm, C., Haegler, K., Müller, N.S., Plant, C.: Coco: coding cost for parameter-free outlier detection. In: Proceedings of the 15th ACM SIGKDD IC on Knowledge Discovery and Data Mining, KDD 2009, pp. 149–158. Association for Computing Machinery, New York (2009). ISBN 9781605584959. https://doi.org/10.1145/1557019.1557042
35. Davis, J., Goadrich, M.: The relationship between precision-recall and ROC curves. In: Proceedings of the 23rd International Conference on Machine Learning, pp. 233–240 (2006)
36. Snoek, J., Larochelle, H., Adams, R.P.: Practical Bayesian optimization of machine learning algorithms. In: Pereira, F., Burges, C.J.C., Bottou, L., Weinberger, K.Q. (eds.) Advances in Neural Information Processing Systems 25, pp. 2951–2959. Curran Associates Inc. (2012). http://papers.nips.cc/paper/4522-practical-bayesian-optimization-of-machine-learning-algorithms.pdf

Image Extrapolation Based on Perceptual Loss and Style Loss

Yongpeng Ren[1], Xian Zhang[1], Hongping Ren[1], Lutao Wang[1], Guanrao Huang[2], Taisong Xiong[1], and Xiaojie Li[1](✉)

[1] College of Computer Science, Chengdu University of Information Technology, Chengdu 610103, China
lixj@cuit.edu.cn

[2] Chengdu Shengdaren Technology Co. Ltd., Chengdu 610000, China

Abstract. In recent years, deep learning-based image extrapolation has achieved remarkable improvements. Image extrapolation utilizes the structural and semantic information from the known area of an image to extrapolate the unknown area. In addition, these extrapolative parts not only maintain the consistency of spatial information and structural information with the known area, but also achieve a clear, beautiful, natural and harmonious visual effect. In view of the shortcomings of traditional image extrapolation methods, this paper proposes an image extrapolation method which is based on perceptual loss and style loss. In the paper, we use the perceptual loss and style loss to restrain the generation of the texture and style of images, which improves the distorted and fuzzy structure generated by traditional methods. The perceptual loss and style loss capture the semantic information and the overall style of the known area respectively, which is helpful for the network to grasp the texture and style of images. The experiments on the Places2 and Paris StreetView dataset show that our approach could produce better results.

Keywords: Image extrapolation · Perceptual loss · Style loss

1 Introduction

In computer vision tasks, deep learning-based image completion methods are widely used. Especially in recent years, image completion has made significant achieve-ments. Image completion is a special task, which actually falls between image editing and image generating. Traditional image completion methods can be divided into texture diffusion-based [3, 4, 16], distribution-based [13, 17] and generator-based models [8, 9, 14, 15]. Texture diffusion-based methods simply collect the similar pixels from the known area to fill the missing area. Due to directly searching similar pixels from a known area to complete an unknown area, the completed results usually have unnatural images and fuzzy boundaries. Based on the idea of data driving, distribution-based methods learn the relevant distribution information from large data to generate plausible structures.

X. Wu et al. (Eds.): QShine 2020, LNICST 381, pp. 179–191, 2021.
https://doi.org/10.1007/978-3-030-77569-8_13

However, these methods produce rough and fuzzy results. Generator-based methods generally employ neural network to extract high-level spatial features of the image and finally generate plausible structures for the missing regions. Since the semantic information of the image can be captured, these methods usually generate coherent, clear and authentic contents.

Image extrapolation, a specific application of image completion, utilizes the fragments of images to infer extensional parts, and finally generating the whole picture. It can be mainly used on texture synthesis, panorama synthesis and video expansion. However, it remains a challenging problem than image inpainting due to two main issues. First, fewer image proximity information can be used to infer the unknown regions. Second, the extrapolative results must have the realistic visual effect and natural structures. A mainstream deep learning-based technology for image extrapolation is the generative adversarial network (GAN) [2]. It as generative model performs remarkably on the unsupervised learning of complex distribution. The generator and discriminator networks are trained jointly with opposite goals: the former minimizes the objective function and the latter maximizes the objective function by adversarial training. This competitiveness helps them to mimic any distribution of data. In this way, generator can capture the data distribution once trained successfully.

How to make the network generate visually-realistic and semantically-plausible contents? In this paper, a new GAN-based method is proposed. In the proposed method, we use the perceptual loss [12] to extract features from both original images and generative images, as a result the network could obtain lower-level details and high-level abstract information. This finally helps network to produce clear and nature contents. Moreover, we utilize the style loss [11] to calculate Gram matrix of the extracted features for analysing the correlation of pairwise features. This eventually catches the overall style of images. Furthermore, the general reconstruction loss has some limitations, since it directly measures differences of the pixels from both original images and generative images. The general reconstruction results indeed have the higher signal-to-noise ratio, but it contains less high-frequency information. This would lead to blurry and distorted images. While perceptual loss acquires semantic information of images by extracting feature, which means that network is in the perception of the images. Owing to using low-level pixel information and high-level abstract features of images, which can restrain both texture and style of data, our proposed method could generate real and reasonable contents for the missing regions.

Our main contributions are as follows:

1. Perceptual loss is used to extract the details and abstract features of the images, ensuring the consistency of the semantic information between the known area and unknown area. Eventually, visually eliminating the blurry of boundary and generating natural and real results.
2. Style loss is used to obtain the overall style of the images through feature extractor and Gram matrix, ensuring the consistency of texture and style between the known area and unknown area. Finally, promoting the generation of real and natural style.

2 Related Work

Patch-based and diffusion-based methods are main non-deep-learning image completion approaches, which aim at attaining non-learning statistics information to complete images. They borrow similar pixels from undamaged images to fill missing parts. It usually generates implausible texture as it fails to understand high-level semantic information of images.

Context encoders (CE) [1] is an early deep learning-based method for image completion. It uses an encoder-decoder network architecture. Specially, the encoder maps the masked image to a low-dimensional feature space, then the decoder utilizes the features to reconstruct a complete image. Moreover, the encoder and decoder are connected by channel-wise fully-connected layer. Based on unsupervised learning method, CE utilizes the known feature information to complete the missing areas. However, it usually produces blurry textures and distorted structures, since the limitation of channel-wise fully-connected layer. Thus, CE needs to be improved in the network architecture.

Then, paper [5] could produce fine complete results by using multi-scale neural patch synthesis. Its network architecture consists of two branches. The first branch is responsible for generating the contents of missing regions, which utilizes the information from known areas. The second is used to produce textures for missing regions, which considers the differences of textures from both original images and generative images. Moreover, it employs the pre-trained VGG network [6] to extract texture features, then generates fine textures by restraining the gap of the feature maps. Due to fully considering the differences of textures, the method made a great improvement. However, owing to a large memory is required and only learning patch information from an image rather than a dataset, the method has some limitations.

Recently, Mark Sabini et al. [7] proposed the image outpainting with GANs (IOGnet). At present, the state-of-the-art approaches mainly apply GAN and CNN. Therefore, IOGnet employed CNN-based GAN to extrapolate the both sides of images, namely completing the regions beyond the border of images. In addition, it could produce a panorama through recursive extrapolation. In order to improve stability of the DCGAN, it changed the traditional training procedure. The training procedure is divided into three stages. The first stage is to train the generator by using mean square error. The second stage is to train the discriminator. The last stage is to train the generator and the discriminator at the same time. However, its extrapolative results are still much vague. Thus, the method also needs further improvements.

More recently, the pluralistic image completion (PICnet) proposed by Chuanxia Zheng et al. [10] achieved good complete results. Most image completion methods generate only one output result for a missing image. In this paper, they creatively proposed a pluralistic image completion method. Namely, they generated diverse and reasonable complete results for one input. However, the pluralistic image completion faced a great challenge. Sampling from CVAE would lead to the minimal diversity, due to a ground truth only provides one instance label in the training dataset. In order to solve this problem, PICnet proposed a probabilistic theory-based framework to maintain the diversity of sampling. Moreover, it used two parallel paths to train the model. One is the reconstructive path, which uses the instance label to obtain the prior distribution of missing regions and finally it reconstructs the image by the prior distribution. Another is the generative path, whose distribution is close to the prior distribution of the reconstructive path. The network achieves the trade-off between the reconstruction of original data and the variance of conditional distribution. Since utilizing the prior information of missing parts to guide the process of image completion, the approach could produce excellently clear and realistic results. In addition, the diversity of output results provides a sufficient condition to select high-quality complete images.

3 Image Extrapolation Based on Perceptual Loss and Style Loss

The encoder-decoder network architecture is used in our method (see Fig. 1). Specially, the encoder extracts high-level abstract features of the image, and the decoder utilizes the abstract features to up-sample a reconstructive image. Moreover, both encoder and decoder are connected by residual block to realize the inference network function. This could generate the mean and variance of VAE for the decoder's sampling in the latent space. To improve the stability of training procedure, we apply LSGAN that uses the least squares loss. In addition, the structure of discriminator is similar to that of the encoder, and we use the global discriminator to score the whole image. This could grasp the overall quality of the image.

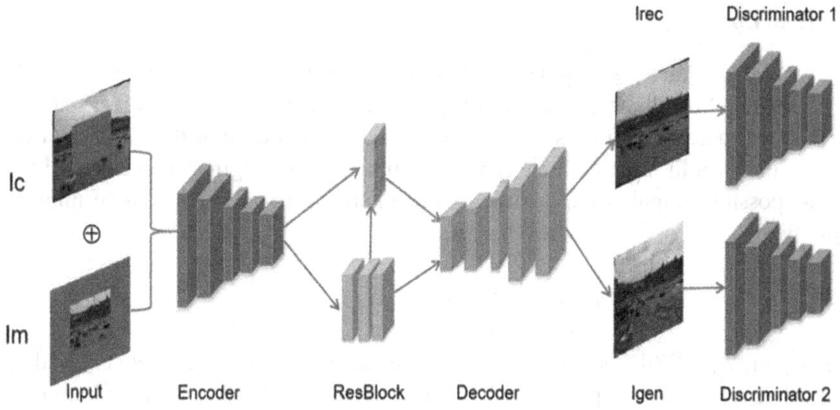

Fig. 1. Overview of our architecture.

In Fig. 1, Ic is the complement of the masked image, and Im is the masked image. ⊕ means to concatenate Ic and Im. The convolution kernels of Encoder, ResBlock, Decoder and Discriminator are all 3 × 3, and the stride is 1 × 1. During training, Ic and Im are concatenated and input into the Encoder. The mean and variance of VAE is generated by the Resblock, then utilizing the mean and variance to up-sample images. Thus, the prior information of reconstructive path is used to guide the process of image generation. Furthermore, the decoder generates the reconstructive image Irec and the generative image Igen respectively, finally sending the Irec and Igen to their own discriminators.

3.1 Perceptual Loss

The perceptual loss extracts the features from both generative images and original images through the pre-trained VGG network, and restrains the features of these images by the L1 norm. Since forcing the generative results to perceptually resemble these labels from pre-trained network, the perceptual loss could improve the quality of extrapolative areas. Formally,

$$L_p = E\left[\left\|\phi_j(I_{gt}) - \phi_j(I_{gen})\right\|_1\right] \tag{1}$$

where I_{gt} and I_{gen} are the original image and the generative image, respectively. $\phi_j(.)$ is the j-th layer feature map extracted by VGG network. Perceptual loss is to compare the features from the convolution of original images and generative images in the VGG network. This aims to make the extracted high-level feature information (such as the contents and structures of images) as close as possible. It also means that the network is perceiving the image. In the training of GAN network, perceptual loss can make the feature map of the generative image close to that of the original image, finally assists the image generation and improves the quality of generative images.

3.2 Style Loss

Style loss is similar to perceptual loss. We also use the pre-trained VGG network to extract features from both generative images and original images. However, the Gram matrix of extracted features is further calculated. As restraining the Gram matrix of features, the overall style of both generative images and original images could be as close as possible. Finally, the quality of the generative images also can be improved. Formally,

$$L_s = E\Big[\Big\|G_{\phi_i}(I_{gt}) - G_{\phi_i}(I_{gen})\Big\|_1\Big] \tag{2}$$

where I_{gt} and I_{gen} are the original image and the generative image respectively, and $\phi_i(.)$ is the i-th layer feature map extracted by VGG network. $G(.)$ denotes the Gram matrix corresponding to the feature maps. Gram matrix calculates the inner product of any k vectors in the n-dimensional Euclidean space. It can be regarded as covariance matrix without subtracting mean between different feature maps. In the convolution network, the shallow layer network extracts low-level features of images, while the deep layer network extracts high-level abstract features. These low-level and high-level features are more like the overall style of an image, which determines the real attribute of an image. By calculating the Gram matrix of these feature maps, the correlation between pairwise eigenvectors can be estimated. In the training procedure, owing to the style of generative images can be gradually close to that of original images, the quality of generative images could be improved.

3.3 Other Loss

In addition, we use the loss of PICnet:

$$L = \alpha_{KL}(L^r_{KL} + L^g_{KL}) + \alpha_{app}(L^r_{app} + L^g_{app}) + \alpha_{ad}(L^r_{ad} + L^g_{ad}) \tag{3}$$

where the superscripts r and g denote the loss of reconstructive path and generative path respectively. L_{KL} is used to constrain the distribution of hidden layer and L_{app} is the reconstruction loss of images. L_{ad} is the adversarial loss.

Finally, we add the two losses to Eq. (3):

$$L_{total} = L + \lambda_1 L_p + \lambda_2 L_s \tag{4}$$

In the experiment, we set $\lambda_1 = 0.1$, $\lambda_2 = 250$.

3.4 Improvement of IOGnet

Furthermore, we also apply perceptual loss and style loss to the improvement of the IOGnet method. The loss of the original IOGnet is as follows:

$$L_{MSE} = ||M \odot (G(I_p) - I_n)||_2^2 \tag{5}$$

$$L_D = -[\log D(I_n) + \log(1 - D(G(I_p)))] \tag{6}$$

$$L_G = L_{MSE} - \gamma \log D(G(I_p)) \tag{7}$$

where M is the mask, I_n is ground truth, and I_p is concatenation of the masked I_n and the mask. D and G are the discriminator and generator, respectively. The training procedure is divided into three stages. The first stage is through Eq. (5) to train the generator. In the second stage, the discriminator is trained through Eq. (6). In the last stage, the generator and discriminator are trained at the same time through Eq. (7).

We modify the losses of Eq. (5) and Eq. (7) in IOGnet as follows:

$$L_r = L_{MSE} + \alpha L_p + \beta L_s \tag{8}$$

$$L_G = L_r - \gamma \log D(G(I_p)) \tag{9}$$

Namely, the perceptual loss and style loss are added in the first and third stages of training, and we set $\alpha = 10$, $\beta = 100$, and $\gamma = 0.0004$ in the experiment.

4 Experiment Results

The experimental metrics can be divided into the qualitative and quantitative comparison. The qualitative comparison is visually estimated for the quality of generative results. In addition, we measure the quantitative comparison by employing the following metrics: 1) Inception Score (IS) and Frechet Inception Distance (FID) are commonly used to evaluate the quality of the generative model, which can be used to measure the diversity and clarity of generative images; 2) peak signal-to-noise ratio (PSNR) is also a widely-used full-reference metric for objective estimation; 3) other metrics include ℓ_1 loss, root mean square error (RMSE) and structural similarity (SSIM). These metrics are based on pixel-wise independence.

$$IS = \exp(E_{x \sim pg} D_{KL}(p(y|x) \| p(y))) \tag{10}$$

where x is the generative image, g is the generator, and y is the label predicted by pre-trained Inception-V3 model.

$$FID = \|\mu_x - \mu_g\|_2^2 + Tr(\sum_x + \sum_g - 2(\sum_x \sum_g)^{1/2}) \tag{11}$$

where the superscripts x and g denote the ground truth and generative image respectively. μ is the mean of eigenvectors, and Σ is the covariance matrix of eigenvectors

$$PSNR = 10 \cdot \log_{10} \frac{MAX_I^2}{MSE} \tag{12}$$

where MAX_I^2 is the maximum pixel-value of the image, and MSE is mean square error.

The experiment is implemented in Python 3.6.9, PyTorch 1.2.0, and Ubuntu 16.04. The GPU is NVIDIA Geforce RTX 2080 Ti. In addition, the batch-size is 64, and the Adam optimizer is used to update the network parameters. The fixed learning rate is 10^{-4}. Furthermore, we train the network in the end-to-end style. The LSGAN is applied to make the training procedure more stable, and updating the discriminator once then updating the generator once. Our training procedure costs 100 epochs in total.

We evaluate the proposed model on the image dataset Paris StreetView [18] and Place2 [19]. All the images are resized to 128 × 128. The test input is masked images with the 64 × 64 center area. The experimental comparison method is the baseline of PICnet and IOGnet. Because of the multiple output results of PICnet, we choose the image with the highest score of discriminator for comparison.

4.1 Qualitative Comparison

Firstly, we evaluate our model on the Paris StreetView dataset. Figure 2 shows the extrapolative results generated by PIC and our method, which could be used to visually estimate these results. In Fig. 2(c), the PIC generates the results with distorted structures and even residual shadows in the extrapolative regions. Moreover, the results of PIC also exhibit slight blurriness and unnaturalness. In Fig. 2(d) and (e), we add style loss alone or add style loss and perceptual loss at the same time into PIC. As a result, the generative results have some improvements. Our model could basically eliminate the parts with residual shadows in PIC, and force the generative images to be closer to the style of ground truth.

Then, we also show the experimental results on the Places2 in Fig. 3. We can find the similar influence on generative images after adding the style loss and perceptual loss. By and large, the visual effects of Fig. 3(d) and (e) are better than Fig. 3(c). Compared to the original PIC, our model could produce more natural and more realistic images. As a result, the experimental results show that our method can improve the quality of image extrapolation.

In order to further evaluate the effectiveness of our method, we also implement the experiments compared with the IOG method. Figure 4 and Fig. 5 show the visual effects of IOG method and our method on the Paris StreetView and Places2 dataset. Similarly, the generative results can be improved after adding the style loss and perceptual loss. Compared with the original IOG in Fig. 4(c) and Fig. 5(c), our model could smooth the coarseness of extrapolative parts, meanwhile enhance the clarity and aesthetics of generative images. On the whole, the images generated by our model are more visually-realistic and more plausible than the original IOG.

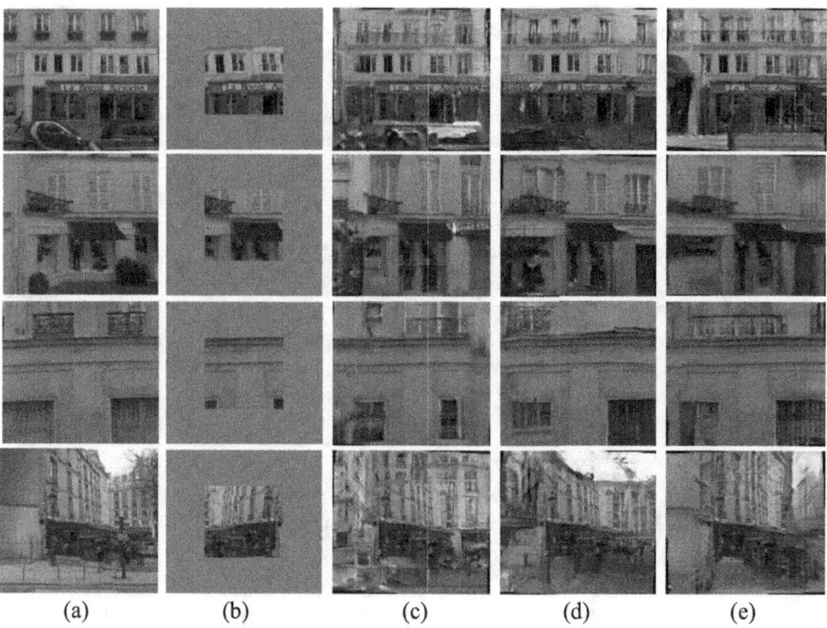

Fig. 2. Qualitative results of different methods on the Paris StreetView dataset. (a) ground truth, (b) input, (c) PIC, (d) PIC+style loss, and (e) PIC+style loss+perceptual loss.

In conclusion, we compared with the original PIC and IOG on the Paris StreetView and Place2 for qualitative comparison. The experiments prove that the style loss and perceptual loss can contribute to the improvement of image extrapolation, especially for these exhibit fuzzyness and unnaturalness.

4.2 Quantitative Comparison

Table 1 and 2 show the quantitative results of different methods on Places2 and Paris StreetView. These results of quantitative metrics show that after adding style loss and perceptual loss the generative images could be improved in some degrees. Table 1 is the quantitative metrics of 20000 test images on the Place2. The IS metric increases by 0.09, and FID decreases by 2.94. Thus, it shows that our methods could effectively improve the diversity and clarity of the generative images. Meanwhile, RMSE and ℓ_1 loss metrics decrease by 1.45 and 0.65 respectively, indicating that the overall pixel differences between the generative images and the original images are smaller. However, owing to the limitation of the 100 test images of Paris StreetView, we only evaluate the SSIM and RMSE. Furthermore, the SSIM and RMSE on the Paris StreetView are also better.

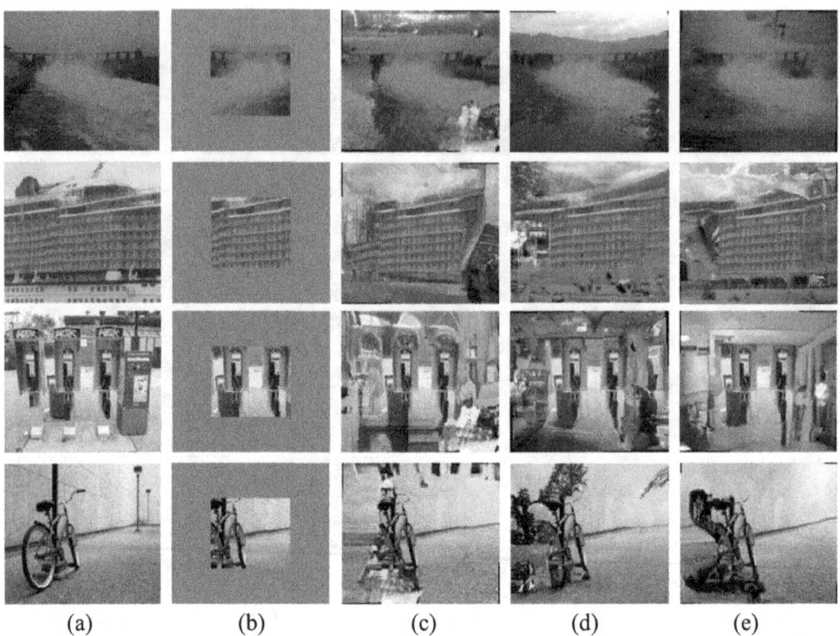

(a) (b) (c) (d) (e)

Fig. 3. Qualitative results of different methods on the Places2 dataset. (a) ground truth, (b) input, (c) PIC, (d) PIC+style loss, and (e) PIC+style loss+perceptual loss.

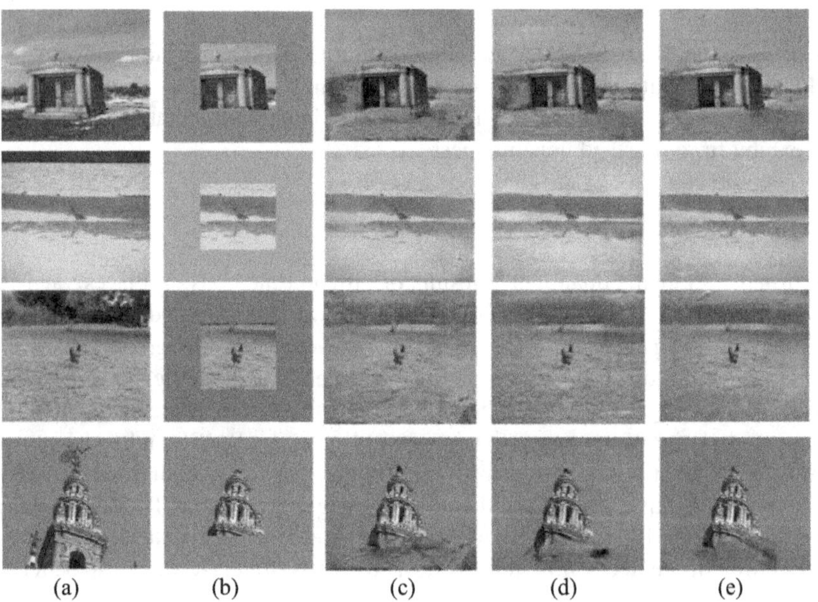

(a) (b) (c) (d) (e)

Fig. 4. Qualitative results of different methods on the Places2 dataset. (a) ground truth, (b) input, (c) IOG, (d) IOG+style loss, and (e) IOG+style loss+perceptual loss.

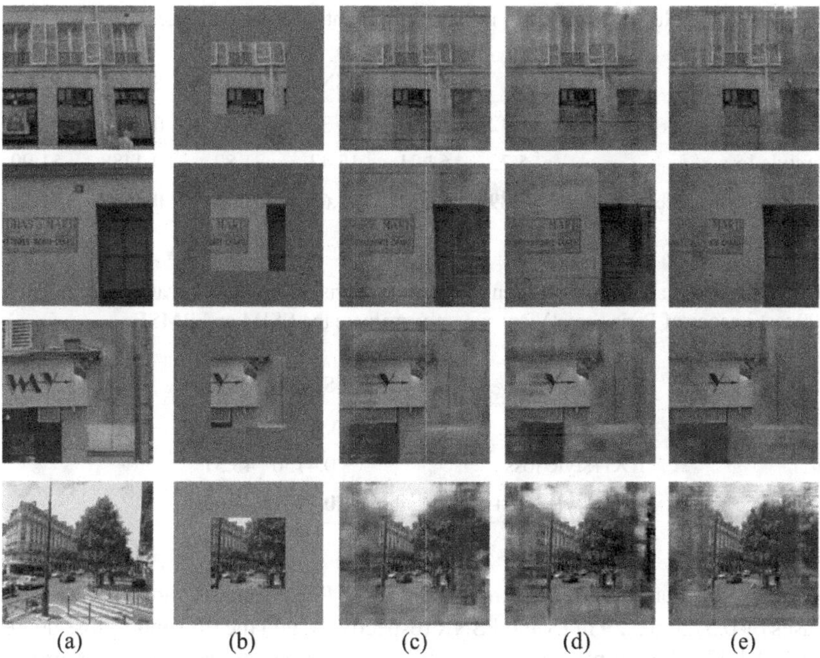

(a) (b) (c) (d) (e)

Fig. 5. Qualitative results of different methods on the Paris StreetView dataset. (a) ground truth, (b) input, (c) IOG, (d) IOG+style loss, and (e) IOG+style loss+perceptual loss.

Table 1. Quantitative results of different methods on the Places2 dataset.

Method	IS	FID	PSNR	ℓ_1 loss	SSIM	RMSE
PIC	5.60	34.81	12.95	38.29	0.4116	70.21
PIC+style loss	5.49	**31.87**	12.92	38.21	0.4109	70.54
PIC+style loss+perceptual loss	**5.69**	32.39	**13.11**	**37.64**	**0.4140**	**68.76**

Table 2. Quantitative results of different methods on Paris StreetView. Because the limitation of the 100 test images of Paris StreetView, we only evaluate the SSIM and RMSE.

Method	SSIM	RMSE
PIC	0.4248	55.63
PIC+style loss	**0.4441**	**55.58**
PIC+style loss+perceptual loss	0.4367	55.89

In addition, we have also implemented the comparative experiments on the IOG method. Table 3 and 4 show the quantitative results of different methods on Place2 and

Table 3. Quantitative results of different methods on Places2.

Method	IS	FID	PSNR	ℓ_1 loss	SSIM	RMSE
IOG	5.32	68.04	15.32	32.43	0.4455	52.56
IOG+style loss	5.72	64.71	15.43	31.82	0.4488	51.90
IOG+style loss+perceptual loss	**5.89**	**62.34**	**15.61**	**31.11**	**0.4660**	**51.14**

Table 4. Quantitative results of different methods on Paris StreetView. Because the limitation of the 100 test images of Paris StreetView, we only evaluate the SSIM and RMSE.

Method	SSIM	RMSE
IOG	0.4670	43.11
IOG+style loss	0.4730	43.37
IOG+style loss+perceptual loss	**0.4731**	**42.60**

Paris Streetview dataset. The experimental results on Place2 dataset show that our method increases by 0.57 and 0.29 in IS and PSNR respectively, and the FID and ℓ_1 loss decrease by 5.7 and 1.32. Thus, it proves that the style loss and perceptual loss can improve the quality of the generative images, and it can also improve the clarity and naturalness of the extrapolative results. Moreover, SSIM increases by 0.02 and RMSE decreases 1.42, indicating that our method could improve the quality of generative images. In addition, SSIM and RMSE metrics on the Paris StreetView dataset are also improved.

5 Conclusion

We proposed a new image extrapolation method. It combined the perceptual loss with style loss. After training and testing on the commonly-used image extrapolation dataset, experimental results show that our model can produce fine textures and natural contents for the missing images. In addition, no matter qualitative or quantitative comparison, our results could exhibit better than these methods. In the future, we will continue to explore the relevant aspects of image extrapolation, and further improve the quality of generative images in the field.

Acknowledgment. This work was supported by the National Key Research and Development Program of China (No. 2017YFC1502203), and the Sichuan Science and Technology program (2019JDJQ0002, 18MZGC0060, 2018RZ0072, and 2018GZ0184), the major Project of Education Department in Sichuan (17ZA0063).

References

1. Deepak, P., Philipp, K., Jeff, D., Trevor, D., Alexei, A.E.: Context encoders: feature learning by inpainting. In: Proceedings of the IEEE Conference on Computer Vision and Pattern Recognition, pp. 2536–2544 (2016)

2. Goodfellow, I., Pouget-Abadie, J., Mirza, M.: Generative adversarial nets. In: Proceedings of the IEEE Conference on Computer Vision and Pattern Recognition, pp. 654–656 (2014)
3. Coloma, B., Marcelo, B., Vient, C., Guillermo, S., Joan, V.: Filling-in by joint interpolation of vector fields and gray levels. In: Proceedings of the IEEE Conference on Computer Vision and Pattern Recognition, pp. 1200–1211 (2001)
4. Sun, J., Yuan, L., Jia, J., Shum, H.-Y.: Image completion with structure propagation. ACM Trans. Graph. **24**, 861–868 (2005)
5. Chao, Y., Xin, L., Zhe, L.: High-Resolution image inpainting using multi-scale neural patch synthesis. In: Proceedings of the IEEE Conference on Computer Vision and Pattern Recognition, pp. 1457–1460 (2017)
6. Simonyan, K., Zisserman, A.: Very deep convolutional networks for large-scale image recognition. In: Proceedings of the IEEE Conference on Computer Vision and Pattern Recognition, pp. 1409–1412 (2014)
7. Mark, S., Gili, R.: Painting outside the box: image outpainting with GANs. In: Proceedings of the IEEE Conference on Computer Vision and Pattern Recognition, pp. 421–420 (2018)
8. Iizuka, S., Simo-Serra, E., Ishikawa, H.: Globally and locally consistent image completion. ACM Trans. Graph. **36**, 1–14 (2017)
9. Wang, C., Liu, M.Y., Zhu, J.Y., Tao, A., Kautz, J.: High-resolution image synthesis and semantic manipulation with conditional GANs. In: Proceedings of the IEEE Conference on Computer Vision and Pattern Recognition, pp. 5–10 (2018)
10. Chuanxia, Z., Tat-Jen, C., Jianfei, C.: Pluralistic image completion. In: Proceedings of the IEEE Conference on Computer Vision and Pattern Recognition, pp. 1–5 (2019)
11. Gatys, L.A., Ecker, A.S., Bethge, M.: Image style transfer using convolutional neural networks. In: Proceedings of the IEEE Conference on Computer Vision and Pattern Recognition, pp. 2414–2423 (2016)
12. Johnson, J., Alahi, A., Fei-Fei, L.: Perceptual losses for real-time style transfer and super-resolution. In: Leibe, B., Matas, J., Sebe, N., Welling, M. (eds.) Computer Vision – ECCV 2016: 14th European Conference, Amsterdam, The Netherlands, October 11-14, 2016, Proceedings, Part II, pp. 694–711. Springer, Cham (2016). https://doi.org/10.1007/978-3-319-46475-6_43
13. Bertalmio, M., Sapiro, G., Caselles, V., Ballester, C.: Image inpainting. In: Proceedings of the 27th Annual Conference on Computer Graphics and Interactive Techniques, pp. 417–424 (2000)
14. Kamyar, N., Eric, B., Tony, J., Faisal, Z., Qureshi, M.: EdgeConnect: generative image inpainting with adversarial edge learning. In: Proceedings of the IEEE Conference on Computer Vision and Pattern Recognition, pp. 5–7 (2019)
15. Dolhansky, B., Ferrer, C.C.: Eye in-painting with exemplar generative adversarial networks. In: Proceedings of the IEEE Conference on Computer Vision and Pattern Recognition, pp. 7902–7911 (2018)
16. Ballester, C., Bertalmio, M., Caselles, V., Sapiro, G., Verdera, J.: Filling-in by joint interpolation of vector fields and gray levels. In: Proceedings of the IEEE Conference on Computer Vision and Pattern Recognition, pp. 1200–1211 (2001)
17. Hays, J., Efros, A.: Scene completion using millions of photographs. ACM Trans. Graph. **26**(3), 4 (2007)
18. Doersch, C., Singh, S., Gupta, A., Sivic, J., Efros, A.: What makes Paris look like Paris? ACM Trans. Graph. **31**(4), 1–9 (2012). https://doi.org/10.1145/2185520.2185597
19. Zhou, B., Lapedriza, A., Khosla, A., Oliva, A., Torralba, A.: Places: A 10 million image database for scene recognition. IEEE Trans. Pattern Anal. Mach. Intell. **40**(6), 1452–1464 (2018)

Comparison of Two Fourier Transform Methods in Modulation Measurement Profilometry

Min Zhong[1(✉)], Feng Chen[1], Chao Xiao[1], Peng Duan[1], Min Li[1], and Wenzao Li[2]

[1] College of Optoelectronic Engineering, Chengdu University of Information Technology, Chengdu 610225, China
zm1013@cuit.edu.cn

[2] College of Communication Engineering, Chengdu University of Information Technology, Chengdu 610225, China

Abstract. The modulation measurement profilometry encoding the spatial distribution information of specimen surface into the fringe defocus can realize the reconstruction of the specimen with complex surface shape. For this technique, the imaging axis of the CCD camera is coaxial with the projecting direction thanks to the application of a beam splitter mounting in the projection optical path. Without doing the phase unwrapping operation, it can accomplish shadow-free measurement for the specimen by extracting modulation values of the fringe pattern. The paper makes a comparison of the modulation retrieval in the conditions of fringe patterns with different surface profiles. Two Fourier transform methods are implemented in our computer simulation and practical experiment to show their performance in demodulating the modulation information from fringe patterns in optical 3D shape measurement.

Keywords: Surface measurements · Fourier transform · Image analysis · Modulation

1 Introduction

Three-dimensional surface shape measurement with the virtues of non-destructive, non-contact, high-resolution, high speed and ease of automation has become attractive to a profusion of industrial metrology, machine vision, robot simulation and automated manufacturing [1,2]. Among the existing optical three-dimensional shape measurement based on trigonometric measurement principle, Phase-shifting profilometry [3,4] is displayed more prominently thanks to its high measurement accuracy and low environmental vulnerability which is superior to that of other optical three-dimensional surface shape measurement technologies such as Four transform profilometry [5,6], Wavelet transform profilometry [7–9] and S-Transform profilometry [10–13]. However, as a multi-frame

X. Wu et al. (Eds.): QShine 2020, LNICST 381, pp. 192–208, 2021.
https://doi.org/10.1007/978-3-030-77569-8_14

fringes analysis technology, Phase shifting profilometry requires the projection of multiple patterns to reconstruct the shape surface of the specimen that speed may be sacrificed in favor of the resolution, and it's not competent for dynamic measurement. Therefore, the single-frame fringe analysis technology will be more widely utilized for measuring dynamically deformable shapes because only one single fringe projection is required to complete instantaneous deformation analysis. For this category of technique, the inherent problem of shadow and occlusion is inexorable because there is a certain angle between the projection and acquisition axes.

Some vertical measurement techniques [14–18] have been proposed in a bid to address this nontrivial problem, including modulation measurement profilometry and three-dimensional surface profilometry based on fringe contrast analysis, as distinct from the optical three-dimensional shape measurement based on the trigonometric measurement principle, applies a configuration with a feature that the projection axis coincides with the acquisition axis or the acquisition direction is the same as the projection direction. It not only avoids the effect of shadows and shutoff, but also solves the issue of phase truncating and spatial discontinuities. For this technology, the spatial distribution information of the specimen is encoded into the fringe defocus, so modulation values instead of phase information in the fringe pattern is required to be analyzed to accomplish the reconstruction.

There are two categories of the modulation measurement systems. For the first category, the grating moves in a direction parallel to the projected axis during the measurement process, and modulation values are extracted from the perspective of the entire two-dimensional fringe pattern. Phase-shifting method and two-dimensional Fourier transform method can be applied to realize the modulation calculation. For another category, a certain angle exists between the direction of the movement for the grating and that of the projection axis. Not only the above approaches, but also the one-dimensional Fourier transform method can be used to complete modulation retrieval. However, this method calculates the modulation information from an intensity curve formed by eponymous pixels from the series of the captured fringe patterns instead of each two-dimensional fringe pattern, and it is not applicable to the first type of system measurement. The one-dimensional Fourier transform method can be regarded as a kind of point-to-point operation instead of a global operation based on each entire fringe pattern analysis utilizing two-dimensional Fourier transform, which can retain more details of the specimen.

This paper gives a brief review of the theory for the modulation measurement in the second case (an angle exists), both one-dimensional Fourier transform method and two-dimensional Fourier transform method are used to extracting the fringe modulation distribution from two different viewpoints. Comparisons are made in both computer simulation and actual experiment to show their performance in demodulating the modulation information from fringe patterns in optical 3D shape measurement.

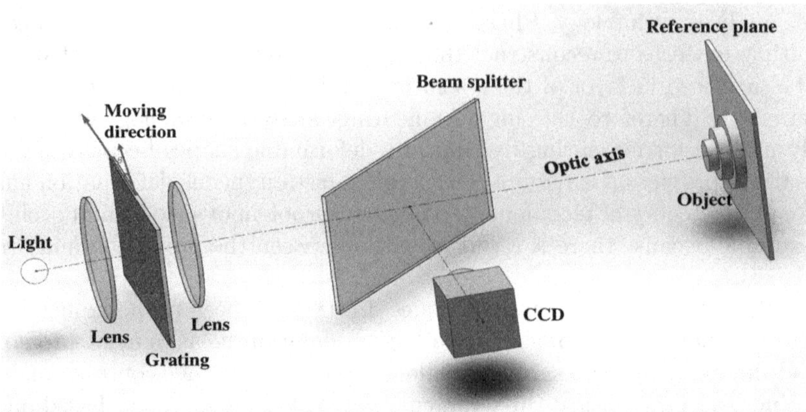

Fig. 1. The system configuration.

2 Principle

Figure 1 shows the system configuration of the modulation measurement profilometry. In this configuration, the grating is fixed on the translation platform and the normal of the grating plane is parallel to the optical axis. There is an angle (90°-β) existing between the bearings of the electric translation table and the projection optical axis. While the direction of detector is coaxial to the optic axis of projector. As such, this system offers a vertical measurement method against the problem of shadow and occlusion caused in triangulation system, which makes it possible to measure the specimen with complex surface. In the process of the measurement, the grating is driven by 1D precision translation platform, whose image will continually scan the specimen. Simultaneously, a CCD camera acquires the corresponding fringe patterns encoding the spatial

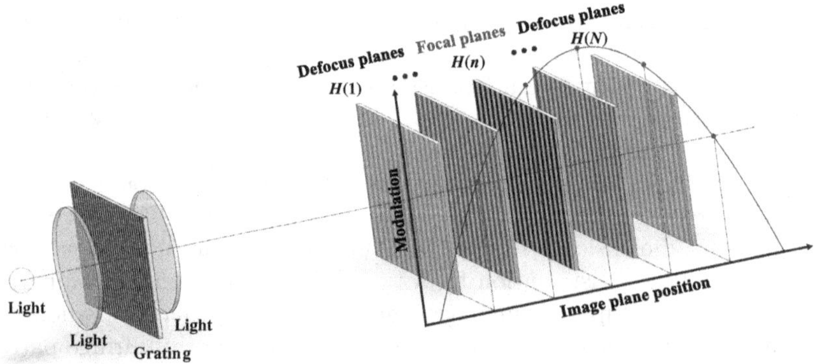

Fig. 2. Imaging principle.

distribution information of the specimen synchronously. Due to the existence of the angle (90°-β) between the direction of grating motion and the optical axis, a fixed phase shifting interval ($2\pi/N$, $N \geq 3$) appears for any two adjacent captured grating images. According to the imaging principle shown in Fig. 2, when project the image of the grating onto the specimen surface, the clearest fringe pattern with the largest modulation value observed by CCD camera will be produced on the focal plane. As the distance from the image plane to the focal plane increases, the image becomes more and more blurred. Therefore, a relationship between the modulation value and the image plane position can be formed as depicted in Fig. 2.

As shown in Fig. 1, in the actual measurement process, two sets of pulses with different time intervals will be sent by a controller. One set pulses are used to control the stepper motor to drive the grating to scan the specimen in succession, another set pulses are applied to trigger CCD camera to synchronously capture the grating image produced on the surface of the specimen that image acquisition will be synchronized with grating projection. Assuming that variable t represents the serial number of the collected image. Actually, variable t contains several other implicit meanings. When the CCD camera captures the tth frame pattern, the tth set pulses have been generated. Also, the grating derived by the 1D precision translation platform has been moved t times at an equal interval, the phase interval between the first grating image and the tth grating image will be t times of $2\pi/N$ that the total phase shift is $t \cdot 2\pi/N$. The algorithm for the tth frame fringe can be mathematically described as [16]

$$I_f(x, y, t) = \frac{R(x, y)I_0(x, y)}{M^2} + \frac{R(x, y)C_0(x, y)}{M^2} \cos\left[2\pi f_0 x + \Phi_0(x, y) + \frac{2\pi(t-1)}{N}\right] \tag{1}$$

$I_f(x, y, t)$ represents the light energy distribution of the tth fringe. $R(x, y)$ is the reflectivity of the specimen. M is the transverse magnification. $I_0(x, y)$ and $C_0(x, y)$ respectively are the background intensity and projected fringe contrast. f_0 represents the grating frequency. $\Phi_0(x, y)$ is the initial phase. N represents the total phase shift numbers for one period. The value of t ranges from 1 to $T(t = 1, 2, 3, \ldots, T)$ that the grating moves T times throughout the whole scanning process.

$I_f(x, y, t)$ represents a clear image on the focal plane. While the image on the defocused plane with a distance γ from the focusing plane can be described as

$$I_d(x, y, t; \gamma) = h(x, y; \gamma) * I_f(x, y, t) \tag{2}$$

Where $I_d(x, y, t; \gamma)$ represents the blurred image. Symbol * is the convolution operation. The expression of $h(x, y; \gamma)$ is

$$h(x, y; \gamma) = \frac{1}{2\pi\sigma_h^2} e^{-\frac{x^2+y^2}{2\sigma_h^2}} \tag{3}$$

Where $\sigma_h = cr$. The value of c is a system parameter, and it often takes the value of [19].

Substitute Eq. 1 and Eq. 3 into Eq. 2, the out-of-focus image can be written to be

$$\begin{aligned} I_d(x,y,t;\gamma) = & \frac{R(x,y)I_0(x,y)}{M^2} + \frac{R(x,y)C_0(x,y)}{M^2} \\ & \bullet e^{-\frac{f_0{}^2\sigma_h^2}{2}} \cos\left[2\pi f_0 x + \Phi_0(x,y) + \frac{2\pi(t-1)}{N}\right] \end{aligned} \tag{4}$$

The corresponding modulation distribution for the image with different degree of defocusing level can be defined as

$$\begin{aligned} M(x,y,t;\gamma) &= \frac{R(x,y)}{M^2} C_0(x,y) e^{-\frac{f_0{}^2\sigma_h^2}{2}} \\ &= R(x,y) M_0(x,y) e^{-\frac{f_0{}^2\sigma_h^2}{2}} \end{aligned} \tag{5}$$

Where $M_0(x, y)$ represents the modulation value on the focusing plane, which will be larger than that at any other defocusing plane.

3 Extraction of Modulation Values

3.1 Two-Dimensional Fourier Transform Method

When applying the two-dimensional Fourier transform method to calculate the modulation values, the result is obtained from the perspective of the entire two-dimensional image from the series of the captured fringe patterns. In mathematics, the convolution theorem states that under suitable conditions the Fourier transform of a convolution of two signals is the pointwise product of their Fourier transforms. Therefore, Eq. 2 can be written in the product form that

$$I_D(u,v,t;\gamma) = H(u,v;\gamma) \bullet I_F(u,v,t) \tag{6}$$

Where

$$\begin{aligned} H(u,v;\gamma) &= e^{-\frac{u^2+v^2}{2\sigma_h^2}} \\ I_F(u,v,t) &= \frac{R(u,v)I_0(u,v)}{M^2}\delta(u,v) \\ &+ \frac{R(u,v)C(u,v)}{2M^2}\delta(u-f_0,0)e^{i\{\Phi_0(u,v)+\frac{2\pi(t-1)}{N}\}} \\ &+ \frac{R(u,v)C(u,v)}{2M^2}\delta(u+f_0,0)e^{-i\{\Phi_0(u,v)+\frac{2\pi(t-1)}{N}\}} \end{aligned} \tag{7}$$

According to Eq. 6 and Eq. 7, the out-of-focus image in the frequency domain can be expressed as

$$I_D(u,v,t;\gamma) = I_{D(0)}(u,v,t;\gamma) + I_{D(1)}(u,v,t;\gamma) + I_{D(-1)}(u,v,t;\gamma) \tag{8}$$

Where

$$
\begin{aligned}
I_{D(0)}(u,v,t;\gamma) &= \frac{R(u,v)I_0(u,v)}{M^2}\delta(u,v) \\
I_{D(1)}(u,v,t;\gamma) &= \frac{R(u,v)C(u,v)}{2M^2}\delta(u-f_0,0)e^{i\{\Phi_0(u,v)+\frac{2\pi(t-1)}{N}\}}e^{-\frac{{f_0}^2\sigma_h^2}{2}} \\
I_{D(-1)}(u,v,t;\gamma) &= \frac{R(u,v)C(u,v)}{2M^2}\delta(u+f_0,0)e^{-i\{\Phi_0(u,v)+\frac{2\pi(t-1)}{N}\}}e^{-\frac{{f_0}^2\sigma_h^2}{2}}
\end{aligned}
\tag{9}
$$

Where $I_{D(0)}(u,\ v,\ t;\ \gamma)$ and $I_{D(1)}(u,\ v,\ t;\ \gamma)$ respectively represent the zero and the fundamental frequency component. $I_{D(-1)}(u,\ v,\ t;\ \gamma)$ is the conjugate of the $I_{D(1)}(u,\ v,\ t;\ \gamma)$. Select an appropriate filter window to extract the fundamental frequency part and then apply the inverse Fourier transform. The modulation value can be obtained by taking the absolute value of the obtained result.

$$
\begin{aligned}
M_{2DFFT}(x,y,t;\gamma) &= \left|\frac{R(x,y)}{2M^2}C_0(x,y)e^{i\{\Phi_0(x,y)+2\pi(t-1)/N\}}e^{-\frac{{f_0}^2\sigma_h^2}{2}}\right| \\
&= \frac{R(x,y)}{2M^2}C_0(x,y)e^{-\frac{{f_0}^2\sigma_h^2}{2}} \\
&= \frac{1}{2}R(x,y)M_0(x,y)e^{-\frac{{f_0}^2\sigma_h^2}{2}}
\end{aligned}
\tag{10}
$$

When making a comparison between Eq. 10 and Eq. 5, it becomes apparent that there is only one difference (constant 1/2), which means that the modulation distribution of out-of-focus image can be acquired by multiplying constant 2.

Two-dimensional Fourier transform method is a global analysis method. The modulation value of any pixel in the fringe pattern is extracted by taking advantage of the information of pixels in the entire two-dimensional fringe pattern. Therefore, even if a pixel in the fringe has no data, a modulation value for this pixel can be estimated based on the information of surrounding pixels, which implies that the sensitivity of this method is not high enough in detection of minor defects. Besides, the application of the filtering operation makes this approach fail to retain the details of the specimen, which will smooth the steep edges and corners, bring down a steep slope.

3.2 One-Dimensional Fourier Transform Method

When utilizing the one-dimensional Fourier transform method to realize the modulation acquisition, the calculation result is completed from an angle of an intensity curve formed by eponymous pixels from the captured images. Therefore, it is a kind of point-to-point operation instead of a global operation based on each entire fringe pattern analysis.

As shown in Fig. 3, points at the same position $(x,\ y)$ are extracted from this series of images, and finally the curve $I(t)(x,\ y)$ (the blue curve, the red one represents its outline) shown in the right figure can be formed. Based on Eq. 1, curve $I(t)(x,\ y)$ from the definite point $(x,\ y)$ of the fringes can be simply expressed as

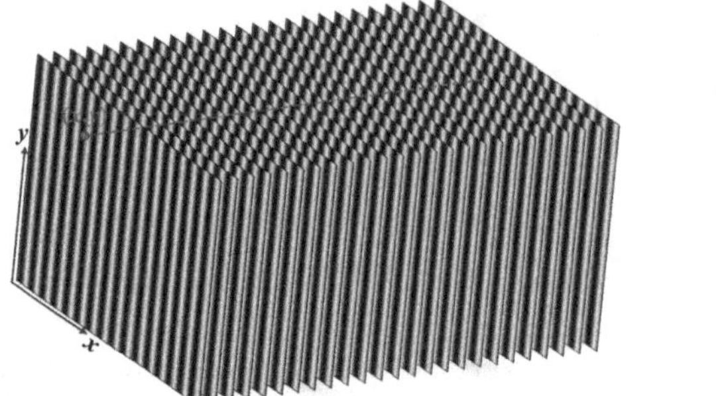

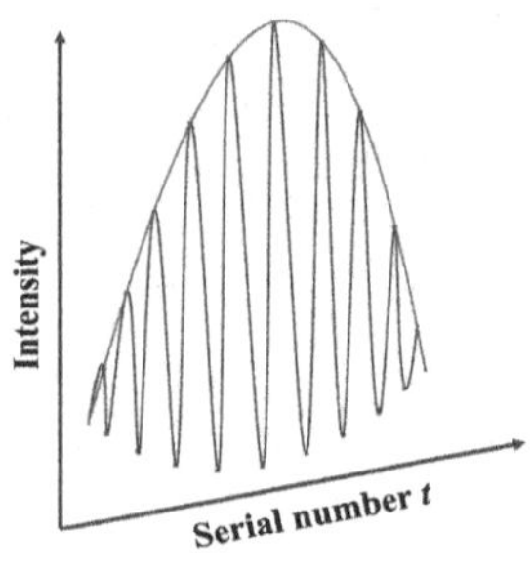

Fig. 3. The curve $I(t)(x,\ y)$. (Color figure online)

$$I_d(t)|_{(x,y)} = \frac{R}{M^2}\left\{I_0 + C_0 e^{-\frac{f_0{}^2\sigma_h^2}{2}}\cos\left[\Phi + \frac{2\pi(t-1)}{N}\right]\right\}\Bigg|_{(x,y)} \tag{11}$$

For the same position $(x,\ y)$ on the fringe patterns,$\Phi = 2\pi f_0 x + \Phi_0$ is a constant. Do Fourier transform operation on Eq. 11 that

$$I_{D'}(t)|_{(u,v)} = I_{D'(0)}(t)\Big|_{(u,v)} + I_{D'(1)}(t)\Big|_{(u,v)} + I_{D'(-1)}(t)\Big|_{(u,v)} \tag{12}$$

Where

$$\begin{aligned} I_{D'(0)}(t)\Big|_{(u,v)} &= \frac{RI_0}{M^2}\delta(u,v) \\ I_{D'(1)}(t)\Big|_{(u,v)} &= \frac{RC}{2M^2}\delta(u-f_0,0)e^{i\{\Phi+\frac{2\pi(t-1)}{N}\}}e^{-\frac{f_0{}^2\sigma_h^2}{2}} \\ I_{D'(-1)}(t)\Big|_{(u,v)} &= \frac{RC}{2M^2}\delta(u+f_0,0)e^{-i\{\Phi+\frac{2\pi(t-1)}{N}\}}e^{-\frac{f_0{}^2\sigma_h^2}{2}} \end{aligned} \tag{13}$$

$I_{D'(0)}(t)\Big|_{(u,v)}$ denotes the zero-spectrum component of the curve, $I_{D'(1)}(t)\Big|_{(u,v)}$ and $I_{D'(-1)}(t)\Big|_{(u,v)}$ respectively represent the fundamental spectrum component of the curve. The utilization of a proper filter can make the acquisition of useful fundamental component come true, and then doing inverse Fourier transform. Modulation value for each fringe at position $(x,\ y)$ can be finally obtained by taking the absolute value of the result.

$$\begin{aligned} M_{1DFFT}(t)|_{(x,y)} &= \left|\frac{R}{2M^2}C_0 e^{i\{\Phi+\frac{2\pi(t-1)}{N}\}}e^{-\frac{f_0{}^2\sigma_h^2}{2}}\Bigg|_{(x,y)}\right| \\ &= \frac{R}{2M^2}C_0 e^{-\frac{f_0{}^2\sigma_h^2}{2}}\Bigg|_{(x,y)} \\ &= \frac{1}{2}RM_0 e^{-\frac{f_0{}^2\sigma_h^2}{2}}\Bigg|_{(x,y)} \end{aligned} \tag{14}$$

Comparing Eq. 14 with Eq. 5, modulation distribution of out-of-focus image for point (x, y) can be calculated by multiplying constant 2. While the modulation maps of the whole images can be obtained by repeatedly doing the above operation for every point in the fringe patterns. For this method in modulation retrieval, the calculation result for each pixel in a fringe pattern in effect terminates the influence from the information of its neighbor pixels. That is, the operation of each pixel in the entire fringe pattern is independent of each other. The frequency spectrum of a curve by one-dimensional Fourier transform method is simpler than that of an image by two-dimensional Fourier transform method. Moreover, the generation of high-frequency components can be effectively avoided by appropriately selecting the scanning range, which refrains from spectrum aliasing between fundamental frequency and high order frequency. Obviously, due to the simple spectrum, it is easy to extract useful fundamental frequency information to retain more details of the specimen.

4 Simulation

In the actual application, the surface of the tested object is unpredictable. This section will make a comparison of the performance of the two Fourier transform methods in the reconstruction of two different surface profiles.

The main parameters are set as following. Both the background intensity and the fringe contrast are 0.5 $(I_0(x, y) = 0.5, C_0(x, y)) = 0.5$. The reflectivity factor is set to be $R(x, y) = 0$, The frequency of the grating is $f_0 = 1/6$ pixel^{-1}, the focal length and the diameter of the lens respectively are $f = 58$ mm and $d = 40$. In the scanning process, a total of 160 frames of fringes were collected by CCD that $T = 160$. The size of each image is 264 × 264 pixels. To match reality more exactly, random noise of 3% fringe intensity is added in each image. All the simulations are performed on MATLAB platform.

4.1 Comparison of Smooth Surface

The first tested object we used is the PEAKS function. It has an absolute height of 60 mm as shown in Fig. 4(a). Figure 4 (b) shows the 60th frame of captured images. Both one-dimensional Fourier transform method (1DFT) and Two-dimensional Fourier Transform method (2DFT) are used to calculating the modulation values of the images, and the performance for the 60th frame of captured images obtained by the two methods are respectively shown in Figs. 4 (c) and (d). It is palpable that there are many small spots in Fig. 4 (c), while the edges in Fig. 4 (d) are blurred. The reconstruction result and error distribution for the two approaches are respectively shown in Figs. 5 (a)–(d). The standard deviation errors (mean square error RMS) are 0.353 mm by 1DFT and 0.144 mm by 2DFT. Obviously, for the tested object with smooth surfaces, there are many burrs on the surface shape obtained by 1DFT. While for 2DFT, it softens the steep area even though the smoothness of the surface is preserved. In order to illustrate features clearly, Figs. 6(a)–(c) show a small area (rows:

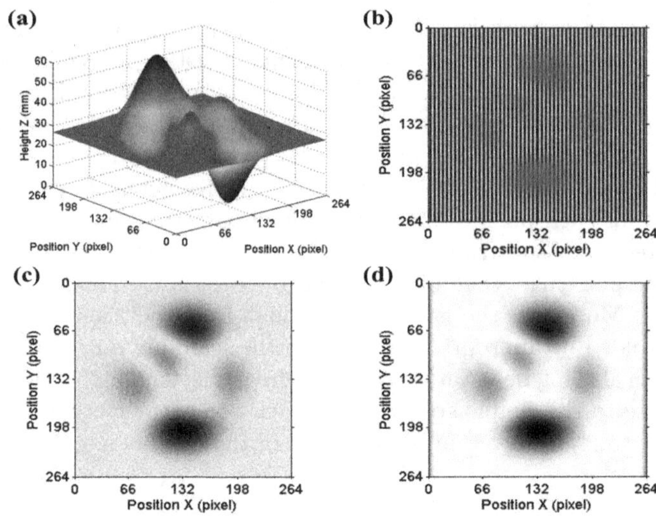

Fig. 4. Simulation: (a) The simulated object; (b) The 60th frame of fringe pattern; (c) Modulation values obtained by 1DFT; (d) Modulation values obtained by 2DFT.

185–220; columns: 115–150) of the simulated object and the reconstructions of the same area by the two methods. Figure 6(d) shows the part from the 170th column to the 225th column in the 100th row of the simulated object, the same part of the reconstruction by 1DFT and that by 2DFT. For the 1DFT method

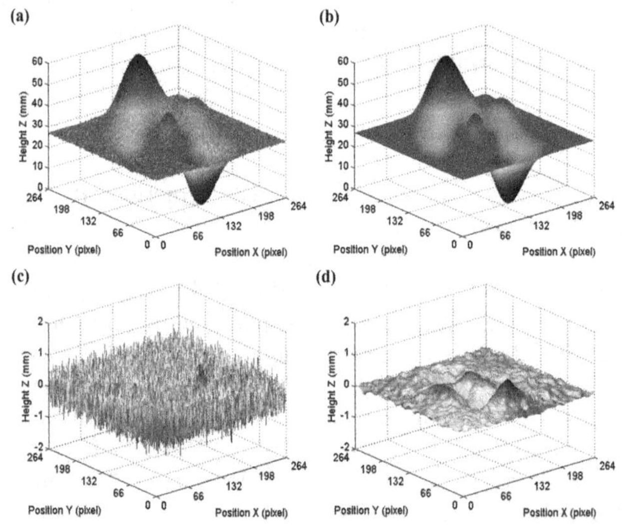

Fig. 5. Simulation results: (a) Reconstruction by 1DFT; (b) Reconstruction by 2DFT; (c) Error distribution by 1DFT; (d) Error distribution by 2DFT.

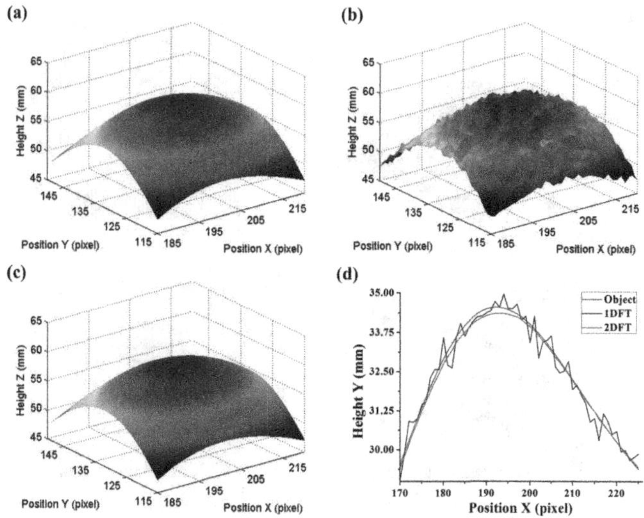

Fig. 6. Comparisons: (a) Part of the simulated object; (b) Part of the reconstruction by 1DFT; (c) Part of the reconstruction by 2DFT; (d) the 170th column to the 225th column in the 100th row of the simulated object, the same part of the reconstruction by 1DFT and that by 2DFT.

in modulation retrieval, calculation result for one pixel in a fringe pattern is extracted from a curve produced by retrieving the same coordinate from the captured fringe patterns, that is, there is no relationship among adjacent pixels, and a slight difference in the modulation values can result in a large difference in height values. This existence of the difference leads to a reconstruction for the measured object with a coarse or irregular surface. However, 2DFT method is a global analysis method. The modulation value of any pixel in the fringe pattern is extracted by taking advantage of the information of pixels in the entire fringe pattern. Besides, the application of the filtering operation brings down steep slopes even though the character of smoothness for the surface shape remains.

4.2 Comparison of Step Surface

To furtherly make a comparison, another computer simulation is used in this section. The simulated object is shown in Fig. 7(a) has a dramatic change in height that there are four discontinuous height steps (10 mm, 30 mm, 50 mm, 70 mm) from the bottom to the top.

In this simulation, the system parameters are the same as those mentioned above. Figure 7 (b) shows the 64th frame of captured images. Figures 7 (c) and (d) are respectively the modulation values of Fig. 7 (b) obtained by 1DFT and 2DFT. Even the measured object has rapid height variation on the shape surface, the application of 1DFT method can obtain accurate modulation value (the edges of the steps are clear) in the areas where height variation is steep. While for the

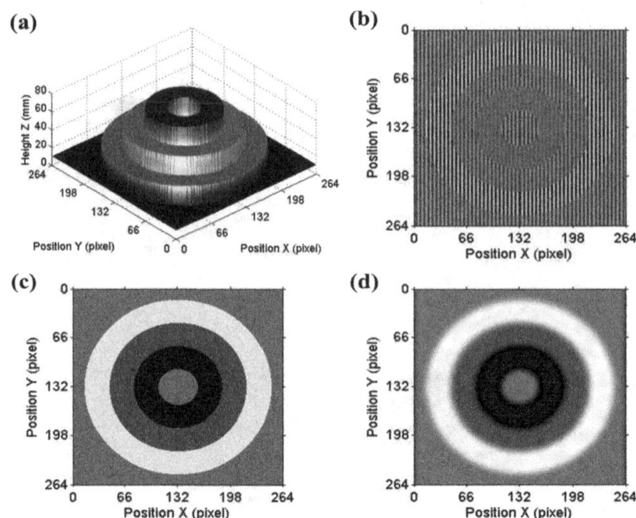

Fig. 7. Simulation: (a) The simulated object; (b) The 60th frame of fringe pattern; (c) Modulation values obtained by 1DFT; (d) Modulation values obtained by 2DFT.

2DFT method, the edge between two steps with different heights is blurred. The reconstructions obtained by the two methods are respectively shown in Figs. 8 (a) and (b). Figures 8 (c) and (d) are the corresponding error distributions. The standard deviation errors are 0.351 mm by 1DFT method and 3.543 mm by 2DFT method. For 2DFT method, due to the filtering operation applied in the analysis of each fringe, high-frequency component containing the details of the measured object is filtered out, the sharp edge of the step becomes smooth. While the result by 1DFT is much better since the modulation is calculated using point-to-point algorithm which eliminates the influence from the neighbor pixels. For clarity, Figs. 9(a)–(c) shows a small area (rows: 195–210; columns: 125–140) on the second step plane of the tested object, the reconstructed result by 1DFT and that by 2DFT. Figure 9(d) shows the part from the 192nd column to the 242nd column in the 152nd row of the simulated object, the same part of the reconstruction by 1DFT and that by 2DFT. For the 1DFT method, the calculation of any point for the height value has no relationship with others that the edge of any two steps can be reconstructed correctly. However, it is the independence of points that the reconstructed plane is not flat (shown in Fig. 9(b)). While for the 2DFT method, the utilization of the filtering operation leads to the result that the high frequency component containing the details of the measured object is filtered out, the sharp edge of the step becomes smooth.

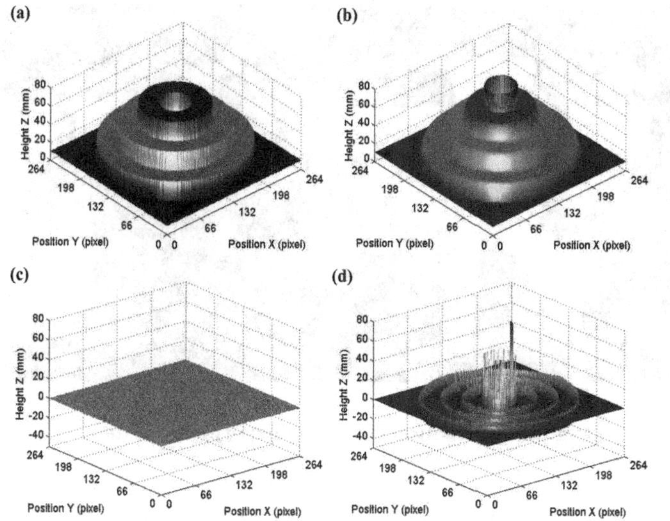

Fig. 8. Simulation results: (a) Reconstruction by 1DFT; (b) Reconstruction by 2DFT; (c) Error distribution by 1DFT; (d) Error distribution by 2DFT.

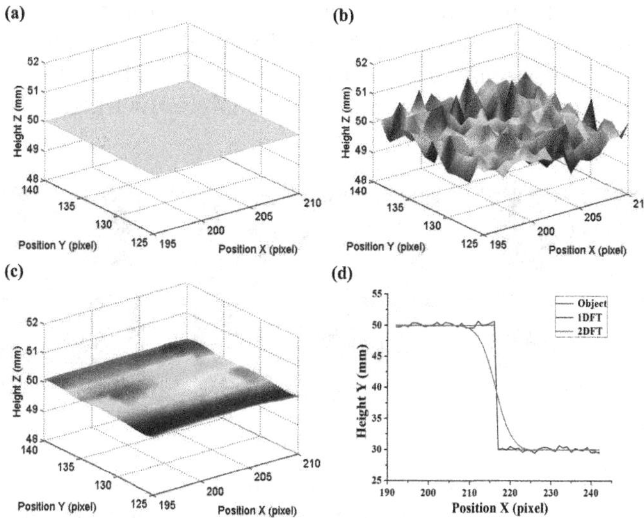

Fig. 9. Comparisons: (a) Part of the simulated object; (b) Part of the reconstruction by 1DFT; (c) Part of the reconstruction by 2DFT; (d) The 192nd column to the 242nd column in the 152nd row of the simulated object, the same part of the reconstruction by 1DFT and that by 2DFT.

5 Experiment

In order to verify the conclusions in the simulation, experiment is carried out to confirm the results. The measurement system is shown in Fig. 10. The grating is 2

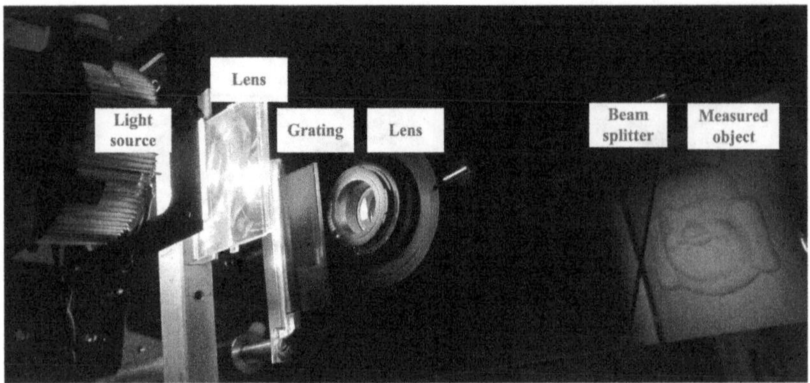

Fig. 10. Configuration of the experiment system.

lines/mm. The face of a Maitreya shown in Fig. 11(a) is utilized as the measured object. In the process of measurement, 471 frames of the fringe patterns are captured by the CCD camera (BASLER A504k). To show the variety of focus plane, Figs. 11(a)–(c) respectively show the 200th, the 300th, the 400th frame of the fringe patterns. It illustrates that the focus plane of the projector changes from top to bottom of the measured object. To reduce the computational work, the size of the captured images is cut to be 880 × 1030 pixels. Both the 1DFT method and the 2DFT method are applied to analyze the fringes.

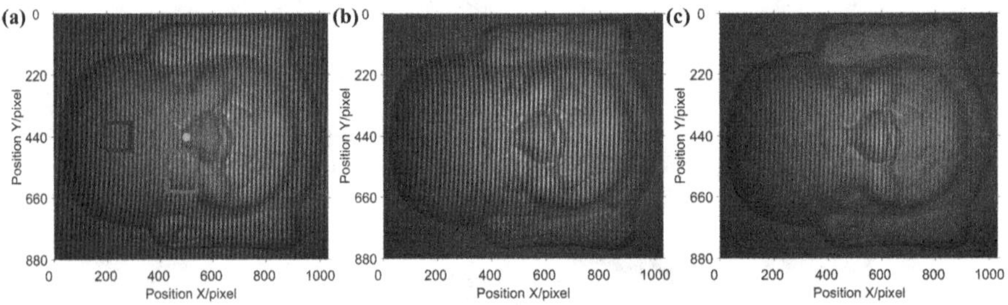

Fig. 11. Fringe patterns: (a) The 200th frame of fringe pattern; (b) The 300th frame of fringe pattern; (c) The 400th frame of fringe pattern. (Color figure online)

For the 1DFT method, the extraction of modulation is done from an angle of an intensity curve formed by eponymous pixels from the captured images. Take the center pixel (441, 515) (marked in Fig. 11(a) with an orange dot) of the cropped image as an example, its intensity cure shows in Fig. 12(a), whose spectrum is shown in Fig. 12(b). Obviously, the spectrum of a cure is very simple, and it's easy for one to filter the useful fundamental frequency information to obtain the envelope of this curve. When the envelope for each pixel is extracted,

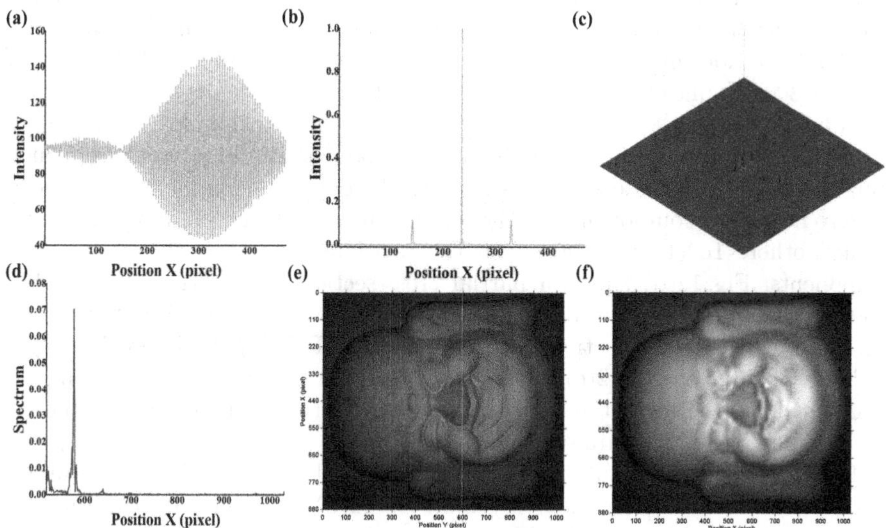

Fig. 12. Modulation retrieval: (a) Intensity curve formed by eponymous pixels from the captured images; (b) The spectrum of (a); (c) The spectrum of Fig. 11(a); (d) Partial cross-section of (c); (e) Modulation values obtained by 1DFT; (f) Modulation values obtained by 2DFT.

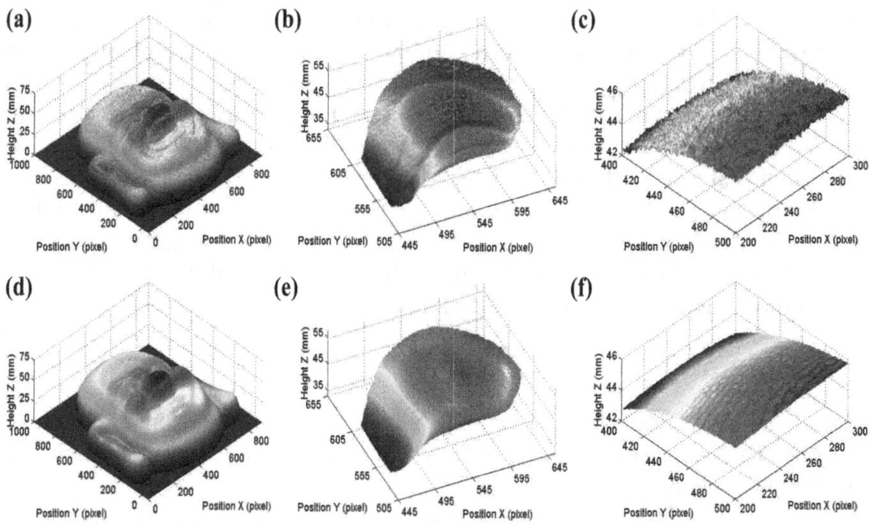

Fig. 13. Experiment results: (a) The reconstruction result obtained by 1DFT; (b) The reconstruction result of the left eye by 1DFT; (c) Enlarged picture of the forehead by 1DFT; (d) The reconstruction result obtained by 2DFT; (e) The reconstruction result of the left eye by 2DFT; (f) Enlarged picture of the forehead by 2DFT.

the modulation distribution for any $t^{\text{th}}(t = 1, 2, 3 \ldots 471)$ fringe can be obtained by extracting the modulation value of the tth point from each envelope line. For the 2DFT method, modulation retrieval is completed from the whole 2D image. Take the 300th frame of images as an example, its spectrum is shown in Fig. 12(c). It is apparent that the spectrum of a 2D image is more complex than that of a curve. What's worse, the zero-frequency component and high-order frequency component are very close to the fundamental frequency component, and both the zero frequency component and the fundamental frequency component extend to each other. To clearly distinguish the relationship among these frequency components, Fig. 12(d) shows a partial cross-section (row: 441, column: 518–1030) of Fig. 12(c). Figures 12(e) and (f) are the modulation distributions of Fig. 11(b), which are respectively obtained by the 1DFT method and the 2DFT method. Obviously, Fig. 12(e) is clearer than Fig. 12(f). Even for the complex regions such as the nose, the mouth and the eyes, the modulation distribution map can be obtained accurately by 1DFT.

The reconstruction results by the 1DFT method and the 2DFT method are respectively shown in Figs. 12(a)–(c) and Figs. 12(d)–(f). Figures. 12(a) and (d) show the whole surface of the Maitreya by the two methods. For the result obtained by the 1DFT method, the details of the shape surface for the object can be well preserved such as the tiny changes in the eyes, mouth and nose. However, the application of the filtering operation makes the 2DFT method fail to retain the characteristics of these areas with small changes in height. For clarity, Figs. 12(b) and Fig. 12(e) respectively show the height distribution of the left eye (marked in Figs. 11(b) with red rectangle) reconstructed by the two methods. The position of the eyelids can be clearly identified in Fig. 12(b), while Fig. 12(e) just shows a smooth cambered surface that completely misses the details. Figures 12(c) and Fig. 13(f) show the reconstructions of a flat area (part of the forehead marked in Figs. 11(b) with blue rectangle, rows: 401–500; columns: 201–300) calculated by the two methods. Due to the application of point-to-point algorithm for 1DFT, there is no relationship for the calculation of any two adjacent pixels. A slight difference in the modulation values can result in a large difference in height values. Therefore, there are many burrs on the surface shape obtained by 1DFT. While, as shown in Fig. 12(f), 2DFT can reconstruct the corresponding region with a smooth surface.

6 Discussion

In the actual application for the two methods, to get more efficient modulation or improve the measurement accuracy, one can set up the experimental system from the following aspects:

- *Period of the grating.* The period of the grating should be small enough, so that the modulation distribution for each pixel point will be narrow, which helps extract the maximum modulation value and the corresponding serial number more accurately. At the same time, one should take the Nyquist–Shannon sampling theorem into consideration that the sampling period for the fringe pattern should be at last more than 4 points.

- *Pitch and range of grating movement.* The distance of each movement for the grating should be appropriately small to ensure that the modulation curve is sampled properly and smoothly. The entire range of grating movement should be large enough to ensure that the modulation curve includes the maximum modulation value and at least one minimum modulation value adjacent to either side of it.
- *Translation platform.* The grating is driven by a 1D precision translation platform, and this movement is manually controlled. If a closed-loop step-motor control system is utilized, the measurement accuracy will be improved.

7 Conclusion

Two methods including 1DFT method and 2DFT are utilized to extract the modulation distribution of fringe patterns. For 1DFT method, the calculation is completed from an angle of an intensity curve formed by eponymous pixels from a series of captured images, which terminates the influence from the information of other neighbor pixels and can better retain the details of the object under test. While the surface of the reconstruction result will be coarse due to the differences of the modulation values for any adjacent pixels. For 2DFT method, the result is obtained from the perspective of each complete two-dimensional image from the series of the captured fringe patterns, and the application of the filtering operation will smooth the steep edges and corners, bring down a steep slope and fail to retain more details of the measured object.

Acknowledgments. The authors would like to acknowledge the support of the National Natural Science Foundation of China (NSFC) (61801057), Sichuan education department project 18ZB0124, College Students' innovation and entrepreneurship training programs S201910621117 and 202010621108.

References

1. Zhang, S.: Recent progresses on real-time 3D shape measurement using digital fringe projection techniques. Opt. Lasers Eng. **48**(2), 149–158 (2010)
2. Zhang, S.: Absolute phase retrieval methods for digital fringe projection profilometry: a review. Opt. Lasers Eng. **107**, 28–37 (2018)
3. Jiao, D., Liu, Z., Shi, W., Xie, H.: Temperature fringe method with phase-shift for the 3D shape measurement. Opt. Lasers Eng. **112**, 93–102 (2019)
4. Zuo, C., Feng, S., Huang, L., Tao, T., Yin, W., Chen, Q.: Phase shifting algorithms for fringe projection profilometry: a review. Opt. Lasers Eng. **109**, 23–59 (2018)
5. Su, X., Chen, W.: Fourier transform profilometry: a review. Opt. Lasers Eng. **35**(5), 263–284 (2001)
6. Zhang, H., Zhang, Q., Li, Y., Liu, Y.: High speed 3D shape measurement with temporal Fourier transform profilometry. Appl. Sci. **9**(19), 4123 (2019)
7. Zhang, Z., Zhong, J.: Applicability analysis of wavelet-transform profilometry. Opt. Express **21**(16), 18777–18796 (2013)
8. Zhong, J., Weng, J.: Phase retrieval of optical fringe patterns from the ridge of a wavelet transform. Opt. Lett. **30**(19), 2560–2562 (2005)

9. Kocahan, O.: Determination of height profile from a two-dimensional fringe signal using a two-dimensional continuous wavelet transform. Turk. J. Phys. **41**, 81–89 (2017)
10. Stockwell, R.G., Mansinha, L., Lowe, R.P.: Localization of the complex spectrum: the s transform. IEEE Trans. Signal Process. **44**(4), 998–1001 (1996)
11. Zhong, M., Chen, W., Jiang, M.: Application of s-transform profilometry in eliminating nonlinearity in fringe pattern. Appl. Opt. **51**(5), 577–587 (2012)
12. Zhong, M., Chen, W., Wang, T., Su, X.: Application of two-dimensional s-transform in fringe pattern analysis. Opt. Lasers Eng. **51**(10), 1138–1142 (2013)
13. Zhong, M., Chen, W., Su, X., Zheng, Y., Shen, Q.: Optical 3D shape measurement profilometry based on 2D s-transform filtering method. Opt. Commun. **300**, 129–136 (2013)
14. Su, X., Su, L., Li, W.: New Fourier transform profilometry based on modulation measurement. In: Proceedings of SPIE - The International Society for Optical Engineering, vol. 3749 (1999)
15. Su, L., Su, X., Li, W., Xiang, L.: Application of modulation measurement profilometry to objects with surface holes. Appl. Opt. **38**(7), 1153–1158 (1999)
16. Zhong, M., Su, X., Chen, W., You, Z., Lu, M., Jing, H.: Modulation measuring profilometry with auto-synchronous phase shifting and vertical scanning. Opt. Express **22**(26), 31620 (2014)
17. Su, L., Li, W., Su, X., Xiang, L.: Phase-shift error calibration in modulation measurement profilometry. In: 18th Congress of the International Commission for Optics (1999)
18. Mizutani, Y., Kuwano, R., Otani, Y., Umeda, N., Yoshizawa, T.: Three-dimensional shape measurement using focus method by using liquid crystal grating and liquid varifocus lens. In: Proceedings of SPIE - The International Society for Optical Engineering, vol. 6000, pp. 336–431 (2005)
19. Subbarao, M., Gurumoorthy, N.: Depth recovery from blurred edges, pp. 498–503 (1988)

Research on Image Enhancement Model Based on Variable Order Fractional Differential CLAHE

Guo Huang[1], Li Xu[2,3(✉)], Qing-li Chen[1], Xiu-qiong Zhang[1], Tao Men[1], and Hong-ying Qin[1]

[1] Sichuan Province University Key Laboratory of Internet Natural Language Intelligent Processing, Leshan Normal University, Leshan 614000, China

[2] School of Electronics and Materials Engineering, Leshan Normal University, Leshan 614000, China

[3] School of Computer Science, Sichuan University, Chengdu 610064, China

Abstract. Image visual effects can be enhanced primarily through edge and texture enhancement or contrast enhancement. Image enhancement based on fractional differential can effectively enhance image details such as edge and texture using the weak derivative property of the 0–1-order fractional differential operator. Image enhancement based on gray statistics involves the redistribution of light and dark pixels to enhance the overall contrast of the enhanced image as well as the enlargement of the gray-level dynamic range, thereby improving the visual effect of the image effectively. To enhance the edge and texture information of the image, enhance the contrast of the image effectively, and then improve the visual effect of the image, an image enhancement model based on contrast limited adaptive histogram equalization incorporating a fractional differential operator is proposed. The image enhancement model incorporates a fractional differential operator into the adaptive limited contrast image enhancement model, which can enhance the image contrast and effectively enhance the edge and texture details of the image simultaneously. Experimental results show that the proposed variable-order fractional differential contrast-limited adaptive histogram equalization image enhancement model can significantly improve the contrast of the image compared with the traditional fractional differential image enhancement model; additionally, it can effectively enhance the edge and texture details of the image compared with the traditional image enhancement model, which is based on statistical methods.

Keywords: Fractional calculus · Image enhancement · Fractional gradient · Variable order · Histogram enhancement

1 Introduction

Image enhancement mainly emphasizes the local and non-local feature information of an image to achieve clearer images or highlight the edge texture and other important

X. Wu et al. (Eds.): QShine 2020, LNICST 381, pp. 209–226, 2021.
https://doi.org/10.1007/978-3-030-77569-8_15

features of the image such that the image contrast can be enhanced; hence, the visual effect of the image will be enhanced for later applications in specific occasions. Two image enhancement methods exist: frequency and spatial methods. In the frequency method, an image is transformed into a two-dimensional discrete Fourier transform or cosine transform, and high-frequency information such as edge and texture is enhanced using a high-pass filter, rendering the enhanced image visual effects better or easier to be processed later [1, 2]. The spatial domain method can be classified into image enhancement based on a difference operator and that based on gray-level statistics [3, 4]. The difference operator uses the gray-level difference of neighboring pixels to extract the edge and texture features of the image and then adds the weighted sum to the original image to enhance the important features of the image, such as edges and textures, thereby achieving a better image visual effect. The gray-level statistical method can improve the dynamic range of gray values and image contrasts by redistributing the probability of light and dark pixel values in the image and hence improve the visual effects of the image.

In recent years, research focus on difference image enhancement has expanded from the traditional integer-order differential operator to the fractional-order sub-differential operator. Pu proposed a classical image enhancement model based on a fractional differential operator using the "weak derivative" and "nonlocal" characteristics of the fractional differential operator; as such, the fractional differential operator can enhance the high-frequency component of the image and retain the low- and medium-frequency information of the signal nonlinearly. Therefore, the application of a low-order fractional differential operator to image enhancement can enrich the weak edge and texture of the enhanced image; additionally, the detailed features of the smooth region of the image can be enhanced accordingly [5–7]. On this basis, scholars have proposed many improved image enhancement models based on fractional differential to solve the problems of fractional differential image enhancement models [8–13]. However, the enhancement effect of the fractional differential operator on low-contrast images is insignificant. This is because the fractional differential operator is essentially a difference operator. It mainly enhances the gray value of pixels in abrupt gray areas, such as edges and textures, and does not redistribute the local or nonlocal gray value distribution of the image. Therefore, the enhancement effect of the low-contrast image is poor. The image enhancement method based on pixel gray value statistics is a histogram enhancement technology, among which the histogram equalization method is widely used because of its simplicity and high efficiency [14–18]. The traditional histogram equalization image enhancement is a global image enhancement method that enhances an entire image in a unified scale. Therefore, this method can not only enhance the contrast of the background image, but also reduce the contrast of useful signals, and the gray level of the transformed image will be reduced owing to tradeoffs. Consequently, some details in the image will disappear, and the hierarchical sense of the processed image will be poor. To solve the problems in traditional histogram enhancement technology, Kare proposed a self-adaptive contrast-limited adaptive histogram equalization (CLAHE) image enhancement model [19], which uses block processing to perform histogram equalization to different degrees in different contrast regions of the image. To overcome noise in enlarged images during the enhancement process, a contrast is set.

A threshold is used to control the effects of noise, and a better contrast enhancement effect is obtained. On this basis, scholars have proposed many improvement methods [20, 21]. However, the effect of the CLAHE image enhancement model on images with rich texture details is insignificant. This is because the image enhancement model is based on the pixel statistical redistribution method, which cannot directly enhance the edge and texture details of the image; hence, the image enhancement ability is limited.

In summary, the fractional differential image enhancement model based on the difference method and the CLAHE image enhancement model based on statistical methods have their own advantages and disadvantages. The image enhancement model based on fractional differential can effectively enhance the edge and texture details of the image, but the effect of the model on low-contrast image enhancement is insignificant; hence, it cannot effectively improve the overall and local contrast of the image. Furthermore, the visual effect of the enhanced image is general. Meanwhile, the CLAHE operator can effectively improve the local and nonlocal contrast of an image; additionally, a contrast threshold is introduced to suppress noise amplification during image enhancement. However, the effect of the CLAHE operator on image enhancement with rich texture details is general, and the enhancement of image clarity is insignificant. To enhance the overall and local contrast of an image and effectively enhance the edge and texture details of the image, this study attempts to integrate a fractional-order differential operator into the contrast-constrained adaptive histogram enhancement model; additionally, a CLAHE image enhancement model based on a fractional differential-order local variable is proposed. The image model offers not only the advantages of the classical fractional differential enhancement operator, but also exhibits the ability of the CLAHE model to enhance the local and nonlocal enhancement effect of low-contrast images.

2 Theoretical Background

2.1 Fractional Calculus Theory

Fractional calculus theory is an extension of integral-order calculus theory. Leibniz initially established fractional calculus theory at the end of the 16th century. Subsequently, the development of fractional calculus theory and its application lagged behind. It was not until Riemann Liouville [22] introduced fractional calculus theory into Brownian motion analysis that fractional calculus theory was initially applied in practice. In recent years, fractional calculus theory has been widely used in signal processing and analysis, fractal theory, and fractional order PID controllers. To date, a unified definition for fractional calculus theory does not exist. Generally, different definitions are used based on applications. The definition of fractional Grümwald–Letnikov (G–L) is realized using a difference scheme; therefore, it is highly suitable for processing the gray values of discrete pixels in digital image processing. Furthermore, the fractional G–L definition is conducive to numerical calculations and can be regarded as the extension of the limit form of the integer-order differential.

According to the definition of the integer-order derivative for continuous functions $f(x)$, the first derivative is defined as shown in Eq. (1).

$$f'(x) = \lim_{\Delta x \to 0} \frac{f(x + \Delta x) - f(x)}{\Delta x} \tag{1}$$

According to a similar definition, the definition of the n-order derivative of a continuous function $f(x)$ can be deduced, as shown in Eq. (2).

$$f^{n}(x)=\lim_{\Delta x\to 0}\frac{1}{(\Delta x)^{n}}\sum_{j=0}^{n}(-1)^{j}\binom{n}{j}f(x-j\Delta x) \tag{2}$$

According to the mathematical properties of the classical gamma function $\Gamma(x)$, $\binom{n}{j}=\frac{n!}{j!(n-j)!}=\frac{\Gamma(n+1)}{\Gamma(j+1)\Gamma(n-j+1)}$. Assuming that any real number v replaces a positive integer n and considering the special case of the fractional G–L definition in digital image processing, i.e., the distance between adjacent pixels is $\Delta x=1$, a fractional G–L differential definition suitable for two-dimensional discrete signal processing is obtained. Equation (3) can be simplified to Eq. (4), where $g(i)=(-1)^{i}\frac{\Gamma(v+1)}{\Gamma(i+1)\Gamma(v-i+1)}$, and the symbol "*" implies a convolution operation.

$$f^{v}(x)=\lim_{n\to\infty}\sum_{i=0}^{n}(-1)^{i}\frac{\Gamma(v+1)}{\Gamma(i+1)\Gamma(v-i+1)}f(x-i) \tag{3}$$

$$f^{v}(x)=\lim_{n\to\infty}\sum_{i=0}^{n}g(i)f(x-i)=g(x)*f(x) \tag{4}$$

2.2 Amplitude Frequency Characteristics of Fractional Calculus Operators

To analyze the amplitude–frequency characteristics of the one-dimensional signal fractional calculus operator, the function $g(x)$ is transformed into a frequency space. Both sides of Eq. (4) are Fourier transformed to obtain Eq. (5), where the frequency domain function of the fractional calculus is $G(\omega)$, as shown in Eq. (6).

$$FT\left(f^{v}(x)\right)=FT\left(g(x)*f(x)\right)=G(\omega)\times F(\omega) \tag{5}$$

$$G(\omega)=FT\left((-1)^{i}\frac{\Gamma(v+1)}{\Gamma(i+1)\Gamma(v-i+1)}\right)=|\omega|^{v}e^{i\frac{v\pi}{2}\mathrm{sgn}(\omega)} \tag{6}$$

Figure 1 shows the amplitude–frequency characteristic curves of the fractional differential operator and the fractional integral operator based on the fractional G–L definition. Subgraph (a) represents the amplitude–frequency characteristic curve corresponding to the differential order defined by the fractional-order G–L in the range of $v\in\{0.0,\ 0.5,\ 0.8,\ 1.0,\ 1.5,\ 2.0\}$. Direct observation shows that the fractional-order G–L calculus operator with a positive order can enhance the high-frequency part of the signal ($\omega>1$). Furthermore, the enhancement amplitude will increase rapidly with the increase in the differential order, but the enhancement amplitude is not as significant as that of the high-order integer differential. The subpositive fractional G–L differential operator can enhance the middle- and low-frequency parts of the signal ($\omega<1$) to a certain extent, but the high-order sub-integer differential operator has a certain degree of attenuation effect on the low- and medium-frequency information. Subgraph (b) shows that the fractional-order range $v \in \{0.0, \ -0.5, \ -0.8, \ -1.0, \ -1.5, \ -2.0\}$ defined by G–L is the corresponding amplitude–frequency characteristic curve. Direct observation shows that the fractional-order G–L calculus operator with a negative order had a certain degree of nonlinear attenuation on the high-frequency part of the signal ($\omega > 1$), and the attenuation amplitude decreased gradually with the decrease in the differential order. Compared with the high-order integral operator, the fractional-order G–L calculus operator can retain more high-frequency information. The fractional-order G–L calculus operator with a negative order had a certain degree of nonlinear enhancement on the middle- and low-frequency parts of the signal ($\omega < 1$), and the enhancement amplitude increased significantly with the decrease in the differential order.

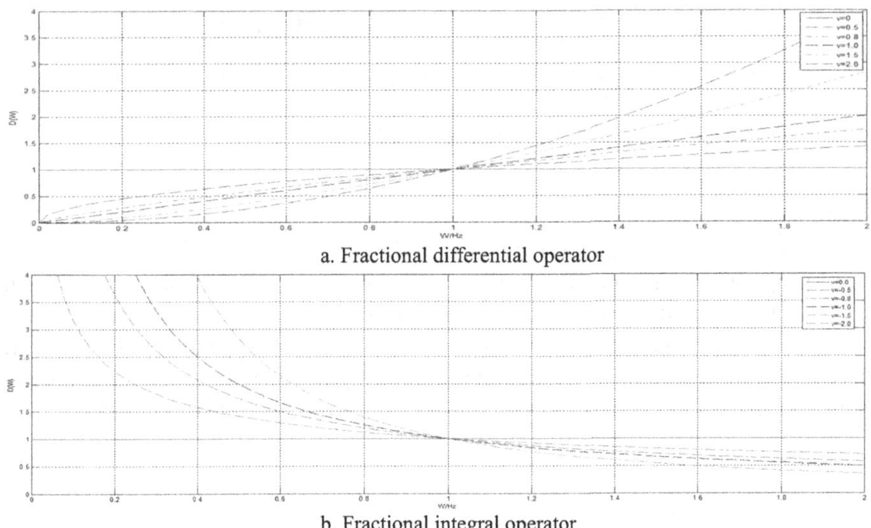

a. Fractional differential operator

b. Fractional integral operator

Fig. 1. Amplitude–frequency characteristic curve of fractional order G–L calculus operator

2.3 CLAHE Model

To solve the block effect and real-time problem of the traditional histogram enhancement model, Karel proposed a contrast-constrained adaptive histogram image enhancement algorithm. The core idea of the CLAHE algorithm is to use the image segmentation mechanism. Before calculating the cumulative histogram of each image block, a clipping threshold is determined to increase the amplitude of the original image cumulative histogram. The clipping part cannot be omitted but is redistributed to the original image histogram according to certain rules. Finally, bilinear interpolation is used to improve the timeliness of the CLAHE algorithm, and the block effect is eliminated. The core steps of the CLAHE algorithm are as follows.

Step 1. The image is divided into n * n image non-overlapping sub-blocks of the same size, as shown in in Fig. 2(a).

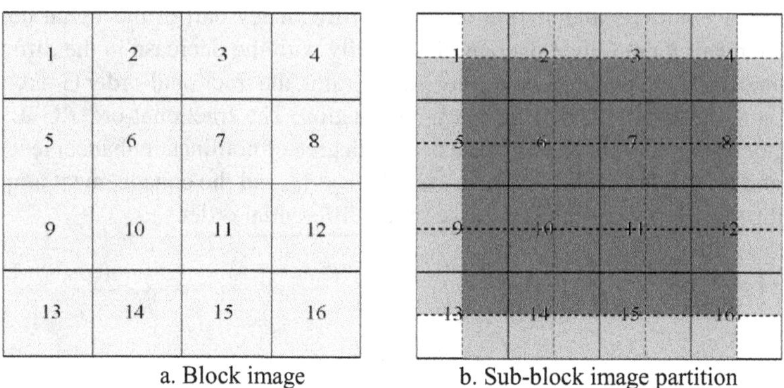

a. Block image b. Sub-block image partition

Fig. 2. Image block diagram for n = 4

Step 2. The histogram of the current image sub-block is calculated, and the clipped image histogram is determined according to the image gray threshold, as shown in Fig. 3. The number of pixels in the image block whose gray histogram is greater than the histogram clipping threshold is accumulated and stored in the exceeding threshold vector. Subsequently, the histogram is redistributed repeatedly to satisfy the clipping threshold. Finally, the image sub-block is processed using traditional histogram equalization.

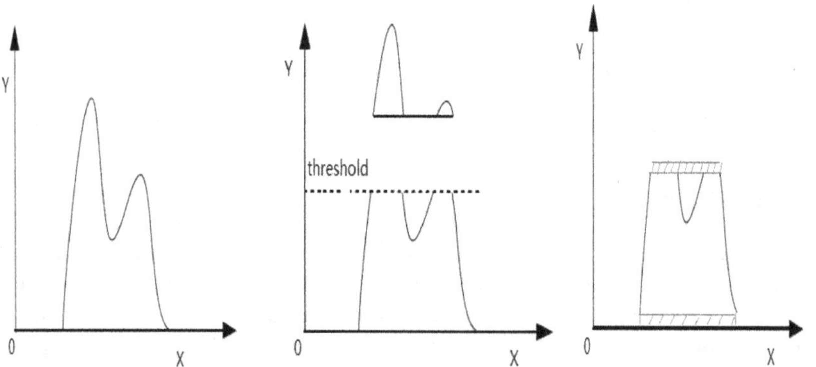

a. Histogram of original image b. Histogram after clipping c. Histogram after reallocation

Fig. 3. Schematic diagram of image sub block histogram clipping process

Step 3. According to the area divided by subgraph (b) in Fig. 2, interpolation calculation is performed using different methods. Specifically, as shown in Fig. 4, for the pixels in the dark region, bilinear interpolation was performed according to the gray value of the pixel in the four neighborhood regions. For the pixel in the light color region, the gray value was linearly interpolated according to the gray value of the pixel in the photographic neighborhood. The gray value of the pixel in the colorless region was determined according to the histogram mapping of the sub-block. The specific interpolation method is expressed in Eq. (7).

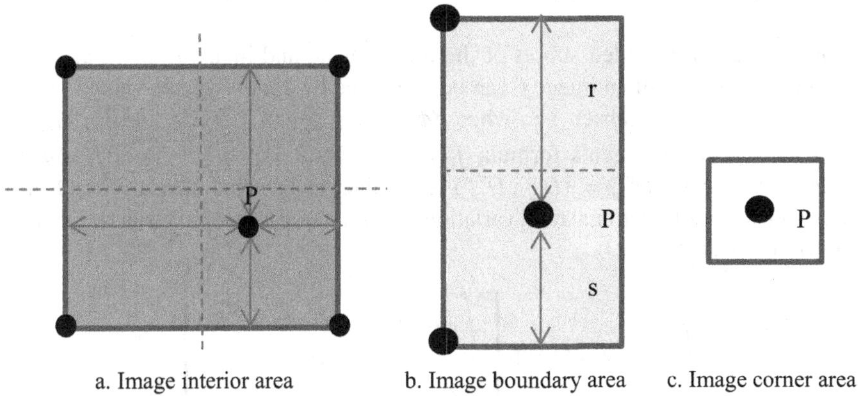

Fig. 4. Schematic diagram of pixel interpolation method in different regions

$$\mathrm{P_{new}} = \begin{cases} \frac{s}{s+r}\left(\frac{y}{x+y}f_{i,j}(P_{old}) + \frac{x}{x+y}f_{i+1,j}(P_{old})\right) + \frac{r}{s+r}\left(\frac{y}{x+y}f_{i,j+1}(P_{old}) + \frac{x}{x+y}f_{i+1,j+1}(P_{old})\right) \\ \frac{s}{s+r}f_{i,j}(P_{old}) + \frac{r}{s+r}f_{i,j}(P_{old}) \\ \mathrm{P_{old}} \end{cases} \tag{7}$$

3 Fractional Differential Contrast-Limited Adaptive Histogram Equalization (FCLAHE) Model

To enhance the edge and texture details of the image, effectively enhance the image contrast, and improve the visual effect of the image, a model combining the advantages of the fractional-order differential operator and the CLAHE model is proposed herein, known as the variable-order FCLAHE image enhancement model.

3.1 Image Detail Measure

(1) Fractional gradient modulus of the image

The image gradient is described as a measure in a bounded variation function or distribution space, denoted as $BV(\Omega)$, in which discontinuous jump features such as edges and textures are allowed. Therefore, the space $BV(\Omega)$ is often used to describe the global characteristics of images in an image processing model based on a variational method [23]. In this study, based on the total variation and AAA $BV(\Omega)$ AA space and combined with fractional calculus theory, the fractional total variation and fractional $BV(\Omega)$ space were extended.

Suppose Ω is a bounded subset of the image plane and image $f \in L^1(\Omega)$. If the distribution derivative of an image f can be expressed by a finite-vector-valued random measure on a bounded subset, i.e., when $\forall\phi = (\phi_1, \phi_2) \in C_0^1(\Omega)^2$ and $\left|\phi_{L^\infty(\Omega)}\right| \leq 1$, then the fractional Green's formula $\int_\Omega f \times div^v \phi d\Omega == \int_\Omega \overline{(-1)^v} D^v f \cdot \phi d\Omega$ is satisfied. In this case, $D^v f = (D_x^v f, D_y^v f)$ is a finite vector value measure on Ω, from which expressions of fractional total variation and fractional step degree can be obtained, as shown in Eqs. (8) and (9).

$$\int_\Omega \left|D^v f\right| d\Omega = \sup\left\{ \begin{array}{l} \int_\Omega f div^v \phi d\Omega : \phi = (\phi_1, \phi_2) \\ \in C_0^1(\Omega)^2, \left|\phi_{L^\infty(\Omega)}\right| \leq 1 \end{array} \right\} \tag{8}$$

$$D^v f = \nabla^v f = \left(\frac{\partial^v f}{\partial x_1^v}, \frac{\partial^v}{\partial x_2^v}\right) \tag{9}$$

Based on Eq. (9) and considering the "eight-neighborhood" of the image, the fractional ladder degree modulus function of the image can be obtained, as shown in Eq. (10).

$$\left|\nabla^v f\right| = \sqrt{\left(\begin{array}{l} \left(\frac{\partial^v f}{\partial x_+^v}\right)^2 + \left(\frac{\partial^v f}{\partial x_-^v}\right)^2 + \left(\frac{\partial^v f}{\partial y_+^v}\right)^2 + \left(\frac{\partial^v f}{\partial y_-^v}\right)^2 + \\ \left(\frac{\partial^v f}{\partial x_{45^\circ}^v}\right)^2 + \left(\frac{\partial^v f}{\partial x_{135^\circ}^v}\right)^2 + \left(\frac{\partial^v f}{\partial x_{275^\circ}^v}\right)^2 + \left(\frac{\partial^v f}{\partial x_{315^\circ}^v}\right)^2 \end{array} \right)} \tag{10}$$

Based on Eq. (10), the fractional gradient modulus of the "sub-block image" can be obtained, as shown in Eq. (11).

$$FGM_{I_k} = \sum_{(i,j)\in I_k} \left|\nabla^v f_{i,j}\right| \tag{11}$$

(2) Image texture measurements

The autocorrelation function $x(t)$ is typically used to represent the relationship between random signals at any time point. It is generally used to describe the cross-correlation of specific signals and the correlation degree at different times in the same sequence. Furthermore, it is typically used to search for repetitive patterns. Because the image texture roughness is proportional to the autocorrelation function, an autocorrelation function is introduced herein to mathematically describe the image texture feature value; hence, the image texture measurement can be obtained, as shown in Eq. (12).

$$ITM_{\varepsilon,\eta}(f_k) = \frac{\sum_{(i,j)\in f_k} f(i,j)f(i-\varepsilon,j-\eta)}{\sum_{(i,j)\in f_k} [f(i,j)]^2} \quad \varepsilon,\eta \in [1,D] \text{ D is the offset distance} \tag{12}$$

3.2 Variable-Order Weight Function

As shown by the amplitude–frequency characteristic curve of the fractional differential operator, the fractional differential operator can more effectively enhance the "weak texture" details of the image than the integer-order differential operator in the low-frequency part of the image; additionally, it can nonlinearly enhance the "strong texture" and edge information of the image in the high-frequency part of the image, but the lifting amplitude is lower than that of the high-order integer differential operator. Therefore, a coincidence score must be constructed for the weight function of the properties of the differential-order image enhancement operator. In Eq. (13), using the special properties of the special function, $g(x)$ as shown in Fig. 5(a), the boundary of function $g(x)$ is approximately $x = k$, the definition domain is $x \in [0,\ 1]$, and the value domain is $g(x) \in [0,\ 1]$. In the definition domain, the function $g(x)$ decreased with the increase in x. In this study, the normalized image eigenvalues were in the range of 0 to 1, and the corresponding fractional differential-order values were between 0 and 2 because the fractional differential operator in this interval can enhance the image details more reasonably. Therefore, the range of the function $g(x)$ was extended to 0–2, and the symmetric function $x = k$ about $g'(x)$ was derived, as shown in Eq. (14). As shown in Fig. 5(b), the function $g'(x)$ is related to $x = k$ (the boundary), definition domain $x \in [0,\ 1]$, and value domain $g'(x) \in [0,\ 2]$. Within the definition domain, the value of function $g'(x)$ increased with x, and the increase range of the function value was controlled by parameter r.

$$g(x) = \frac{1}{1 + (x/k)^r} \tag{13}$$

$$g'(x) = \frac{2}{1 + ((2k - x)/k)^r} \tag{14}$$

3.3 Description of FCLAHE Algorithm

Step 1: The Image is Classified into Two Levels. Based on the CLAHE algorithm, the image is divided into N*N sub-blocks, and the sub-block image is further divided into M*M block "secondary sub-block images" by combining the template scale of the

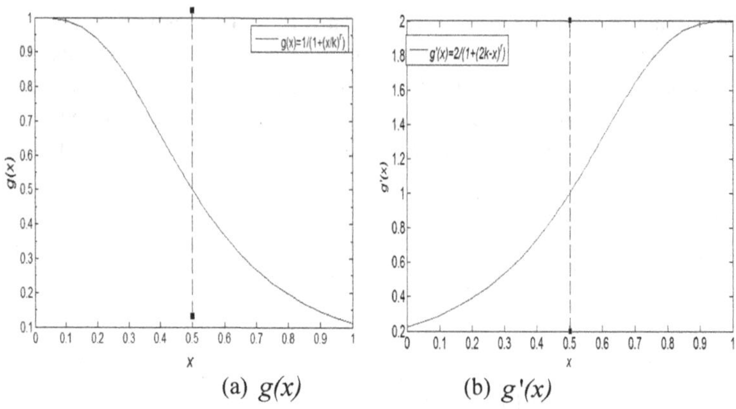

(a) $g(x)$ (b) $g'(x)$

Fig. 5. Schematic diagram of weight function within $x \in [0,\ 1]$ range

traditional fractional differential mask operator. If the domain Ω is a bounded open subset of the real space R^2, then the image can be represented as $f = \{f_\Omega : \Omega \in R^2 \to R\}$. As shown in Fig. 6, the entire image f_Ω in the definition domain Ω is blocked, and it is numbered in the line order to obtain the line vector set $[f_1, f_2, f_3 \ldots f_{N*N}]$ of N* N * N "sub-block images." Hence, the row vector set $[f_{k1}, f_{k2}, f_{k3} \cdots f_{kM*M}]$ of M * M "secondary sub-block images" can be obtained after the current "sub-block image" f_k, $k \in 1, N*N$ is divided into two levels (Fig. 7).

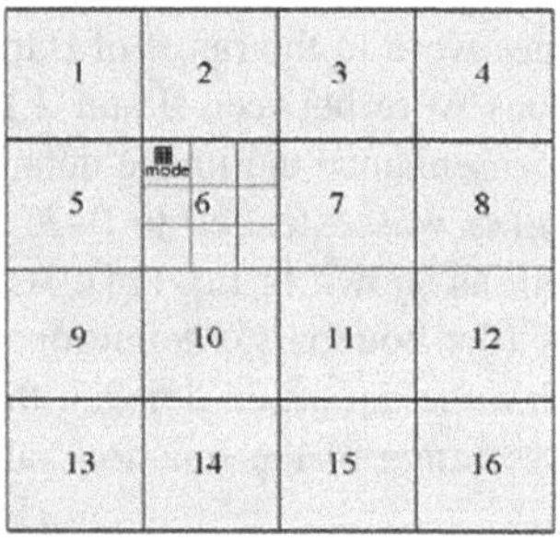

Fig. 6. N = 4, M = 3 block diagram

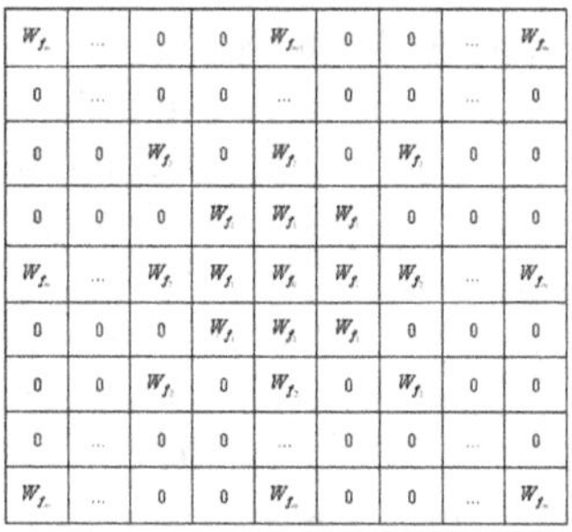

Fig. 7. The fractional order G - L operator

Step 2: Calculate the "Sub-Block Image" Eigenvalue. The fractional step modulus value of each "sub-block image" was calculated using Eq. (11) and then stored into the fractional ladder modulus value vector FGM_{blk} of the "sub-block image." Using the image texture formula show in Eq. (12), the texture measure value of each "sub-block image" was calculated and then stored into the texture measure vector ITM_{blk} of the "sub-block image." Subsequently, the fractional ladder degree modulus vector FGM_{blk} and image texture measure vector ITM_{blk} were normalized by a sigmoid function in the global range, as shown in Eq. (15). Finally, the weighted sum of the two types of image eigenvalues was calculated to obtain the "sub-block image" eigenvalue vector SIF_{blk}, as shown in Eq. (16).

$$\begin{cases} FGM_{blk} = \frac{1}{1+e^{-FGM_{blk}}} \\ ITM_{blk} = \frac{1}{1+e^{-ITM_{blk}}} \end{cases} \tag{15}$$

$$\begin{aligned} &SIF_{blk} = k_1 FGM_{blk} + k_2 ITM_{blk}\ k_1 + k_2 = 1 \\ &k_1 \text{ and } k_2 \text{are weight adjustment coefficients} \end{aligned} \tag{16}$$

Step 3: Calculate characteristic value of "secondary sub-block image". As in step 2, the fractional step modulus vector of the "sub-block image" can be obtained using the fractional step modulus formula shown in Eq. (11) and the image texture measurement formula shown in Eq. (12). Subsequently, the same normalization process was performed. Finally, the weighted sum of the two image eigenvalues was calculated to obtain the "secondary sub-block image" eigenvalue vector SIF_{sub_blk}.

Step 4: Determine the Order of Variable-Order Fractional Differential. The "sub-block image" eigenvalue vector SIF_{blk} and the "secondary sub-block image" eigenvalue vector SIF_{sub_blk} are weighted and summed again; consequently, the local and nonlocal image eigenvalues SIF of the current processing module can be obtained, as shown in Eq. (17). Using the $g'(x)$ property of the weight function deduced in this study, i.e., it is an increasing function in the real number field, and that the enhancement amplitude increases with the independent variable, the fractional differential-order function with variable order can be obtained, as shown in Eq. (18).

$$SIF = k_3 SIF_{blk} + k_4 SIF_{sub\_blk}\ k_3 \text{ and } k_4 \text{ are weight coefficients}'\ k_3 + k_4 = 1 \tag{17}$$

$$v = \frac{2}{1 + ((2k - SIF)/k)^r} k \text{ and } r \text{ are adjustment parameters} \tag{18}$$

Step 5: Variable-Order Fractional Differential Enhancement. Using the fractional G–L differential operators in [6], as shown in Eqs. (19) and (20), and extending them to eight directions in the image, we can obtain the fractional G–L differential mask operator based on the fractional-order G–L differential mask operator, as shown in Fig. 7, where the coefficient of the mask operator is expressed by Eq. (21). Finally, the current sub-block image f_k and variable-order mask operator are convoluted to obtain the edge and texture details of the current sub-block image.

$$\begin{aligned} &{}_a^G D_t^v f(x,y)_x \overset{\Delta}{=} f(x,y) + (-v) f(x-1,y) + \frac{-v(-v+1)}{2} f(x-2,y) \\ &+ \frac{-v(-v+1)(-v+2)}{6} f(x-3,y) + \cdots + \frac{\Gamma(-v+m)}{\Gamma(m+1)\Gamma(-v)} f(x-m,y) \end{aligned} \tag{19}$$

$$\begin{aligned} &{}_a^G D_t^v f(x,y)_y \overset{\Delta}{=} f(x,y) + (-v) f(x,y-1) + \frac{-v(-v+1)}{2} f(x,y-2) \\ &+ \frac{-v(-v+1)(-v+2)}{6} f(x,y-3) + \cdots + \frac{\Gamma(-v+n)}{\Gamma(n+1)\Gamma(-v)} f(x,y-n) \end{aligned} \tag{20}$$

$$\begin{cases} W_{f_0} = 1 \\ W_{f_1} = -v \\ W_{f_2} = \frac{v(v-1)}{2} \\ \cdots\cdots \\ W_{f_m} = \frac{v(v-1)(v-2)\ldots(v-m+1)}{m!} \end{cases} \tag{21}$$

Step 6: FCLAHE Image Enhancement. Redistributing the current sub-block image according to certain rules, the histogram distribution of f_k shows the occurrence times of all pixels in the original histogram within the clipping threshold. Subsequently, the redistributed histogram was equalized to obtain a better contrast in the sub-block image. Finally, the image details extracted by the CLAHE method and fractional differential were weighted and summed, as shown in Eq. (22); the function min() ensures that the gray value of the processed image is within the specified range.

$$f_k = \min\left(k_5 \sum_{i \in f_k} \left(f_{k_i} * \frac{W_v}{\max(W_v)}\right) + k_6 CLAHE(f_k), 1\right)$$

k_5 and k_6 are weight adjustment coefficients (22)

4 Experimental Simulation and Comparative Analysis

4.1 Evaluation Parameters

In a subjective evaluation, the human eye directly views the enhanced image, and the enhanced image mainly focuses on the human eye's feeling, to reflect the human visual perception. Hence, the image is more vivid. Because the human eye is sensitive to the texture details and edge parts of the image, direct observation was performed in this study to contrast the difference between visual light and shade, as well as the difference in binary images with edge and texture features. The objective index was evaluated by constructing the relevant evaluation function of important objective image evaluation features based on the subjective feeling of the human eyes. Subsequently, according to the evaluation function, the numerical quantization results based on some image features were obtained. In this study, the edge preserving coefficient, average gradient, contrast, and image entropy were used to evaluate the objective enhancement effect of image enhancement operators.

(1) Edge Preservation Index (EPI)

The edge preserving index indicates that the enhancement operator maintains the horizontal or vertical edge of the image. The higher the EPI value, the better is the edge preserving ability of the operator. The formula for the edge retention coefficient is shown in Eq. (23).

$$EPI = \sum_{i=1}^{row}\sum_{j=1}^{col}\left|\begin{matrix}\Delta_{level} f_{after}(i,j) + \\ \Delta_{vertical} f_{after}(i,j)\end{matrix}\right| \Bigg/ \sum_{i=1}^{row}\sum_{j=1}^{col}\left|\begin{matrix}\Delta_{level} f_{befor}(i,j) + \\ \Delta_{vertical} f_{befor}(i,j)\end{matrix}\right| \tag{23}$$

(2) Average Gradient (AG)
The AG value of the image can describe detailed contrast and texture changes in the image as well as reflect the clarity of the image to a certain extent. The formula to calculate the AG value is shown in Eq. (24).

$$AG = \frac{1}{M * N}\sum_{i=1}^{row}\sum_{j=1}^{col}\sqrt{\Delta_{level}f(i,j)^2 + \Delta_{vertical}f(i,j)^2} \tag{24}$$

(3) Entropy (E)
The E of an image describes the average amount of information contained in the image. The higher the E value, the more information it contains, and the richer are the edge texture details of the image. The formula to calculate the image E is as follows:

$$Entropy(f_{i,j}) = -\sum_{L=1}^{NUM\_GL} P(f_{i,j})\log P(f_{i,j}) \tag{25}$$

p is the probability function of the image pixel

(4) Contrast (C)
C represents the ratio of black to white, i.e., the gradient from black to white. The larger the ratio, the more gradients exist from black to white, and the better is the texture level of the detail. The formula for C is as follows:

$$C = \left|\sum_{i=1}^{row}\sum_{j=1}^{col}\Delta f(i,j)\right| \Big/ Number \tag{26}$$

Here, “number” indicates the number of differences between the gray values of eight adjacent regions of the image.

4.2 Experimental Results Analysis

The variable-order fractional order CLAHE image enhancement model proposed herein involves many parameters. The weighted coefficient or threshold parameter of the model can be determined using the empirical value or average proportion after many experiments. In this study, the number of first-order blocks was N = 4, the number of second-order blocks was M = 3, the offset distance of the image texture was D = 5, and the weights of the fractional-step eigenvalue and texture eigenvalue of the “sub-block image" and "secondary sub-block image" were $k_1 = 0.5$ and $k_2 = 0.5$, respectively. For the variable-order fractional differential, the weights of the two-level sub-block image eigenvalues were $k_3 = 0.5$ and $k_4 = 0.5$, and the variable-order fractional differential function $k = 0.5$, $r = 2$. The first four coefficients were used as the estimated values of the fractional differential. The weights of the FCLAHE contrast enhancement image and fractional differential detail image were $k_5 = 1.1$ and $k_6 = 1$, respectively.

To verify the advantages of the FCLAHE image enhancement model proposed herein, the current classic image enhancement methods were compared, including the Laplace image enhancement (Laplace), traditional fractional-order G–L image enhancement, histogram equalization (HE), and CLAHE methods. The experimental results show that superior performance of the proposed method based on comparing the images obtained from different image enhancement methods and the texture characteristics of the image. The experimental results show that the proposed FCLAHE image enhancement method had a higher contrast and edge texture than those of existing image enhancement methods, as indicated by the calculated quantitative values of objective indicators of the image after different enhancement methods, including the EPI, AG, E, and C. Furthermore, it demonstrated clearer and better visual effects.

Figure 8 shows the contrast of different enhancement models to enhance the Barbara image. Direct observation of the red circle part of the Barbara image shows that the Laplace operator, fractional G–L operator, and FCLAHE image enhancement model proposed herein significantly improved the texture details of the rattan chair information of the Barbara image compared with the HE and CLAHE image enhancement methods. Because the Laplace operator is a second-order differential operator, the edge and texture amplitude of the enhanced image improved significantly. However, if the differential order of the Laplace operator is extremely large, the strong edge information in the image will be enhanced excessively, thereby resulting in the false "bright line" phenomenon on the arm edge of the character image, which may cause image distortion. Direct observation of the Barbara blue circle shows that the CLAHE and FCLAHE image enhancement models can improve the local contrast information of the image compared with the Laplace, fractional-order G–L, and HE methods because it considers the local and nonlocal statistical information of the image, resulting in better bright-contrast visual effects and more visible low-contrast details. Figure 9 shows the contrast effect of the texture features of image Barbara enhanced by different enhancement models. Direct observation shows that the method proposed herein can enhance the detailed texture features of the image more effectively than other image enhancement methods, as well as effectively improve the local and overall contrasts of the image.

Figure 10 shows the contrast curves of the AG, edge retention coefficient, C, and E values of the two images above enhanced using the Laplace, fractional-order G–L, HE, CLAHE, and FCLAHE methods. Direct observation shows that the FCLAHE method improved the low-frequency weak edge and texture details of the image more effectively

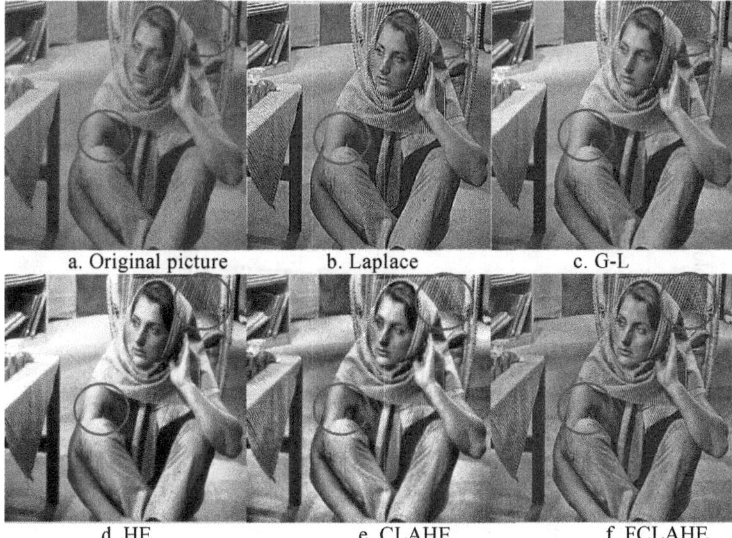

Fig. 8. Pictures obtained using different image enhancement models

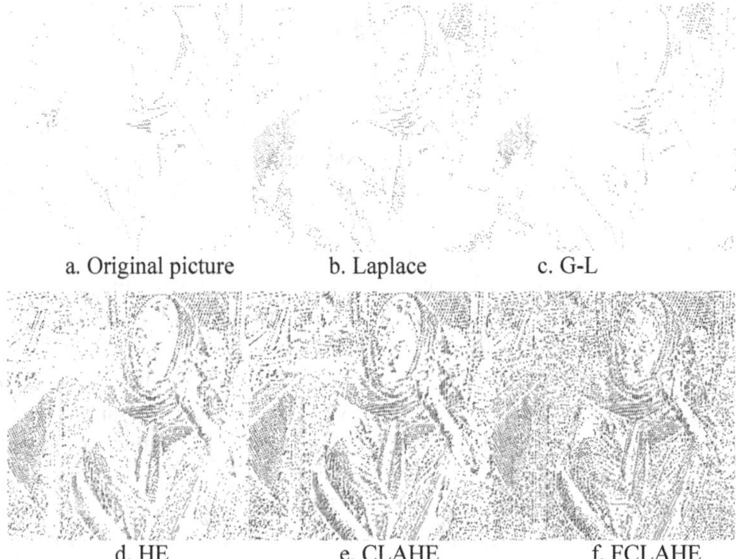

Fig. 9. Texture feature map of different image enhancement methods

compared with the difference or statistical image enhancement methods. Moreover, it exhibited the ability of the nonlocal adaptive histogram operator to improve the local and nonlocal contrasts of the image; consequently, the evaluation index of the enhanced image was better than those of other classical image enhancement methods.

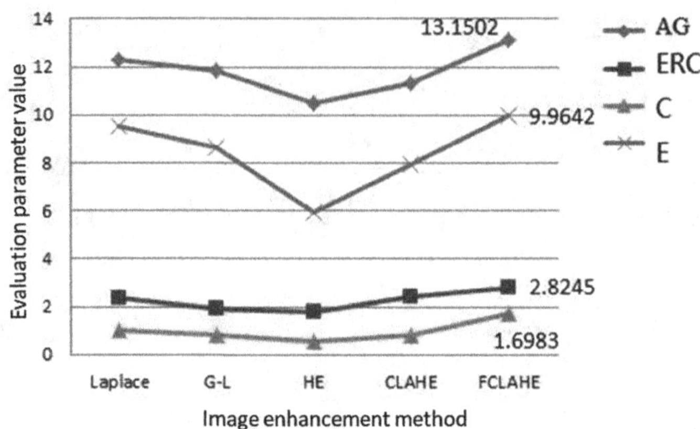

Fig. 10. Evaluation data corresponding to different image enhancement models

5 Conclusion

The fractional differential image enhancement model and its improved model typically utilizes the "weak derivative" and "nonlocal" properties of the fractional differential operator; hence, it can effectively enhance the edge and texture information of an image. However, the effect of this model on low-contrast image enhancement is not ideal. The CLAHE image enhancement model utilizes the probability redistribution of light and dark pixels to enhance the overall contrast of an enhanced image and expands the dynamic range of the gray level, which can effectively improve the light and dark contrasts of the image. However, the effect of the CLAHE model on images with rich texture details is insignificant. Considering the advantages and disadvantages of the difference and statistical methods for image enhancement, a variable-order fractional order CLAHE image enhancement model based on the existing fractional-order differential image enhancement model and the classical CLAHE model was proposed. The model uses the blocking mechanism of the CLAHE model, in which the current sub-block image is divided into two levels. The order of the fractional differential operator is determined by the linear weighted value of the fractional ladder degree modulus and the image texture measure of the current image sub-block and secondary block images. The simulation results show that the proposed FCLAHE method afforded different degrees of improvement compared with other classical image enhancement models owing to the combination of the fractional differential operator and the CLAHE operator's excellent characteristics, whether in subjective direct observations or based on objective evaluation data.

References

1. Grigoryan, A.M., Jenkinson, J., Agaian, S.S.: Quaternion Fourier transform based alpha-rooting method for color image measurement and enhancement. Signal Process. **109**, 269–289 (2015)
2. Tiwari, M., Lamba, S.S., Gupta, B.: A software supported image enhancement approach based on DCT and quantile dependent enhancement with a total control on enhancement

level. Multimed. Tools Appl. **78**(12), 16563–16574 (2018). https://doi.org/10.1007/s11042-018-7056-4

3. Qing, S., Cosman, P.C.: Luminance enhancement and detail preservation of images and videos adapted to ambient illumination. IEEE Trans. Image Process. **27**, 4901–4915 (2018)
4. Kandhway, P., Bhandari, A.K.: An optimal adaptive thresholding based sub-histogram equalization for brightness preserving image contrast enhancement. Multidimension. Syst. Signal Process. **30**(4), 1859–1894 (2019). https://doi.org/10.1007/s11045-019-00633-y
5. Pu, Y.F., Zhou, J.L., Yuan, X.: Fractional differential mask: a fractional differential-based approach for multiscale texture enhancement. IEEE Trans. Image Process. **19**(2), 491–511 (2010)
6. Pu, Y.-F., Wang, W., Zhou, J.-L., et al.: Fractional-order derivative detection of texture of image and the realize of fractional-order derivative filtering. Sci. China (Ser. E) **38**(12), 2252–2272 (2008)
7. Pu, Y.-F., Siarry, P., Chatterjee, A., et al.: A fractional-order variational framework for retinex: fractional-order partial differential equation-based formulation for multi-scale nonlocal contrast enhancement with texture preserving. IEEE Trans. Image Process. **27**(3), 1214–1229 (2017)
8. Chen, Q., Huang, G., Zhang, X., et al.: A caputo fractional differential approach to image enhancement. J. Comput.-Aided Design Comput. Graph. **25**(04), 519–525 (2013)
9. Cao, T., Wang, W.: Depth image enhancement and detection on NSCT and fractional differential. Wirel. Pers. Commun. **103**(1), 1025–1035 (2018). https://doi.org/10.1007/s11277-018-5494-y
10. Yu, Q., Liu, F., Turner, I., et al.: The use of a Riesz fractional differential-based approach for texture enhancement in image processing. Anziam J. **54**(C), C590–C607 (2013)
11. Gao, C.B., Zhou, J.L., Hu, J.R., Lang, F.N.: Edge detection of colour image based on quaternion fractional differential. IET Image Proc. **5**(3), 261–272 (2011)
12. Si, S., Hu, F., Fu, B., et al.: An algorithm for texture enhancement based on fractional differential mask using adaptive non-integer step. J. Comput. -Aided Design Comput. Graph. **26**(09), 1438–1449 (2014)
13. Nandal, A., et al.: Image edge detection using fractional calculus with feature and contrast enhancement. Circuits Syst. Signal Process. **37**(9), 3946–3972 (2018). https://doi.org/10.1007/s00034-018-0751-6
14. Li, S., Jin, W., Li, L., et al.: An improved contrast enhancement algorithm for infrared images based on adaptive double plateaus histogram equalization. Infrared Phys. Technol. **90**, 164–174 (2018)
15. Singh, P., Mukundan, R., De Ryke, R.: Feature enhancement in medical ultrasound videos using contrast-limited adaptive histogram equalization. J. Digit. Imaging **33**(1), 273–285 (2019). https://doi.org/10.1007/s10278-019-00211-5
16. Shakeri, M., Dezfoulian, M.H., Khotanlou, H., et al.: Image contrast enhancement using fuzzy clustering with adaptive cluster parameter and sub-histogram equalization. Digit. Signal Process. **62**, 224–237 (2017)
17. Wong, C.Y., Liu, S., Liu, S.C., et al.: Image contrast enhancement using histogram equalization with maximum intensity coverage. J. Mod. Opt. **16**, 1–12 (2016)
18. Shanmugavadivu, P., Balasubramanian, K.: Thresholded and Optimized Histogram Equalization for contrast enhancement of images. Comput. Electr. Eng. **40**(3), 757–768 (2014)
19. Zuiderveld, K.: Contrast limited adaptive histogram equalization. In: Graphics Gems, pp. 474–485 (1994)
20. Reza, A.M.: Realization of the contrast limited adaptive histogram equalization (CLAHE) for real-time image enhancement. J. VLSI Signal Process. Syst. Signal Image Video Technol. **38**(1), 35–44 (2004). https://doi.org/10.1023/B:VLSI.0000028532.53893.82

21. Wang, Q., Zhu, Y., Li, H.: Imaging model for the scintillator and its application to digital radiography image enhancement. Opt. Exp. **23**(26), 33753–33776 (2015)
22. Lim, S.C.: Fractional Brownian motion and multifractional Brownian motion of Riemann-Liouville type. J. Phys. A: Gen. Phys. **34**(7), 1301–1310 (2001)
23. Huang, G., Li, X., Chen, Q., et al.: Research on Image denoising based on space fractional partial differential equations. J. Sichuan Univ. (Eng. Sci. Ed.) **44**(2), 94–101 (2012)

Optimum Parameter Estimation Under Additive Cauchy-Gaussian Mixture Noise

Yuan Chen[1], Dingfan Zhang[2], and Longting Huang[3](✉)

[1] University of Science and Technology Beijing, Beijing, China
[2] Beihang University, Beijing, China
zdf1999@buaa.edu.cn
[3] Wuhan University of Technology, Wuhan, China
huanglt08@whut.edu.cn

Abstract. In this paper, a mixture process is proposed for modelling the summation of Cauchy and Gaussian random variables. The probability density function (PDF) of the mixture can be derived as the Voigt profile. To further study the noise, the estimation of the constant model is taken as an illustration. Here the scenarios of both known and unknown density parameters are considered. The maximum likelihood estimator (MLE) with Voigt function is first employed to devised the optimal estimator. Then an M-estimator with pseudo-Voigt function is developed to improve the computational complexity of MLE. Simulation results indicate the superior of both proposals, which can attain the Cramér-Rao lower bound.

Keywords: Cauchy distribution · Gaussian distribution · Additive mixture noise · Maximum likelihood estimation · Voigt function · M-estimation · Pseudo-Voigt function

1 Introduction

Impulsive noise is commonly encountered in many applications such as wireless communications and image processing [1]. Differ with Gaussian noise, it belongs to a family of heavy-tailed noise depicting varying characteristics. Typical impulsive noise models in the literature are α-stable distribution, and generalized Gaussian distribution (GGD) [2]–[3]. Unfortunately, these models cannot cover all variety of impulsive noise types in the real-world, especially for noise obtained at receivers both in radio communication network [4] and binary transmission [5], where the noise is modeled as the sum of two random variables following different distribution. It is also discussed in [6] that in astrophysical images, cosmic microwave background radiation was contaminated with Gaussian noise from the satellite beam and the radiation from galaxies which can be modelled as an α-stable distribution. Here the noise can be expressed as the sum of α-stable and Gaussian variables.

X. Wu et al. (Eds.): QShine 2020, LNICST 381, pp. 227–236, 2021.
https://doi.org/10.1007/978-3-030-77569-8_16

In this paper, to further study the mixture noise, the sum of symmetric Cauchy distribution ($\alpha = 1$) with median γ and zero mean Gaussian distribution with variance σ^2 is taken as an example. The probability density function (PDF) of the mixture noise is called Voigt function, which is derived by the convolution of the PDF of Cauchy and Gaussian's PDF [7]. When the density parameters, namely, γ and σ^2 are known, the PDF of the mixture is determined as a Voigt function, and hence maximum likelihood estimator (MLE) on the PDF can be directly applied. However, when γ and σ^2 are unknown, the expression of PDF is specified. Therefore, they should be estimated prior to employing MLE, with the use of the relationship between the empirical characteristic function (ECF) and characteristic function (CF). Since the Voigt function has a complicated expression with complex integral, MLE suffers from a high computational complexity. To reduce the complexity of MLE, the M-estimator is utilized, whose main idea is finding the minimum of the logarithm of a loss function [8]. Here the loss function can be selected as an arbitrary function. Particularly, the M-estimator becomes the MLE in the case that the loss function is the logarithm of likelihood function. In order to keep the high performance of the MLE, the loss function is chosen as the pseudo-Voigt function, which is the approximation of the Voigt function [9], which is referred to as MEPV. Similar to the MLE, both the known and unknown γ and σ^2 are considered. While they should be estimated firstly, in the unknown density parameters case.

The rest of this paper is organized as follows. Sections 2 and 3 devise the details of the proposed methods, refered to as MLE and MEPV, respectively, where both the known and unknown density parameters cases are investigated individually. Section 4 presented the derivation of the Cramér-Rao lower bound (CRLB) of unknown parameters are derived. Computer simulations are conducted in Sect. 5 to evaluate the accuracy and complexity of proposals. Finally, conclusions are drawn in Sect. 6.

2 Maximum Likelihood Estimator

The observed data $\mathbf{y} = [y_0, y_1, \ldots, y_{N-1}]^T$ can be modeled as:

$$y_n = A + e_n, \quad n = 0, 1, \ldots, N-1 \tag{1}$$

where

$$e_n = q_n + p_n \tag{2}$$

and A is the constant of interest, e_n is the sum of independent and identically distributed (IID) Cauchy variable with median γ and IID zero mean Gaussian noise with the variance σ^2. Since the PDFs of Cauchy and Gaussian distributions are:

$$f_q(x; \gamma) = \frac{\gamma}{\pi(x^2 + \gamma^2)} \tag{3}$$

$$f_p(x;\sigma^2) = \frac{1}{\sqrt{2\pi}\sigma} e^{-\frac{x^2}{2\sigma^2}}, \tag{4}$$

the PDF of e_n is calculated by the convolution of f_q and f_p [7], which is

$$f_e(x;\sigma,\gamma) = \int_{-\infty}^{\infty} \frac{\gamma}{\pi((e_n-\tau)^2+\gamma^2)} \frac{1}{\sqrt{2\pi}\sigma} \exp\left(-\frac{\tau^2}{2\sigma^2}\right) d\tau \tag{5}$$

$$= \frac{\mathrm{Re}\{w_n\}}{\sigma\sqrt{2\pi}} \tag{6}$$

where

$$w_n = \exp\left(-\left(\frac{e_n+i\gamma}{\sigma\sqrt{2}}\right)^2\right)\left(1+\frac{2i}{\sqrt{\pi}}\int_0^{\frac{e_n+i\gamma}{\sigma\sqrt{2}}} \exp\left(t^2\right) dt\right) \tag{7}$$

where w_n is referred to as Faddeeva function [10], with $\mathrm{Re}\{w_n\}$ being the real part of w_n.

In the following, we will first consider the scenario of known γ and σ^2, and then extend the study with unknown distribution parameters.

2.1 Scenario I: Known σ^2 and γ

The PDF of the observed data y_n is expressed as:

$$f_e(\mathbf{y}, A) = \prod_{n=0}^{N-1} \frac{\mathrm{Re}\{w_n\}}{\sigma\sqrt{2\pi}} \tag{8}$$

where

$$w_n = e^{-\left(\frac{y_n-A+i\gamma}{\sigma\sqrt{2}}\right)^2}\left(1+\frac{2i}{\sqrt{\pi}}\int_0^{\frac{y_n-A+i\gamma}{\sigma\sqrt{2}}} e^{t^2}\, dt\right). \tag{9}$$

According to the idea of MLE [11], the estimate of A, denoted by $\hat{A}$, can be obtained:

$$\hat{A} = \arg\min_A \ln f_e(\mathbf{y}, A) = \arg\min_A \sum_{n=0}^{N-1} \ln \mathrm{Re}\{w_n\} \tag{10}$$

For a constant, the location range is $(-\infty, \infty)$ and thus the grid search method is not practical. Therefore the Newton's method is utilized:

$$\hat{A}^{(k+1)} = \hat{A}^{(k)} - \left.\frac{\sum_{n=0}^{N-1} f_e^{'}(y_n;A)}{\sum_{n=0}^{N-1} f_e^{''}(y_n;A)}\right|_{A=\hat{A}^{(k)}} \tag{11}$$

where

$$f_e'(y_n;A) = \frac{1}{\mathrm{Re}\{w_n\}}\mathrm{Re}\left\{\frac{\partial w_n}{\partial A}\right\} \tag{12}$$

$$f_e''(y_n;A) = \frac{1}{\mathrm{Re}\{w_n\}}\mathrm{Re}\left\{\frac{\partial^2 w_n}{\partial A^2}\right\} - \left(f_e'(y_n;A)\right)^2 \tag{13}$$

$$\frac{\partial w_n}{\partial A} = \frac{2}{i\sqrt{\pi}} + 2w_n\frac{y_n - A + i\gamma}{\sigma\sqrt{2}} \tag{14}$$

$$\frac{\partial^2 w_n}{\partial A^2} = \left(4(\frac{y_n - A + i\gamma}{\sigma\sqrt{2}})^2 - 2\right)w_n + \frac{4}{i\sqrt{\pi}}\frac{y_n - A + i\gamma}{\sigma\sqrt{2}} \tag{15}$$

and $^{(k)}$ denotes the kth iteration. The (11) is updated until the relative error $|\frac{\hat{A}^{(k+1)}-\hat{A}^{(k)}}{\hat{A}^{(k+1)}}| < \epsilon$ is reached, with ϵ being the tolerance. To simplify the problem, the least absolute deviation is utilized to initial the value of A, which is $\hat{A}^{(0)} = \mathrm{median}\{y_n\}$.

2.2 Scenario II: Unknown σ^2 and γ

If γ and σ^2 are unknown, an exact expression of PDF cannot be derived readily, so γ and σ^2 are estimated using the relationship between CF and ECF first. For the ACG noise, the CF of y_n has the form of

$$\phi(t) = E\{e^{iy_nt}\} = \exp(itA - \gamma|t| - \frac{t^2}{2}\sigma^2) \tag{16}$$

where E is expectation operator and the magnitude of $\phi(t)$ can be written as

$$|\phi(t)| = \exp(-\gamma|t| - \frac{t^2}{2}\sigma^2). \tag{17}$$

Take the logarithm on both sides of (17), we have

$$\Phi(t) = -\ln|\phi(t)| = \gamma|t| + \frac{t^2}{2}\sigma^2. \tag{18}$$

Accordingly, the ECF $\psi(t)$ has the form of

$$\psi(t) = \frac{1}{N}\sum_{n=1}^{N} e^{iy_nt} \tag{19}$$

and $\Psi(t) = -\ln(|\psi(t)|)$. If the interval of t is chosen as $t \in [t_0, t_{M-1}]$, then γ and σ^2 can be estimated as

$$\{\hat{\gamma}, \hat{\sigma}^2\} = \arg\min_{\gamma,\sigma^2} J \tag{20}$$

where $J = ||\Psi - \mathbf{F}\mathbf{x}||_1$ with $\Psi = [\Psi(t_0), \Psi(t_1), \ldots, \Psi(t_{M-1})]^T$, $\mathbf{x} = [\gamma, \sigma^2]^T$ and $\mathbf{F} = [\mathbf{b}_0, \mathbf{b}_1, \mathbf{b}_2, \ldots, \mathbf{b}_{M-1}]^T$ with $\mathbf{b}_n = [|t_n|, \frac{t_n^2}{2}]$. In this study, (20) can be solved with the use of the subgradient method [12]:

$$\hat{\mathbf{x}}^{(k)} = \hat{\mathbf{x}}^{(k)} - \alpha_k g^{(k)} \tag{21}$$

where $g^{(k)} = -\mathbf{F}^T \text{sign}(\Psi - \mathbf{F}\mathbf{x}^{(k)})$ and $\alpha_k = 1/||g^{(k)}||_2$. We employ the least squares solution for minimizing $||\Psi - \mathbf{F}\mathbf{x}||_2^2$ as $\hat{\mathbf{x}}^{(0)}$ and the stopping criterion follows that of scenario I. According to the analysis in [13], the interval of t is set to $[0.1, 1]$. Since γ and σ^2 are estimated by (20), the PDF of e_n is calculated by (8) and then $\hat{A}$ is obtained by (11).

3 *M*-Estimator Using Pseudo-Voigt Function

Even though the MLE has a high accuracy performance, this method suffers from high computational complexity due to the integral in the Faddeeva function, i.e., the PDF of y_n. To reduce the cost, the logarithm of the pseudo-Voigt function is chosen as the loss function, and M-estimator is employed to estimate the constant.

It has been proved that the Voigt function can be approximately described by the sum of the PDF of Cauchy and Gaussian distributions, which is called pseudo-Voigt function [9]:

$$f_P(e_n; \gamma, \sigma^2) = \mu f_1(e_n; \gamma) + (1 - \mu) f_2(e_n; \sigma^2) \tag{22}$$

where f_1 and f_2 are the PDFs of Cauchy and Gaussian distributions, respectively:

$$f_1(e_n; \gamma) = \frac{C\gamma_v}{\sqrt{\pi}(e_n^2 + \gamma_v^2)} \tag{23}$$

$$f_2(e_n; \sigma^2) = \frac{C}{\sqrt{\pi}\gamma_v} \exp\left(-\log(2) \left(\frac{e_n}{\gamma_v}\right)^2 \right) \tag{24}$$

with

$$\mu = \frac{C - \sqrt{g}}{C(1 - \sqrt{\pi g})}, \quad C = b_{1/2}(a) e^{a^2} \left(1 - \text{erf}(a)\right),$$
$$\gamma_v = \sqrt{2}\sigma b_{1/2}(a) \tag{25}$$

$$b_{1/2}(a) = a + \sqrt{g} \exp(-0.605a + 0.072a^2 - 0.005a^3 + 0.00014a^4) \tag{26}$$

and $a = \frac{\gamma}{\sqrt{2}\sigma}$, with $\text{erf}(\cdot)$ denoting the error function and $g = \log(2)$.

3.1 Known γ and σ^2

In this case, the pseudo-Voigt function parameters are known, A is estimated by minimizing

$$\hat{A} = \arg\min_{\tilde{A}} \left\{ -\sum_{n=1}^{N} \log\left(f_P(y_n, A; \gamma, \sigma^2)\right) \right\} \tag{27}$$

where $f_P(y_n, A; \gamma, \sigma^2) = (1 - \mu) f_1(y_n - \mathbf{h}_n A; \gamma) + \mu f_2(y_n - \mathbf{h}_n A; \sigma^2) = (1 - \mu) f_1(y_n, A; \gamma) + \mu f_2(y_n, A; \sigma^2)$.

To solve (27), the Newton's method is employed

$$\hat{A}^{(\ell+1)} = \hat{A}^{(\ell)} - \left(\frac{\partial^2 J}{\partial A^2}\right)^{-1} \left(\frac{\partial J}{\partial A}\right)\Bigg|_{A=\hat{A}^{(\ell)}} \tag{28}$$

where

$$\frac{\partial J}{\partial A} = \begin{bmatrix} \sum_{n=1}^{N} n V_n \\ \sum_{n=1}^{N} V_n \end{bmatrix} \tag{29}$$

$$\frac{\partial^2 J}{\partial A^2} = \begin{bmatrix} \sum_{n=1}^{N} n^2 \left(U_n - V_n^2\right) & \sum_{n=1}^{N} n \left(U_n - V_n^2\right) \\ \sum_{n=1}^{N} n \left(U_n - V_n^2\right) & \sum_{n=1}^{N} \left(U_n - V_n^2\right) \end{bmatrix} \tag{30}$$

with

$$\begin{aligned} V_n =& \frac{1}{f_P(y_n, A; \gamma, \sigma^2)} \Bigg((1 - \mu) \left(-\frac{2\log(2)}{\gamma_v^2}\right) f_2(y_n, A; \sigma^2) \\ & + \mu \left(\frac{-2(y_n, A)}{(y_n, A)^2 + \gamma_v^2}\right) f_1(y_n, A; \gamma) \Bigg) \end{aligned} \tag{31}$$

$$\begin{aligned} U_n =& \frac{1}{f_P(y_n, A; \gamma, \sigma^2)} \Bigg((1 - \mu) \left(-\frac{2\log(2)}{\gamma_v^2}\right)^2 f_2(y_n, A; \sigma^2) \\ & + \mu \left(\frac{2(y_n, A) - 2\gamma_v^2}{\left((y_n, A)^2 + \gamma_v^2\right)^2}\right) f_1(y_n, A; \gamma) \Bigg) \end{aligned} \tag{32}$$

In this method, $\hat{A}$ is updated by (28) and the initial guess and stop criterion are same with those of the MLE.

3.2 Unknown γ and σ^2

Similar to the scenario in the MLE, the density parameters γ and σ^2 should be estimated first, which can be derived from (19). After $\hat{\gamma}$ and $\hat{\sigma}^2$ are obtained, $f_P(y_n, A; \hat{\gamma}, \hat{\sigma}^2)$ can be reconstructed. Then, the unknown parameter vector A can be estimated by updating (28).

4 Cramér-Rao Lower Bound

The CRLB [11] of $\hat{A}$ can be calculated by the diagonal elements of the inverse of the Fisher information matrix $\mathbf{I}$, which has the form of

$$\begin{aligned}\mathbf{I} &= -E\left\{\sum_{n=1}^{N}\frac{\partial^2 \log f_M(y_n, A; \gamma, \sigma^2)}{\partial A^2}\right\}\\ &= E\left\{\sum_{n=1}^{N}\frac{\partial \log f_M(y_n, A; \gamma, \sigma^2)}{\partial A}\left(\frac{\partial \log f_M(y_n, A; \gamma, \sigma^2)}{\partial A}\right)^T\right\}.\end{aligned} \tag{33}$$

With the use of (16), (33) is:

$$\frac{\partial \log f_M(y_n, A; \gamma, \sigma^2)}{\partial A} = \frac{1}{\sigma^2}\frac{\mathrm{Re}\{(y_n - A + i\gamma)w_n\}}{\mathrm{Re}\{w_n\}}. \tag{34}$$

Since it is difficult to derive the closed-form expression of (34), the average of sufficient number of independent runs is utilized to replace the expectation.

For unknown γ and σ^2, we define $\boldsymbol{\alpha} = [A\ \gamma\ \sigma^2]^T$. Then the CRLB of $\hat{A}$ in this case corresponds to the (1, 1) entry of $\mathbf{I}^{-1}$. Then the (k, l) entry of $\mathbf{I}$ is

$$\begin{aligned}&\mathbf{I}_{k,l}\\ &= -E\left\{\sum_{n=1}^{N}\frac{\partial \log f_M(y_n, A; \gamma, \sigma^2)}{\partial \alpha_k}\frac{\partial \log f_M(y_n, A; \gamma, \sigma^2)}{\partial \alpha_l}\right\},\\ &\qquad k, l = 1, 2, 3.\end{aligned} \tag{35}$$

where

$$\begin{aligned}&\frac{\partial \log f_M(y_n, A; \gamma, \sigma^2)}{\partial \boldsymbol{\alpha}}\\ &= \begin{bmatrix} \frac{1}{\sigma^2}\frac{\mathrm{Re}\{(y_n - A + i\gamma)w_n\}}{\mathrm{Re}\{w_n\}} \\ -\frac{\frac{1}{\sigma^2}\mathrm{Re}\{i(y_n - A + i\gamma)w_n\} + \frac{2}{\sqrt{2\pi\sigma^2}}}{\mathrm{Re}\{w_n\}} \\ \frac{\frac{1}{\sigma^2}\mathrm{Re}\{(y_n - A + i\gamma)^2 w_n\} + \frac{\gamma}{\sqrt{2\pi\sigma^2}\sigma^2}}{\mathrm{Re}\{w_n\}} - \frac{1}{2\sigma^2} \end{bmatrix}.\end{aligned} \tag{36}$$

5 Simulation Results

To assess the performance of the proposed methods, computer simulations are provided. The mean square error (MSE), $E\{(\hat{A} - A)^2\}$, is utilized as the performance measure. The constant is defined as $A = 0.5$. In the case of unknown density parameter, the interval of t is chosen as [0.1, 1] with 1000 uniform grid points, according to the analysis in [13]. Due to the complexity of the mixture noise, the signal to noise ratio (SNR) is difficult to define, and hence we set $\gamma = \sigma^2$ and scale σ^2 to produce different noise conditions. Comparison with the

ℓ_1-norm estimator is provided and the CRLB is also included as the benchmark. It is noted that the ℓ_1-norm minimizer is solved by the least absolute deviation (LAD) [14]. All results are based on 1000 independent runs with a data length of $N = 60$.

First of all, the scenario of the known distribution parameters is investigated. It is shown in Figs. 1 and 2 that the MSEs of both MLE and MEPV can attain the CRLB for $\sigma^2 \in [-30, 10]$ dB case, while they are superior to the ℓ_1-norm estimator. Here, the performance of both MLE and MEPV are same to each other, this is because the main idea of them similar. Figure 3 indicates the average computational cost versus the data length N, where σ^2 and γ are set to 10 dB. A stopwatch timer is employed to measure the computation time. It is observed that the complexity of MEPV is significantly reduced compared with the MLE. Furthermore, for large values of N ($N > 2550$), the computational cost of MLE is increasing exponentially, while MEPV grows linearly because it requires less iterations to converge.

Then, the scenario of the unknown γ and σ^2 is studied. Here the ECF and CF are employed to estimate the density parameters γ and σ^2 in prior. It can be seen in Figs. 1 and 3 that the MSE of both MLE and MEPV can achieve the CRLB and they still outperforms the ℓ_1-norm estimator.

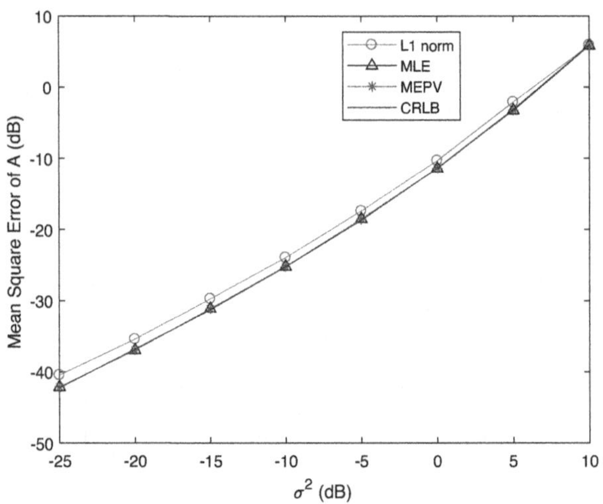

Fig. 1. Mean square error of A versus σ^2 with known γ and σ^2

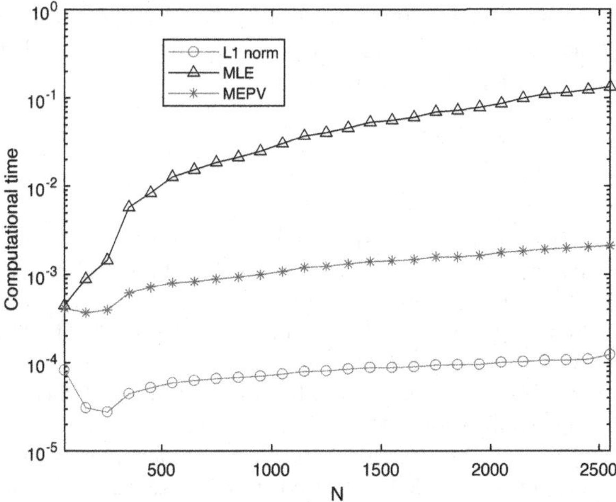

Fig. 2. Computational complexity versus N

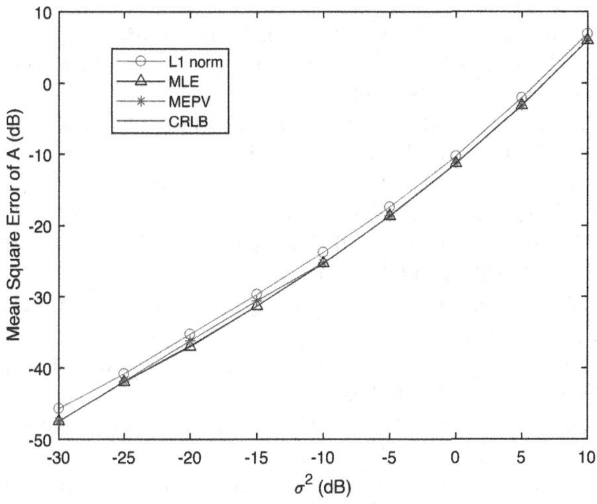

Fig. 3. Mean square error of A versus σ^2 with unknown γ and σ^2

6 Conclusion

A mixture noise modeled as the sum of two popular noise processes, namely, Cauchy and Gaussian distributions is investigated in this paper. To estimate the parameter in this mixture noise, the MLE is first developed for both known and unknown density parameters cases, which is regarded as a special M-estimator. The MEPV is also devised to estimate the parameters with reduced computations. Computer simulation results are provided to show the optimality of the proposed methods.

Funding. The work was financially supported by National Natural Science Foundation of China (Grant No. 61701021) and Fundamental Research Funds for the Central Universities (Grant No. FRF-TP-19-006A3).

References

1. Zoubir, A.M., Koivunen, V., Chakhchoukh, Y., Muma, M.: Robust estimation in signal processing: a tutorial-style treatment of fundamental concepts. IEEE Sig. Process. Mag. **29**(4), 61–80 (2012)
2. Nikias, C.L., Shao, M.: Signal Processing with Alpha-Stable Distribution and Applications. Wiley, New York (1995)
3. Shynk, J.J.: Probability, Random Variables, and Random Processes: Theory and Signal Processing Applications. Wiley, Hoboken (2013)
4. Ilow, J., Hatzinakos, D., Venetsanopoulos, A.N.: Performance of FH SS radio networks with interference modeled as a mixture of Gaussian and alpha-stable noise. IEEE Trans. Commun. **46**(4), 509–520 (1998)
5. Ilow, J., Hatzinakos, D.: Detection for binary transmission in a mixture of Gaussian noise and impulsive noise modeled as an alpha-stable process. IEEE Sig. Process. Lett. **1**(3), 55–57 (1994)
6. Herranz, D., Kuruoglu, E.E., Toffolatti, L.: An α-stable approach to the study of the P(D) distribution of unresolved point sources in CMB sky maps. Astron. Astrophys. **424**(3), 1081–1096 (2004)
7. Olver, F.W.J., Lozier, D.M., Boisvert, R.F.: NIST Handbook of Mathematical Functions, pp. 167–168. Cambridge University Press, Cambridge (2010)
8. Huber, P.J.: Robust Statistics. Wiley, New York (1981)
9. Dirocco, H.O., Cruzado, A.: The Voigt profile as a sum of a Gaussian and a Lorentzian functions, when the coefficient depends only on the widths ratio. Acta Physica Polonica A **122**(4), 666–669 (2012)
10. Weideman, J.A.C.: Computation of the complex error function. SIMAM J. Numer. Anal. **31**(5), 1497–1518 (1994)
11. Kay, S.M.: Fundamentals of Statistical Signal Processing: Estimation Theory. Prentice-Hall, Englewood Cliffs (1993)
12. Boyd, S., Vandenberghe, L.: Convex Optimization. Cambridge University Press, New York (2004)
13. Kogon, S.M., Williams, D.B.: On the characterization of impulsive noise with α-stable distributions using Fourier techniques. In: Proceedings of the Asilomar Conference on Signals, Systems and Computers, vol. 2, pp. 787–791, November 1995
14. Li, Y., Arce, G.: A maximum likelihood approach to least absolute deviation regression. EURASIP J. Appl. Sig. Process. **12**, 1762–1769 (2004). https://doi.org/10.1155/S1110865704401139

Face Reconstruction with Specific Weight Mask

Wentao Shi(✉), Tianji Ma, Nangyang Bai, and Lutao Wang

Chengdu University of Information Technology, No. 24, 1st Section, Xuefu Road, Southwest Airlines, Chengdu, Sichuan, China
waglt@cuit.edu.cn

Abstract. Create a 3D face model from a 2D face image, generally extract facial feature points, calculate a 3D deformation model, and perform deformation and stretching on the generated face database. However, this approach is not only time-consuming but also has no calculation errors. Ideally, neural networks' use to obtain deformation model parameters is also affected by factors such as pose, angle, and datasets. 3D face reconstruction methods rely excessively on the accuracy of the labeling and the face detector's accuracy. This article proposes a method that is not affected by pose. We adopt a feature point extractor that can obtain more features, design an hourglass network to get a model, and consider each feature area differently, effectively using the feature point information. Map from two-dimensional coordinates to three-dimensional space to achieve face reconstruction, and obtain a high-precision face model. We do experiments on the three-dimensional face datasets AFLW2000-3D and 300W-3D. The results show that this method can obtain good performance in face multi-angle reconstruction, and the accuracy is also improved.

Keywords: Computer vision · Face alignment · Dense alignment · Face reconstruction

1 Introduction

With the improvement of perception technology and the fermentation of deep learning, studys in face recognition improve rapidly in recent years, and the accuracy of 3D face reconstruction has also been continuously improved. Due to its wide range of applications, 3D face reconstruction has always attracted attention. Obtaining 3D information from 2D images is of great significance to various fields. It plays a crucial character on animation, games, smart shopping, information security, and other fields. However, 3D face reconstruction often relies on expensive capture equipment and professional technicians, and the cost is exceptionally high. For a long time, the loss of in-depth information and lack of prior knowledge has been a problem.

2D image recognition can easily obtain accurate information and has high robustness, while 3D face reconstruction is affected by various factors such as angle, illumination, skin texture, and lack of datasets with real marks. In order to get good results, the input picture has to perform a good angle, and the face of the image cannot be blocked. Also,

X. Wu et al. (Eds.): QShine 2020, LNICST 381, pp. 237–247, 2021.
https://doi.org/10.1007/978-3-030-77569-8_17

the method might be time-consuming, and the results might not work well. Therefore, the transition from two-dimensional to three-dimensional is a challenging topic in the computer vision area.

The face can be regarded as a three-dimensional object containing texture information and structure information. The three-dimensional face reconstruction method is mainly based on the optimization algorithm [1], and the corresponding face structure and texture information is obtained by obtaining the 3D Morphable Model (3DMM) coefficients [2]. 3DMM learns the prior knowledge of 3D face utilizing statistical analysis and obtains the required face model by controlling the average face database model's deformation. Reference [9] proposed an idea to separate the four areas of the face, find the best-fitting model in each area, and further deform and combine to find the best-fitting model corresponding to each area. However, these methods will be computationally time-consuming due to their high computational complexity and rely on prior knowledge, difficulty in initialization, and easy to fall into local optima. Even with these shortcomings in those methods, 3DMM is still proposing solutions to nonlinear regression functions. With the rise of machine learning and deep learning, most of the work is still based on 3DMM. Recently, methods of using CNNs to regress 3DMM coefficients have achieved good results [3], Zhu [4], End to End method [5]. However, many methods are restrained by poses, they need the input data to have a good angle, and the feature regions are not displaying well.

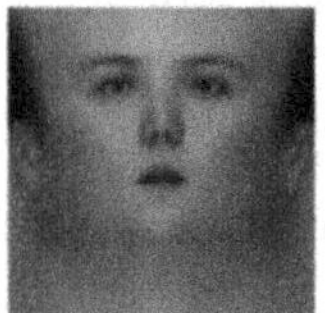
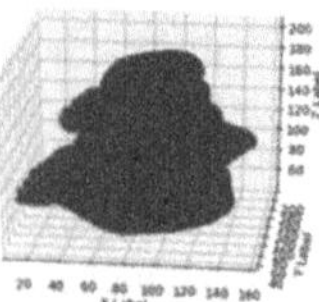

Fig. 1. Obtain 3D point cloud information from a single image

In order to solve the problem of insufficient robustness, limited by the rotation angle, restoration accuracy, we create a two-dimensional coordinate system that carries 3D semantic information and divide the face into different regions on the two-dimensional surface through feature points, and we give different weights to different feature areas. Figure 1 shows an example of obtaining 3D point cloud information from a single image. We achieved good results on different angles and performed robustness in different datasets.

2 Related Works

From 1999 to 2010, Blanz and Vetter [2] proposed a 3D Morphable Model (3DMM), whose method can construct a 3D face model based on 2D images. As pointed out above, the face is divided into the texture part and structure part. The texture coefficient and structure coefficient are shown in the Eq. (1) and Eq. (2), α is the structure coefficient and β is the texture coefficient. These two coefficients control the transformation of the face model. Generate an average face model that can be deformed according to the images, and change the deformable model's coefficients to stretch and deform to obtain the desired result.

$$S_{model} = \bar{S} + \sum_{i=1}^{m-1} \alpha_i S_i \quad (1)$$

$$\mathrm{T_{model}} = \bar{\mathrm{T}} + \sum_{\mathrm{i=1}}^{\mathrm{m-1}} \beta_{\mathrm{i}} \mathrm{T_i} \quad (2)$$

Later, in 2004, Blanz [10] proposed sparse facial feature points for model parameter estimation. Rara [11] and others proposed a model between 2D facial feature points and 3DMM parameters, using principal component regression analysis (PCR) to estimate 3DMM parameters. Due to the facial posture's influence, the possible accuracy of the detection of the detected 2D facial key feature points may be reduced. Dou [12] proposed a dictionary-based method to regress 3D face shape, using sparse coding to estimate model parameters from facial landmarks. Similarly, Zhou [13] also used a dictionary-based method and proposed a convex formula to estimate model parameters. Anbarjafari [9] proposed an end-to-end concept, dividing the face into four parts. All texture maps are distorted to fit the same UV map, and all the parts behind the face are discarded. For the four regions, corresponding facial models are obtained separately, and the four regions obtained by stitching are combined to obtain a complete face. This method is severely affected by noise and hair.

Recently, 3D face reconstruction methods based on deep learning came out. There are methods to add corrections or details to the rough 3DMM prediction [6], Tewari [15], Guo [16]. Many cutting-edge methods use CNNs to regress 3DMM parameters, please see the example, Richardson [6], Tuan[7], Jackson [8], Richardson [14], Feng Y [17], Tewari [20], Piotraschke [21], Huber [22], He K [23]. Figure 2 shows the process of face recognition, which performs excellently in many applications. However, 3D face reconstruction has many issues, such as lack of datasets, rely on the good pose of input data.

Reference [7] solves the problem of the insufficient training set and is more robust than previous methods. The author uses multiple pose photos of the same object to generate a high-accuracy 3D face model [21], and then uses the generated model as a training set and uses a threaded deep convolutional neural network to generate a robust face model. Reference [17] proposed to use 2D pictures with semantic information and non-equivalently consider the weights of different points for evaluation, but the texture information of this method is rough.

3 Network and Loss Function

Our proposed method uses more accurate facial feature extraction to construct a UV map regression 3D face model. The main steps are divided into the extraction of facial feature points, construction of UV maps, and a simple CNN network. In the feature point extraction, we use the face key point extraction method of more key points proposed by Niko. This method belongs to the branch of [22]. Compared with 68 key feature points, 13 key points are added (including forehead area). The difference between the two methods has been shown in Fig. 3 below.

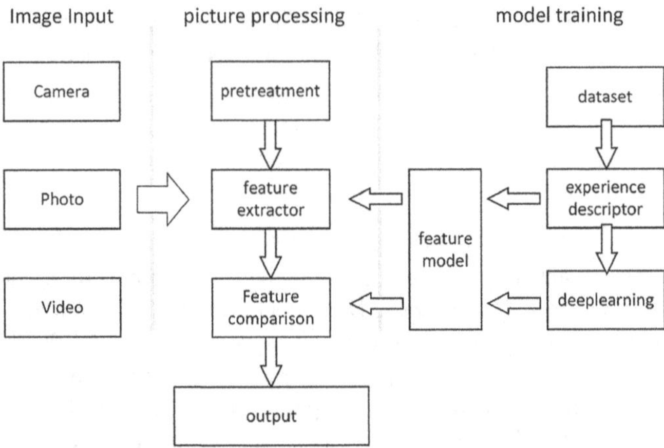

Fig. 2. Application of deep learning in the field of face recognition

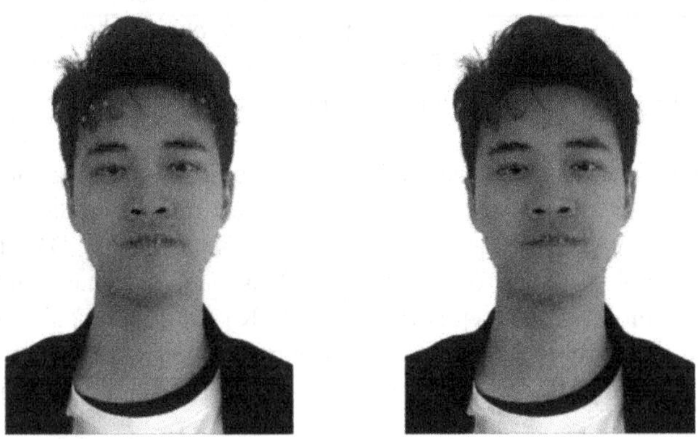

Fig. 3. 81 feature key point extractor (left), 68 feature key point extractor (right)

3.1 Network

We directly regress the parameters of a 3D face from a single 2D face through the CNN network. Meanwhile, we need to emphasize the robustness of the method. Therefore, we need a dataset containing 2D faces and corresponding 3D information and containing yaw angles as our training set. The 300W-3D data set contains large-angle face images, which satisfies our requirements very well. We choose 300W-3D as our training set. Simultaneously, to evaluate the superiority of the method, we manually annotated the 81 feature points of some of the acquired pictures and showed them in the experimental results section.

In order to ensure the effectiveness of the results, we should consider the facial feature regions differently. It is tough to consider features directly from 3D information. Converting a three-dimensional problem to a two-dimensional problem can solve this problem well.

The UV map is a two-dimensional image that records the information of all 3D point clouds. We refer to [19] to propose a method for constructing a UV location map. We design and use an hourglass network, mainly borrowed from the fully convolutional network and the residual network. According to the input color face photo, the UV position map is obtained from a single 2D face image.

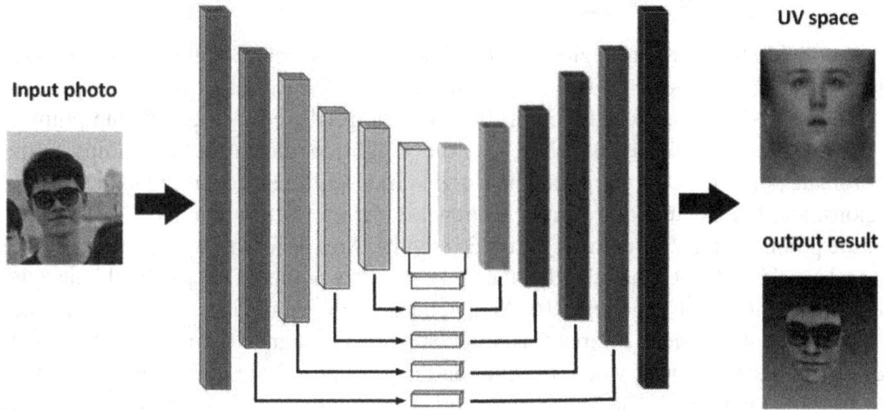

Fig. 4. CNN network, rectangle represents residual block.

Each rectangle is a residual block representing a feature. To ensure the dense geometric structure information of the face, the size of the UV map is 256, which can obtain a high-precision point cloud. The network structure is shown in Fig. 4. The network converts the input RGB image into the corresponding UV image, using encoder-decoder. The coding part is composed of cascaded residual blocks, and each layer is activated by the Relu function and input to the next convolutional layer. Finally, after activation of the Relu function, the output of each residual block structure is obtained, and the picture with the size of $256 \times 256 \times 3$ is reduced to the feature map of $8 \times 8 \times 512$, and the decoding part includes the transposed convolution layer, and the picture is restored to the UV map with the size $256 \times 256 \times 3$. The stride is 1, all kernel size is set to 4, and the Relu function is used to activate.

3.2 Loss Function

To display the information represented by the distinct regions, Anbarjafari [9] proposed an idea, considering the difference in the characteristic information of each region, modeling each region separately, and finally fusing each region.

Realize this vision by creating a UV map. The UV image is a 2D image that contains 3D point cloud information. Reference [9] Use UV maps to represent the texture information of the face. Different from others, Feng Y [17] uses UV space to store 3D point cloud information. The real 3D point cloud information can accurately match the projection of the 2D plane. The picture's RGB information becomes the x, y, z coordinate points in the texture map. Simultaneously, the UV coordinates also contain 3DMM parameters, which carry enough 3D semantic information. We learn from the method they proposed. Since the UV map is obtained through the network, we can divide the face region. According to the difference of the region's feature information, different weights are assigned to each region when calculating the loss. Experiments show that doing so can get better experimental results.

$$\text{Loss} = \sum \|P(x, y) - P(\sim)(x, y)\| \bullet W(x, y) \tag{3}$$

We divide the face into regions in the obtained UV space. Compared with the previous method, we distribute the weights more carefully according to the proportion. Equation (3) shows our loss function, P(x,y) is the predicted face coordinate point, P is the real face coordinate, and W(x,y) is the weight coefficient corresponding to the coordinate point. In our conception, the weight ratio is different according to the divided regions, and the calculation formula is shown in the above formula. Area 1 (81 facial feature points): Area 2 (eyes, nose): Area 3 (mouth): Area 4 (forehead, other areas): Area 5 (neck) = 16:5:4:3:0. Undoubtedly, the 81 key feature points of area 1 should have the highest weight, which is given to 16. Area 3, because the mouth is an important feature, it is assigned 4; the neck of area 5 is an outside area, assigned 0; area 2 and area 4 It belongs to a distinct area, where the eyes and nose of area 2 are iconic features, which are 1 higher than area 3; the forehead and other facial areas of area 4 are relatively less obvious, and 3 is assigned. In this way, we can consider each feature non-equivalently.

4 Experimental Results

Since the dataset requires both 2D pictures and their corresponding 3D semantic information, 300W-3D is selected as the training set because it contains 3DMM coefficients and different angle facial pictures, which enables it to store 3D point cloud information. Compare with 3DDFA [4], DeFA [4], and PRN [17] on the AFLW2000-3D and Florence datasets.

AFLW2000-3D is a dataset used to evaluate the performance of 3D face reconstruction for unconstrained images. This dataset contains the first two thousand face avatars in the AFLW dataset, which can be used for head deflection or head 2D or 3D detection, as well as large-angle faces. The dataset contains two data types. The first is JPG format data, which contains two-dimensional face pictures; MAT format data is a dictionary that contains feature points, 3DMM information, and various image parameters. Since the dataset carries 3DMM coefficients, the three-dimensional information is reconstructed by 3DMM and contains 68 feature points of three-dimensional information. It is the most commonly used dataset to evaluate the performance of facial reconstruction.

300W is a huge face dataset. The dataset has more than thousands of images, each image contains more than one face, but only one face is labeled, mainly used for face alignment. 300W-3D is a dataset that marked 300W data with 3DMM parameters, which can be used to train, test, and evaluate reconstruction performance. We show some examples in Fig. 5. Also, 300W-LP is suitable for our experiment. 300W-LP a subset of 300W samples containing large angles. It can be used to evaluate the robustness of the method to deflection angle rotation.

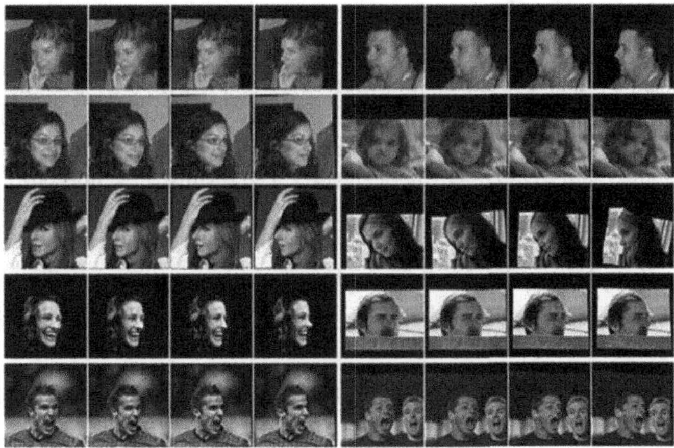

Fig. 5. Some large pose face in 300W-LP.

The Florence-3D dataset contains 53 labeled objects. Every object has different angles. In this section, we compare the performance of some methods, such as 3DDFA and DeFa.

We manually marked 81 key points of faces in some pictures and conducted a separate display experiment. The results have been shown at the rendering part, and we can see the performance intuitively.

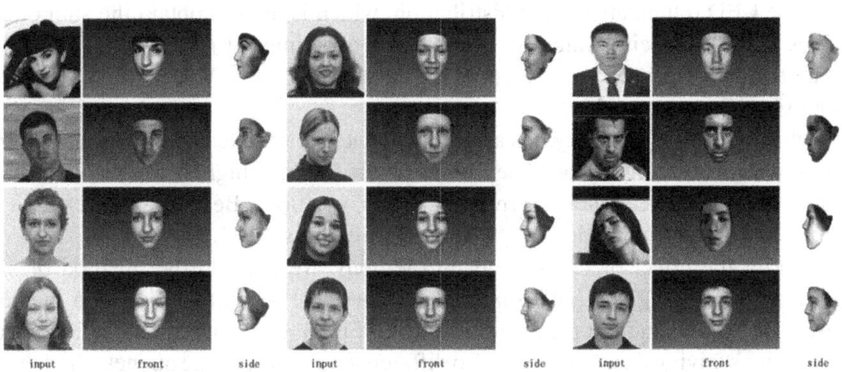

Fig. 6. Output of our method, including the front face and the side face. Some input faces are captured by ourselves

Use the carried 3DMM coefficients to generate the corresponding 3D position map and render it into the UV space. The size of the picture remains unchanged in the UV space, still 256 × 256. Use the hourglass network for training and Adam optimizer operation. The learning rate starts from 0.0001, decays by half every 20 batches, and the input face (rotation) is randomized. Use tensorflow to run. The graphics card used in the experimental hardware is a GTX1070Ti graphics card and I7-8700 central processing unit. Figure 6 shows some examples of our outputs; some input images were captured and manually mark by ourselves.

Although the regularized mean square error (MSE) is generally used as the performance evaluation index, direct use of MSE will cause the loss of key information. To better evaluate the network performance, we evaluate the method at large pose datasets by calculating the regularized mean error (NME). The results show good robustness and well performance. Furthermore, we manually mark 81 key points in some photos and give the results at the rendering display part. In order to further study the performance of the method in this paper under different angles, some data are compared with the NME under small, different, and large deflection angles. The results are shown in Table 1.

Table 1. Performance compared with other methods at different angles (NME).

Method	0° to 30°	30° to 60°	60° to 90°	Mean
3DDFA	3.80	4.55	7.88	5.41
3DSTN	3.13	4.43	5.78	4.45
PRN	2.75	3.50	4.60	3.62
SDM	3.67	4.90	9.67	6.08
Ours	2.69	3.48	4.58	3.58

We use CED (cumulative error distribution, which is used to obtain the sum of all variables below a specific value) to observe the experimental results intuitively. We evaluate the performance of our method on the AFLW2000-3D. The result has been shown in Fig. 7.

We divide the face into 5 regions according to the extracted feature points and assign different weights to each region. The 81 key points have the highest weights to ensure that the network can accurately learn these points' positions. Because the neck area is meaningless for the face model's regression, the weight is zero.

To verify the effect of dividing feature regions, we compare the method of not considering feature regions (all facial regions are equal to 1), the allocation method is shown in Table 2, and the results are shown in Fig. 8. Obviously, considering facial feature regions, giving different weights to different regions can make our network better.

The best way to evaluate the performance of the face model is to observe the rendering result directly. We manually mark 81 key points. We display the outputs below. See Fig. 9. Figure (a) is the reconstruction result of the front face, and figure (b) is the reconstruction result of the side face. Even if the picture uses a picture that includes a rotation angle, we can get satisfactory results.

As shown upon, all facial details, wrinkles, spots are basically restored, and the model will not be influenced by changing the angle. Even we zoom in on the picture, and we still can get a clear vision of the texture details.

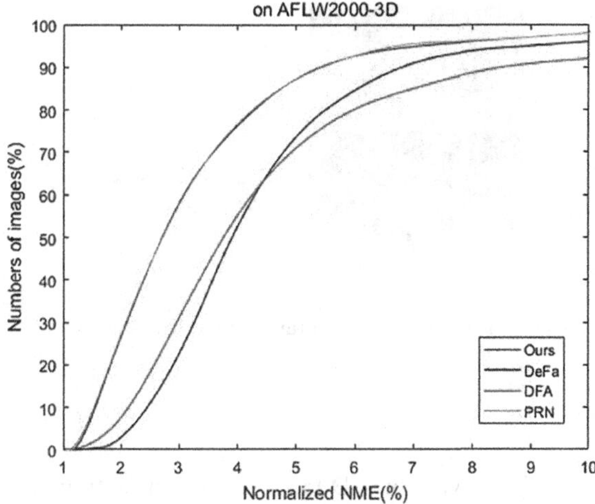

Fig. 7. 3D reconstruction performance comparison, the cumulative error distribution of 4 different methods on the AFLW2000-3D dataset, smooth the curve for better observing

Table 2. Different weight ratio

	Area1	Area2	Area3	Area4	Area5
Weight r1	16	5	4	3	0
Weight r2	1	1	1	1	0

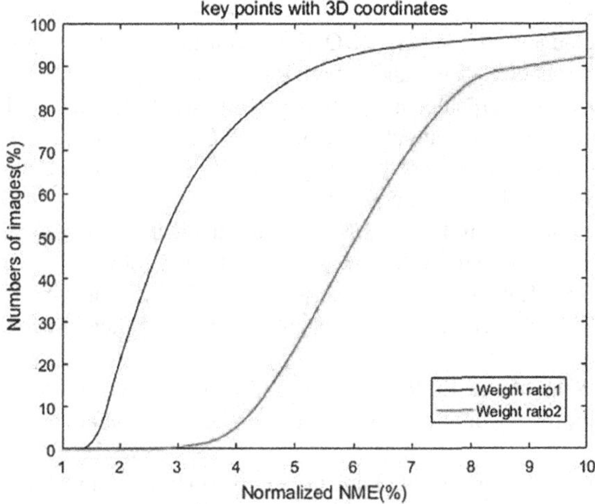

Fig. 8. Comparison of weight ratio 1 and weight ratio 2.

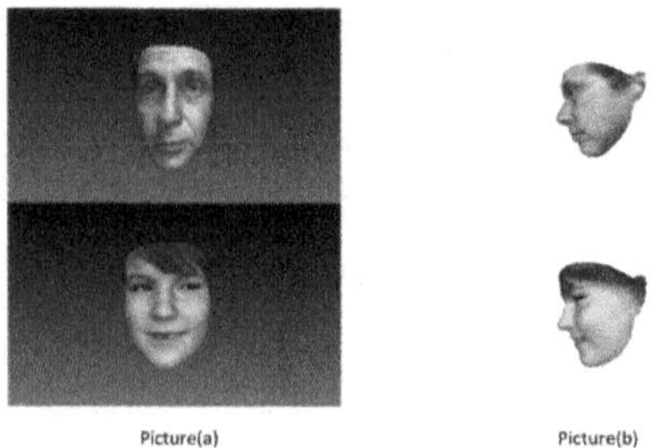

Fig. 9. Outputs of our methods, including both the front face and the side face

5 Conclusion

We provide a new idea for solving the 3D face reconstruction from a single 2D picture. This idea does not rely on expensive hardware equipment and complex networks and can fully return to face features. Experiments show that this method has an excellent performance in 3D face reconstruction.

In this paper, we show the CNN using the hourglass network structure to regress the UV parameters. During the training process, feature points are added as a guide to obtain a robust face image. Meanwhile, an excellent facial feature extractor is introduced to visually display the rendering results. Even if the input image is rotated, excellent results can be obtained without being affected by the rotation. There still has many ways to extend this work, such as applying more accurate feature area weights.

References

1. Amberg, B., Romdhani, S., Vetter, T.: Optimal step nonrigid ICP algorithms for surface registration. In: Computer Vision and Pattern Recognition, pp. 1–8 (2007)
2. Blanz, V., Vetter, T.: A morphable model for the synthesis of 3D faces. In: Proceedings of the 26th Annual Conference on Computer Graphics and Interactive Techniques (1999)
3. Jourabloo, A., Liu, X.: Large-pose face alignment via CNN-based dense 3D model fitting. In: Computer Vision and Pattern Recognition (2016)
4. Zhu, X., Lei, Z., Liu, X., Shi, H., Li, S.Z.: Face alignment across large poses: a 3D solution. In: Computer Vision and Pattern Recognition, pp. 146–155 (2016)
5. Dou, P., Shah, S.K., Kakadiaris, I.A.: End-to-end 3D face reconstruction with deep neural networks. In: IEEE Conference on Computer Vision and Pattern Recognition, pp. 21–26 (2017)
6. Richardson, E., Sela, M., Or-El, R., Kimmel, R.: Learning detailed face reconstruction from a single image. In: IEEE Conference on Computer Vision and Pattern Recognition, pp. 5553–5562 (2017)
7. Tran, A.T., Hassner, T., Masi, I., Medioni, G.: Regressing robust and discriminative 3D morphable models with a very deep neural network. In: IEEE Conference on Computer Vision and Pattern Recognition, pp. 1493–1502 (2017)
8. Jackson, A.S., Bulat, A., Argyriou, V., Tzimiropoulos, G.: Large pose 3D face reconstruction from a single image via direct volumetric CNN regression. In: International Conference on Computer Vision (ICCV), pp. 1031–1039 (2017)

9. Anbarjafari, G., Haamer, R.E., Lusi, I., et al.: 3D face reconstruction with region based best fit blending using mobile phone for virtual reality based social media. arXiv preprint arXiv: 1801.01089 (2017)
10. Blanz, V., Mehl, A., Vetter, T., Seidel, H.-P.: A statistical method for robust 3D surface reconstruction from sparse data. In: Proceedings of the International Symposium on 3D Data Processing Visualization and Transmission, Thessaloniki, Greece, pp. 293–300, 6–9 September 2004 (2004)
11. Rara, H., Farag, A., Davis, T.: Model-based 3D shape recovery from single images of unknown pose and illumination using a small number of feature points. In: Proceedings of the International Joint Conference on Biometrics, Washington, DC, pp. 1–7, 11–13 October 2011 (2011)
12. Dou, P., Wu, Y., Shah, S.K., Kakadiaris, I.A.: Robust 3D facial shape reconstruction from single images via twofold coupled structure learning. In: Proceedings of the British Machine Vision Conference, Nottingham, United Kingdom, pp. 1–13, 1–5 September 2014 (2014)
13. Zhou, X., Leonardos, S., Hu, X., Daniilidis, K.: 3D shape reconstruction from 2D landmarks: a convex formulation. In: Proceedings of the IEEE Conference on Computer Vision and Pattern Recognition, Boston, Massachusetts, pp. 4447–4455, 7–12 June 2015 (2015)
14. Richardson, E., Sela, M., Kimmel, R.: 3D face reconstruction by learning from synthetic data. In: International Conference on 3D Vision (3DV), pp. 460–469 (2016)
15. Tewari, A., Zollhöfer, M., Garrido, P., Bernard, F., Kim, H., Pérez, P., Theobalt, C.: Self-supervised multi-level face model learning for monocular reconstruction at over 250 Hz. In: IEEE Conference on Computer Vision and Pattern Recognition, pp. 2549–2559 (2018)
16. Guo, Y., Zhang, J.Z., Cai, J., Jiang, B., Zheng, J.: CNN-based real-time dense face reconstruction with inverse rendered photo-realistic face images. IEEE Trans. Pattern Anal. Mach. Intell. (TPAMI) **41**, 1294–1307 (2018)
17. Feng, Y., et al.: Joint 3D face reconstruction and dense alignment with position map regression network. In: Proceedings of the European Conference on Computer Vision (2018)
18. Genova, K., Cole, F., Maschinot, A., Sarna, A., Vlasic, D., Freeman, W.T.: Unsupervised training for 3D morphable model regression. In: IEEE Conference on Computer Vision and Pattern Recognition, June 2018
19. Bas, A., Huber, P., Smith, W.A., Awais, M., Kittler, J.: 3D morphable models as spatial transformer networks. In: International Conference on Computer Vision Workshop on Geometry Meets Deep Learning, pp. 904–912 (2017)
20. Tewari, A., Zollhöfer, M., Kim, H., Garrido, P., Bernard, F., Pérez, P., Theobalt, C.: MoFa model-based deep convolutionalface auto encoder for unsupervised monocular reconstruction. In: International Conference on Computer Vision, pp. 1274–1283 (2017)
21. Piotraschke, M., Blanz, V.: Automated 3D face reconstruction from multiple images using quality measures. In: Proceedings of the Conference on Computer Vision Pattern Recognition, June 2016
22. Huber, P., et al.: A multiresolution 3D morphable face model and fitting framework. In: Proceedings of the 11th International Joint Conference on Computer Vision, Imaging and Computer Graphics Theory and Applications (2016)
23. He, K., Zhang, X., Ren, S., Sun, J.: Deep residual learning for image recognition. In: Computer Vision and Pattern Recognition (2016)

Stability Analysis of Quaternion-Valued Neural Network with Non-differentiable Time-Varying Delays and Constant Delays

Hongying Qin[1], Zhenhao Chen[2], Xiaomei Wang[2(✉)], and Guo Huang[1]

[1] School of Artificial Intelligence, Leshan Normal University, Leshan, Sichuan, China
[2] School of Mathematical Science, University of Electronic Science and Technology of China, Chengdu, Sichuan, China
xmwang16@uestc.edu.cn

Abstract. The main goal of this paper is to investigate the problems of the uniqueness of equilibrium and the global μ-stability for the QVNN (quaternion-valued neural network) with leaky constant delay, non-differentiable discrete time-varying delay, distributed constant delay, which is closer to practical application than the QVNN with differentiable time-varying delay. Firstly, we discuss the QVNN as entirety, and prove the equilibrium of the QVNN is unique by using Homeomorphism mapping theorem and quaternion-valued linear matrix inequality. Then a new Lyapunov-Krasovskii functional is derived from the delayed state. The sufficient condition of the global μ-stability is given, while appraising the derivative of the Lyapunov-Krasovskii functional and quaternion-valued linear matrix inequality, this result is new and different from the approaches in available literatures. A quaternion-valued numerical example is presented to illustrate these results.

Keywords: Quaternion-valued neural network · Non-differentiable delays · Constant delays · Global stability

1 Introduction

In recent years, Neural networks has been becoming as foundation of machine learning technology, and used in diverse fields widely, such as engineering control, economy, transportation, psychology, and so on [1–3]. Applications of real-valued

Supported by the Sichuan Province science and technology department application foundation project (2016JY0238), Sichuan Province Education Department Key Projects (18ZA0235), Sichuan Province Education Department General Project (18ZB0268, 18ZB0266), Research Fund of Leshan Normal University (JG2018-1-04, LZD003).

X. Wu et al. (Eds.): QShine 2020, LNICST 381, pp. 248–259, 2021.
https://doi.org/10.1007/978-3-030-77569-8_18

neural networks, complex-valued neural networks have been extensively investigated in past decades. W. Hamilton proposed quaternion comprised of one real part as well as three imaginary parts in 1872 [4]. Quaternion has stronger information storage capacity. In high-dimensional data processing, quaternion-valued neural networks (QVNNs) have irreplaceable advantages compare with real-valued neural networks and complex-valued neural networks, by leveraging the benefits of this [5–9]. In practical applications, time delays may reduce the transmission speed of Neural network, and destabilize the overall stable system [10–12], and there are a lot of non-differentiable time-varying delays in reality, it is important to take non-differentiable time-varying delays into neural networks [13–16]. Hence, the stability of neural networks is primary consideration for ensuring its practicality. In 2007, the definition of the μ-stability is proposed firstly [17], that is recapitulative and could be specialized into six stability states, including power stability, exponential stability, log stability, log-log stability, the global asymptotical stability, the Lyapunov stability, via changing the property of the time delay and μ function [18].

In our research, we investigate the global μ-stability of the QVNN with leaky constant delay, non-differentiable discrete time-varying delay, distributed constant delay. Firstly the equilibrium of the QVNN is unique is proved by using Homeomorphism mapping theorem and quaternion-valued linear matrix inequality. Next, based on the delayed state's feature, we constructed a new Lyapunov-Krasovskii function. The global μ-stability of the QVNN is obtained while appraising the derivative of the Lyapunov-Krasovskii function as well as quaternion-valued linear matrix inequality. The vital contributions of this study are summarized as follows:

(1) We discuss the global μ-stability of the QVNN with non-differentiable discrete time-varying delay, which is closer to practical application than the QVNN with differentiable time-varying delay;
(2) We investigate the QVNN as entirety, not decompose the QVNN into two complex-valued neural networks or four real-valued neural networks, the increasing of data dimension caused by decomposition method is avoided.

2 Preliminaries

x is a quaternion, and can be defined as follow:

$$x = x^R + i \cdot x^I + j \cdot x^J + k \cdot x^K,$$

$x^R, x^I, x^J, x^K \in \mathbb{R}$ all are real coefficient, i, j, k all denote the imaginary units. A quaternion satisfies the Hamilton rule:

$$ij = -ji = k, jk = -kj = i, ki = -ik = j,$$

$$i^2 = j^2 = k^2 = -1.$$

Due to the Hamilton rule, quaternion doesn't meet commutative law of multiplication. In the next content, we use following notations: $\mathbb{Q}^{n\times m}$,$\mathbb{C}^{n\times m}$,$\mathbb{R}^{n\times m}$

represent, respectively, the set of $n \times m$ quaternion, complex, real matrices; $SC_n(\mathbb{Q})$ denotes the set of self-conjugate matrices, $SC_n^{>}(\mathbb{Q})$ denotes the set of positive definite matrices of quaternions [18]; $x \in \mathbb{Q}^n$, x^* is the conjugate transpose of x; $A \in \mathbb{Q}^{n\times m}$, $\bar{A}$, A^*, and $\lambda_{min}(A)$ represent, respectively, the conjugate, the conjugate transpose, and the minimum eigenvalue of A [19].

Considering the QVNN as follow:

$$\frac{dx(t)}{dt} = -Dx(t-\tau_1) + Ag(x(t)) + Bg\big(x(t-\tau(t))\big) + C\int_{t-\tau_2}^{t} g\big(x(s)\big)\,ds + v \tag{1}$$

$x(t) = (x_1(t), x_2(t), ..., x_n(t))^T \in \mathbb{Q}^n$ is the state vector. $D \in \mathbb{R}^{n\times n}$, with $D = diag(d_1, d_2, ..., d_n) \succ 0$ means the self-feedback connection weight matrix. $A, B, C \in \mathbb{Q}^{n\times n}$ mean the connection weight matrix. $\tau_1, \tau_2, \tau(t)$ denote, respectively, the leakage time delay, the distributed constant time delay, and the non-differentiable discrete time-varying delay. $g(x(t)) = (g_1(x_1(t)), g_2(x_2(t)), ..., g_n(x_n(t)))^T \in \mathbb{Q}^n$ denotes activation function. $v = (v_1, v_2, ..., v_n) \in \mathbb{Q}^n$ means the external input vector. $\bar{\lambda}(A)$ is the minimum of eigenvalues matrix A.

Assumption 1. *There are positive constants δ_l, such that:*

$$\| g_l(x) - g_l(x') \| \le \delta_l \| x - x' \|,$$

where $l = 1, 2, ..., n$. We define matrix:

$$\Gamma = diag(\delta_1, \delta_2, ..., \delta_n),$$

Definition 1. *μ-stability: $\mu(t) \ge 0$ is continuous function, when $t \to \infty$, $\mu(t) \to \infty$. If there exists a constant ω, then*

$$\|x(t)\| \le \frac{\omega}{\mu(t)}.$$

Lemma 1 ([20]). *$G = \begin{bmatrix} G_{11} & G_{12} \\ G_{21} & G_{22} \end{bmatrix} \in \mathbb{Q}^{2n\times 2n}$ where $G_{11} = G_{11}^*$, $G_{12} = G_{21}^*$, $G_{22} = G_{22}^*$, then $G \prec 0$ is equivalent two conditions:*

(1) $G_{22} \prec 0$, $G_{11} - G_{12}G_{22}^{-1}G_{12}^* \prec 0$;
(2) $G_{11} \prec 0$, $G_{22} - G_{12}^*G_{11}^{-1}G_{12} \prec 0$.

Lemma 2. *$H(x)$ is specified as a homeomorphism of $\mathbb{Q}^n$ onto itself, if $H(x) : \mathbb{Q}^n \to \mathbb{Q}^n$ is a continuous map accord with two qualifications:*

(1) $H(x)$ is a injective on $\mathbb{Q}^n$;
(2) $\lim_{\|x\|\to\infty} H(x) = \infty$.

Lemma 3 ([3]). *For the Hermitian constant matrix $W \in \mathbb{Q}^{n\times n}$, $W \ge 0$, and scalar function $g : [n, m] \to \mathbb{Q}^n$, $n \le m$, then*

$$\left(\int_n^m g(s)\,ds\right)^* W \left(\int_n^m g(s)\,ds\right) \le (m-n)\int_n^m g^*(s)Wg(s)\,ds.$$

Lemma 4 ([3]). *Let $q, \breve{q} \in \mathbb{Q}^n$, $Q \in SC_n^{>}(\mathbb{Q})$, then*

$$q^*\breve{q} + \breve{q}^*q \leq \frac{1}{\varepsilon}q^*Q^{-1}q + \varepsilon\breve{q}^*Q\breve{q}.$$

3 Main Result

Theorem 1. *If Assumption 1 holds, the equilibrium of the QVNN (1) is unique, when there are positive matrix U and positive matrices J_1, J_2, J_3, such that*

$$\Sigma = \begin{bmatrix} \Sigma_1 & UA & UB & UC \\ * & -J_1 & 0 & 0 \\ * & * & -J_2 & 0 \\ * & * & * & -J_3 \end{bmatrix} \prec 0 \tag{2}$$

where

$$\Sigma_1 = -UD - DU + \Gamma J_1 \Gamma + \Gamma J_2 \Gamma + \tau^2 \Gamma J_3 \Gamma$$

Proof. Establishing a continuous map

$$H(x) = -Dx + Ag(x) + Bg(x) + \tau Cg(x) + V \tag{3}$$

we assume there are two different vectors $x_1, x_2 \in \mathbb{Q}^n$, such that $H(x_1) = H(x_2)$,

$$\begin{aligned} &-(x_1 - x_2)^*(UD + DU)(x_1 - x_2) + (x_1 - x_2)^*U(A + B)\big(g(x_1) - g(x_2)\big)^* J)_1 \\ &\cdot \big(g(x_1) - g(x_2)\big) + \tau(x_1 - x_2)^*UC\big(g(x_1) - g(x_2)\big) + \tau\big(g(x_1) - g(x_2)\big)^* C^*U \\ &\cdot (x_1 - x_2) = 0 \end{aligned} \tag{4}$$

Based on Lemma 4,

$$\begin{aligned} &(x_1 - x_2)^*U(A + B)\big(g(x_1) - g(x_2)\big)^* J)_1\big(g(x_1) - g(x_2)\big) \\ \leq &(x_1 - x_2)^*UAJ_1^{-1}A^*U(x_1 - x_2) + \big(g(x_1) - g(x_2)\big)^* J_1\big(g(x_1) - g(x_2)\big) + (x_1 \\ &- x_2)^*UBJ_2^{-1}B^*U(x_1 - x_2) + \big(g(x_1) - g(x_2)\big)^* J_2\big(g(x_1) - g(x_2)\big) \end{aligned} \tag{5}$$

$$\begin{aligned} &\tau(x_1 - x_2)^*UC\big(g(x_1) - g(x_2)\big) + \tau\big(g(x_1) - g(x_2)\big)^* C^*U(x_1 - x_2) \\ \leq &(x_1 - x_2)^*UCJ_3^{-1}C^*U(x_1 - x_2) + \tau^2\big(g(x_1) - g(x_2)\big)^* J_3\big(g(x_1) - g(x_2)\big) \end{aligned} \tag{6}$$

$$\big(g(x_1) - g(x_2)\big)^* J_l\big(g(x_1) - g(x_2)\big) \leq (x_1 - x_2)^*\Gamma J_l \Gamma(x_1 - x_2) \tag{7}$$

where $l = 1, 2, 3$.

$$\begin{aligned} &(x_1 - x_2)^*(-UD - DU + UAJ_1^{-1}A^*U + UBJ_2^{-1}B^*U \\ &+ UCJ_3^{-1}C^*U + \Gamma J_1 \Gamma + \Gamma J_2 \Gamma + \tau^2 \Gamma J_3 \Gamma)(x_1 - x_2) \geq 0 \end{aligned} \tag{8}$$

According to Lemma 1 and $\Sigma \prec 0$, the following inequality can be obtained

$$\begin{aligned} &-UD - DU + UAJ_1^{-1}A^*U + UBJ_2^{-1}B^*U \\ &+ UCJ_3^{-1}C^*U + \Gamma J_1\Gamma + \Gamma J_2\Gamma + \tau^2\Gamma J_3\Gamma \prec 0, \end{aligned} \tag{9}$$

therefore, $H(x)$ is an injective on $\mathbb{Q}$. Besides,

$$\begin{aligned} &x^*U\big(H(x) - H(0)\big) + \big(H(x) - H(0)\big)^*Ux \\ \leq &x^*(-UD - DU + UAJ_1^{-1}A^*U + UBJ_2^{-1}B^*U + UCJ_3^{-1}C^*U + \Gamma J_1\Gamma + \Gamma J_2\Gamma \\ &+ \tau^2\Gamma J_3\Gamma)x \\ \leq &-\bar{\lambda}(-UD - DU + UAJ_1^{-1}A^*U + UBJ_2^{-1}B^*U + UCJ_3^{-1}C^*U + \Gamma J_1\Gamma + \Gamma J_2 \\ &\cdot \Gamma + \tau^2\Gamma J_3\Gamma)x^*x \\ = &\lambda x^*x \end{aligned} \tag{10}$$

$$\lambda \parallel x \parallel^2 \leq 2 \parallel x \parallel \parallel U \parallel \big(\parallel H(x) \parallel + \parallel H(0) \parallel \big) \tag{11}$$

$$\lambda \parallel x \parallel \leq 2 \parallel U \parallel \big(\parallel H(x) \parallel + \parallel H(0) \parallel \big) \tag{12}$$

Consequently, $\parallel H(x) \parallel \to \infty$, as $\parallel x \parallel \to \infty$, an unique equilibrium point of the QVNN (1) is proved.

Choosing the variable substitution $\widetilde{x}(t) = x(t) - \hat{x}$, $\hat{x}$ is the equilibrium point of the QVNN (1), the following system is obtained from the QVNN (1):

$$\frac{d\widetilde{x}(t)}{dt} = -D\widetilde{x}(t - \tau_1) + Ag(\widetilde{x}(t)) + Bg(\widetilde{x}(t - \tau(t))) + C\int_{t-\tau_2}^{t} g(\widetilde{x}(s))\,ds,$$

Theorem 2. *The QVNN (1) is global $\mu-$stability, if $t \geq T$, $max\{\tau_1, \tau_2, \tau(t)\} \leq \tau$, there are constants α, β, such that $\frac{\dot{\mu}(t)}{\mu(t)} \leq \beta$, $\eta \leq \frac{\mu(t-\tau)}{\mu(t)} \leq \alpha$, and matrices $R_1, R_2, R_3, R_4, R_5 \in SC_n^{>}(Q)$, the following matrix Π exists:*

$$\Pi = \begin{bmatrix} \Pi_1 & R_1 & R_3 & \eta^2R_2 & 0 & 0 & 0 \\ * & \tau_1R_2 & 0 & 0 & 0 & 0 & 0 \\ * & * & 2\beta R_3 & -R_3 & 0 & 0 & 0 \\ * & * & * & -\eta^2R_2 & 0 & 0 & 0 \\ * & * & * & * & R_4 & 0 & 0 \\ * & * & * & * & * & -\eta^2R_4 & 0 \\ * & * & * & * & * & * & -\eta^2R_5 \end{bmatrix} \prec 0 \tag{13}$$

$$\Pi_1 = 2\beta R_1 - \eta^2R_2 + \tau_2R_5$$

Proof. Using Lyapunov-Krasovskii function as below,

$$V(t) = \sum_{q=1}^{5} V_q(t) \tag{14}$$

$$\begin{aligned}
V_1(t) =& \mu^2(t)\widetilde{x}^*(t)R_1\widetilde{x}(t),\\
V_2(t) =& \tau_1 \int_0^{\tau_1}\int_{t-\theta}^{t} \mu^2(s)\dot{\widetilde{x}}^*(s)R_2\dot{\widetilde{x}}(s)dsd\theta,\\
V_3(t) =& \mu^2(t)\Big(\int_{t-\tau_1}^{t} \widetilde{x}(s)ds\Big)^* R_3\Big(\int_{t-\tau_1}^{t} \widetilde{x}(s)ds\Big),\\
V_4(t) =& \int_{t-\tau}^{t} \mu^2(s)g^*\big(\widetilde{x}(s)\big)R_4 g\big(\widetilde{x}(s)\big)ds,\\
V_5(t) =& \tau_2 \int_0^{\tau_2}\int_{t-\theta}^{t} \mu^2(s)\widetilde{x}^*(s)R_5\widetilde{x}(s)dsd\theta.
\end{aligned}$$

The derivative of $V(t)$ along the trajectories of the QVNN (1) is:

$$\begin{aligned}
\dot{V}_1(t) =& 2\mu(t)\dot{\mu}(t)\widetilde{x}^*(t)R_1\widetilde{x}(t) + \mu^2(t)\dot{\widetilde{x}}^*(t)R_1\widetilde{x}(t) + \mu^2(t)\widetilde{x}^*(t)R_1\dot{\widetilde{x}}(t)\\
\leq& 2\beta\mu^2(t)\widetilde{x}^*(t)R_1\widetilde{x}(t) + \mu^2(t)\dot{\widetilde{x}}^*(t)R_1\widetilde{x}(t) + \mu^2(t)\widetilde{x}^*(t)R_1\dot{\widetilde{x}}(t)
\end{aligned} \tag{15}$$

$$\begin{aligned}
\dot{V}_2(t) =& \tau_1 \int_0^{\tau_1} \mu^2(t)\dot{\widetilde{x}}^*(t)R_2\dot{\widetilde{x}}(t) - \mu^2(t-\theta)\cdot\dot{\widetilde{x}}^*(t-\theta)R_2\dot{\widetilde{x}}(t-\theta)d\theta\\
\leq& \tau_1^2\mu^2(t)\dot{\widetilde{x}}^*(t)R_2\dot{\widetilde{x}}(t) - \eta^2\tau_1\mu^2(t)\int_0^{\tau_1}\dot{\widetilde{x}}^*(t-\theta)R_2\dot{\widetilde{x}}(t-\theta)d\theta\\
\leq& \tau_1^2\mu^2(t)\dot{\widetilde{x}}^*(t)R_2\dot{\widetilde{x}}(t) - \eta^2\mu^2(t)\big(\widetilde{x}^*(t)R_2\widetilde{x}(t) - \widetilde{x}^*(t)R_2\widetilde{x}(t-\tau_1) - \widetilde{x}^*(t\\
&-\tau_1)R_2\widetilde{x}(t) + \widetilde{x}^*(t-\tau_1)R_2\widetilde{x}(t-\tau_1)\big)
\end{aligned} \tag{16}$$

$$\begin{aligned}
\dot{V}_3(t) =& \mu^2(t)\big(\widetilde{x}(t)-\widetilde{x}(t-\tau_1)\big)^* R_3\int_{t-\tau_1}^{t}\widetilde{x}(s)ds + 2\mu(t)\dot{\mu}(t)\Big(\int_{t-\tau_1}^{t}\widetilde{x}(s)ds\Big)^* R_3\\
&\cdot\int_{t-\tau_1}^{t}\widetilde{x}(s)ds + \mu^2(t)\Big(\int_{t-\tau_1}^{t}\widetilde{x}(s)ds\Big)^* R_3\big(\widetilde{x}(t)-\widetilde{x}(t-\tau_1)\big)
\end{aligned} \tag{17}$$

$$\dot{V}_4(t) = \mu^2(t)g^*\big(\widetilde{x}(t)\big)R_4 g\big(\widetilde{x}(t)\big) - \mu^2(t-\tau)g^*\big(\widetilde{x}(t-\tau)\big)R_4 g\big(\widetilde{x}(t-\tau)\big) \tag{18}$$

$$\begin{aligned}
\dot{V}_5(t) =& \tau_2 \int_0^{\tau_2} \mu^2(t)\widetilde{x}^*(t)R_5\widetilde{x}(t) - \mu^2(t-\theta)\widetilde{x}^*(t-\theta)R_5\widetilde{x}(t-\theta)d\theta\\
\leq& \mu^2(t)\big(\tau_2^2\widetilde{x}^*(t)R_5\widetilde{x}(t) - \eta^2\tau_2\int_{t-\tau_2}^{t}\mu^2(s)\widetilde{x}^*(s)R_5\widetilde{x}(s)ds\big)\\
\leq& \mu^2(t)\Big(\tau_2\widetilde{x}^*(t)R_5\widetilde{x}(t) - \eta^2\Big(\int_{t-\tau_2}^{t}\widetilde{x}(s)ds\Big)^* R_5\int_{t-\tau_2}^{t}\widetilde{x}(s)ds\Big)
\end{aligned} \tag{19}$$

With (14)–(19),

$$
\begin{aligned}
D^+V(t) \leq & 2\beta\mu^2(t)\widetilde{x}^*(t)R_1\widetilde{x}(t) + \mu^2(t)\dot{\widetilde{x}}^*(t)R_1\widetilde{x}(t) + \mu^2(t)\widetilde{x}^*(t)R_1\dot{\widetilde{x}}(t) + \tau_1^2\mu^2(t) \\
& \cdot \dot{\widetilde{x}}^*(t)R_2\dot{\widetilde{x}}(t) - \eta^2\mu^2(t)\big(\widetilde{x}^*(t)R_2\widetilde{x}(t) - \widetilde{x}^*(t)R_2\widetilde{x}(t-\tau_1) - \widetilde{x}^*(t-\tau_1) \\
& \cdot R_2\widetilde{x}(t) + \widetilde{x}^*(t-\tau_1)R_2\widetilde{x}(t-\tau_1)\big)\mu^2(t)\big(\widetilde{x}(t) - \widetilde{x}(t-\tau_1)\big)^* R_3 \\
& \cdot \int_{t-\tau_1}^{t} \widetilde{x}(s)ds + 2\mu(t)\dot{\mu}(t)\Big(\int_{t-\tau_1}^{t} \widetilde{x}(s)ds\Big)^* R_3 \int_{t-\tau_1}^{t} \widetilde{x}(s)ds + \mu^2(t) \\
& \cdot \Big(\int_{t-\tau_1}^{t} \widetilde{x}(s)ds\Big)^* R_3\big(\widetilde{x}(t) - \widetilde{x}(t-\tau_1)\big)\mu^2(t)g^*\big(\widetilde{x}(t)\big)R_4 g\big(\widetilde{x}(t)\big) - \mu^2(t \\
& - \tau)g^*\big(\widetilde{x}(t-\tau)\big)R_4 g\big(\widetilde{x}(t-\tau)\big) + \mu^2(t)\Big(\tau_2\widetilde{x}^*(t)R_5\widetilde{x}(t) - \eta^2 \\
& \cdot \Big(\int_{t-\tau_2}^{t} \widetilde{x}(s)ds\Big)^* R_5 \int_{t-\tau_2}^{t} \widetilde{x}(s)ds\Big) \\
\leq & \mu^2(t)\Big(2\beta\widetilde{x}^*(t)R_1\widetilde{x}(t) + \dot{\widetilde{x}}^*(t)R_1\widetilde{x}(t) + \widetilde{x}^*(t)R_1\dot{\widetilde{x}}(t) + \tau_1^2\dot{\widetilde{x}}^*(t)R_2\dot{\widetilde{x}}(t) \\
& - \eta^2\big(\widetilde{x}^*(t)R_2\widetilde{x}(t) - \widetilde{x}^*(t)R_2\widetilde{x}(t-\tau_1) - \widetilde{x}^*(t-\tau_1)R_2\widetilde{x}(t) + \widetilde{x}^*(t-\tau_1) \\
& \cdot R_2\widetilde{x}(t-\tau_1)\big)\big(\widetilde{x}(t) - \widetilde{x}(t-\tau_1)\big)^* R_3 \int_{t-\tau_1}^{t} \widetilde{x}(s)ds + 2\beta\Big(\int_{t-\tau_1}^{t} \widetilde{x}(s)ds\Big)^* \\
& \cdot R_3 \int_{t-\tau_1}^{t} \widetilde{x}(s)ds + \Big(\int_{t-\tau_1}^{t} \widetilde{x}(s)ds\Big)^* R_3\big(\widetilde{x}(t) - \widetilde{x}(t-\tau_1)\big)g^*\big(\widetilde{x}(t)\big)R_4 \\
& \cdot g\big(\widetilde{x}(t)\big) - \eta^2 g^*\big(\widetilde{x}(t-\tau)\big)R_4 g\big(\widetilde{x}(t-\tau)\big) + \Big(\tau_2\widetilde{x}^*(t)R_5\widetilde{x}(t) - \eta^2 \\
& \cdot \Big(\int_{t-\tau_2}^{t} \widetilde{x}(s)ds\Big)^* R_5 \int_{t-\tau_2}^{t} \widetilde{x}(s)ds\Big)\Big) \\
= & \mu^2(t)q(t)\Pi q^*(t)
\end{aligned}
\tag{20}
$$

and

$$
\begin{aligned}
q(t) = \Big(& \widetilde{x}^*(t), \dot{\widetilde{x}}(t), \Big(\int_{t-\tau_1}^{t} \widetilde{x}(s)ds\Big)^*, \widetilde{x}^*(t-\tau_1), \\
& g^*\big(\widetilde{x}(t)\big), g^*\big(\widetilde{x}(t-\tau)\big), \Big(\int_{t-\tau_2}^{t} \widetilde{x}(s)ds\Big)^*\Big).
\end{aligned}
$$

With (13) and (20),

$$D^+V(t) \leq 0. \tag{21}$$

For $t \in [T, +\infty)$,

$$\mu^2(t)\bar{\lambda}(R_1)\|\widetilde{x}(t)\|^2 \leq V(t) \leq V_0 = \max_{0<s<T} V(s) \tag{22}$$

Hence,

$$\| \widetilde{x}(t) \| \leq \frac{\omega}{\mu(t)}, \tag{23}$$

with

$$\omega = \sqrt{\frac{V_0}{\overline{\lambda}(R_1)}}.$$

Therefore, the QVNN (1) is global μ-stability.

4 Simulation Example

A numerical simulation example is presented to strengthen the new conclusions above.

Considering the following QVNN:

$$\frac{dx(t)}{dt} = -Dx(t-\tau_1) + Ag(x(t)) + Bg\big(x(t-\tau(t))\big) + C\int_{t-\tau_2}^{t} g\big(x(s)\big)\,ds + v \tag{24}$$

where

$$A = \begin{bmatrix} a_{11} \; a_{12} \\ a_{21} \; a_{22} \end{bmatrix}, B = \begin{bmatrix} b_{11} \; b_{12} \\ b_{21} \; b_{22} \end{bmatrix}, C = \begin{bmatrix} c_{11} \; c_{12} \\ c_{21} \; c_{22} \end{bmatrix},$$

$$D = \begin{bmatrix} 5 \; 0 \\ 0 \; 5 \end{bmatrix}, v = \begin{bmatrix} v_1 \\ v_2 \end{bmatrix},$$

we randomly define quaternion matrices of the QVNN: $a_{11} = 0.27 - 0.35i + 0.043j - 0.18k, a_{12} = -0.15 - 0.22i + 0.28j - 0.04k, a_{21} = -0.31 - 0.293i + 0.2j - 0.06k, a_{22} = 0.175 - 0.2i + 0.18j - 0.165k; b_{11} = -0.38 + 0.223i - 0.57j - 0.139k, b_{12} = 0.16 - 0.121i + 0.031j - 0.09k, b_{21} = 0.93 - 0.701i + 0.02j - 0.27k, b_{22} = 0.0907 - 0.24i + 0.104j + 0.08k; c_{11} = -0.231 + 0.208i - 0.179j + 0.05k, c_{12} = 0.77 + 0.01i + 0.04j - 1.23k, c_{21} = 0.092 - 0.113i + 0.25j - 0.7k, c_{22} = -0.336 + 0.134i - 0.22j + 0.4k; v_1 = 0.052 - 0.14i + 0.08j - 0.12k, v_2 = -0.58 + 0.304i - 0.19j + 0.145k$.

Define $g(x(t)) = tanh(x(t))$, $\tau_1 = 0.4$, $\tau_2 = 0.3$, $\tau(t) = 0.37|sint|$, $\mu = e^{0.8t}$, obviously, $\tau = 0.4, \beta = 0.8, \eta = 0.6, \alpha = 0.8$.

The solutions of $\Sigma \prec 0$ and $\Pi \prec 0$ can be resolved as:

$$\Sigma = \begin{bmatrix} \Sigma_1 & UA & UB & UC \\ * & -J_1 & 0 & 0 \\ * & * & -J_2 & 0 \\ * & * & * & -J_3 \end{bmatrix} \prec 0$$

$$\Pi = \begin{bmatrix} \Pi_1 & R_1 & R_3 & \eta^2 R_2 & 0 & 0 & 0 \\ * & \tau_1 R_2 & 0 & 0 & 0 & 0 & 0 \\ * & * & 2\beta R_3 & -R_3 & 0 & 0 & 0 \\ * & * & * & -\eta^2 R_2 & 0 & 0 & 0 \\ * & * & * & * & R_4 & 0 & 0 \\ * & * & * & * & * & -\eta^2 R_4 & 0 \\ * & * & * & * & * & * & -\eta^2 R_5 \end{bmatrix} \prec 0$$

$$\Pi_1 = 2\beta R_1 - \eta^2 R_2 + \tau_2 R_5$$

where

$$\Sigma_1 = -UD - DU + \Gamma J_1 \Gamma + \Gamma J_2 \Gamma + \tau^2 \Gamma J_3 \Gamma$$

$$U = 10^{-9} diag[0.1335, 0.1155, 0.1357, 0.0649]$$

$$J_1 = 10^{-10} \begin{bmatrix} j_1^{11} & j_1^{12} \\ j_1^{21} & j_1^{22} \end{bmatrix}, J_2 = 10^{-10} \begin{bmatrix} j_2^{11} & j_2^{12} \\ j_2^{21} & j_2^{22} \end{bmatrix}, J_3 = 10^{-14} \begin{bmatrix} j_3^{11} & j_3^{12} \\ j_3^{21} & j_3^{22} \end{bmatrix},$$
$$R_1 = 10^{-11} \begin{bmatrix} r_1^{11} & 0 \\ 0 & r_1^{22} \end{bmatrix}, R_2 = 10^{-10} \begin{bmatrix} r_2^{11} & 0 \\ 0 & r_2^{22} \end{bmatrix}, R_3 = 10^{-11} \begin{bmatrix} r_3^{11} & 0 \\ 0 & r_3^{22} \end{bmatrix},$$
$$R_4 = 10^{-11} \begin{bmatrix} r_4^{11} & 0 \\ 0 & r_4^{22} \end{bmatrix}, R_5 = 10^{-10} \begin{bmatrix} r_5^{11} & 0 \\ 0^{21} & r_5^{22} \end{bmatrix}.$$

$j_1^{11} = -0.1771 - 0.1771j; j_1^{12} = 0.0193 + 0.0111i + 0.0193j + 0.0111k; j_1^{21} = 0.0193 - 0.0111i + 0.0193j - 0.0111k; j_1^{22} = -0.1801 - 0.1801j; j_2^{11} = 0.03846 - 0.4789j; j_2^{12} = 0.00341 + 0.2058i + 0.0035j + 0.0023k; j_2^{21} = 0.00341 - 0.2058i + 0.0035j - 0.0023k; j_2^{22} = -0.7801 - 0.4674j; j_3^{11} = 0.189 + 0.189j; j_3^{12} = 0.0012 + 0.0032i + 0.0012j + 0.0012k; j_3^{21} = 0.0012 - 0.0032i + 0.0012j - 0.0032k; j_3^{22} = 0.2105 + 0.2105j; r_1^{11} = -0.5063 - 0.5063j; r_1^{22} = -0.5063 - 0.5063j; r_2^{11} = 0.3272 + 0.3272j; r_2^{22} = 0.3272 + 0.3272j; r_3^{11} = -0.2932 - 0.2932j; r_3^{22} = -0.2932 - 0.2932j; r_4^{11} = 0.4514 + 0.451j; r_4^{22} = 0.4514 + 0.451j; r_5^{11} = 0.8523 + 0.8523j; r_5^{22} = 0.8523 + 0.8523j.$

Hence the QVNN (24) has unique equilibrium and is global μ-stability under Theorem 1 and Theorem 2.

Figure 1, 2, 3 and 4 respectively are the trajectories of $x^R(t)$, $x^I(t)$, $x^J(t)$, $x^K(t)$ in QVNN (4.1) with original value $x(0) = (-0.5 - 0.32i + 0.28j - 0.541k, 0.69 + 1.05i - 1.1j + 1.2k)$.

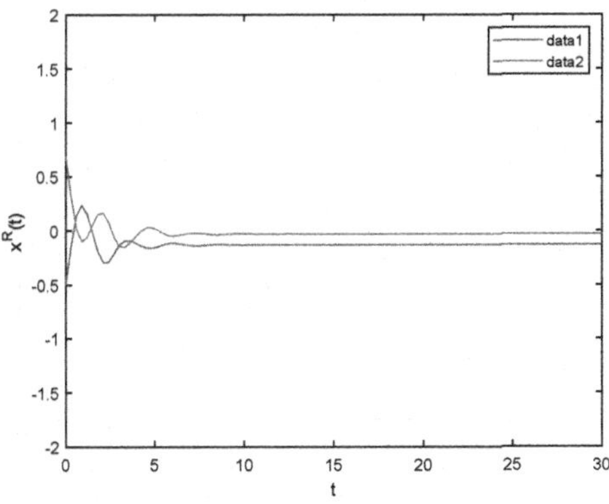

Fig. 1. The trajectories of $x^R(t)$ in the QVNN (24)

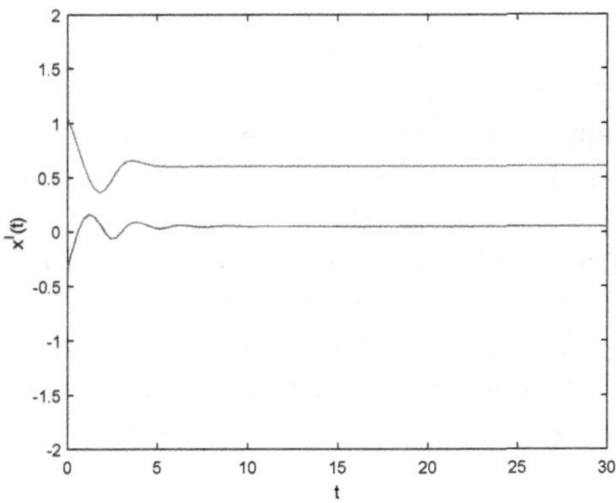

Fig. 2. The trajectories of $x^I(t)$ in the QVNN (24)

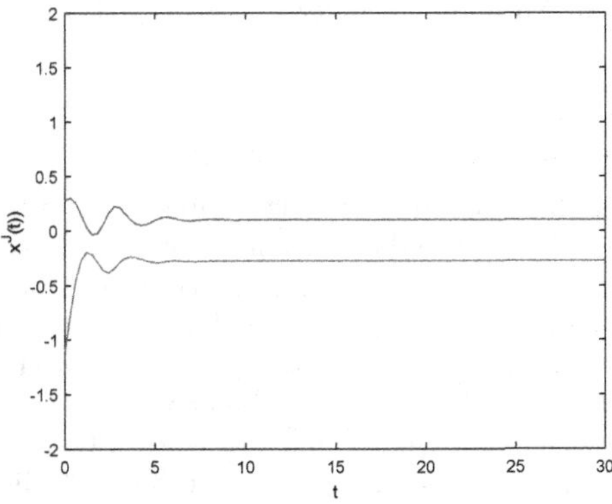

Fig. 3. The trajectories of $x^J(t)$ in the QVNN (24)

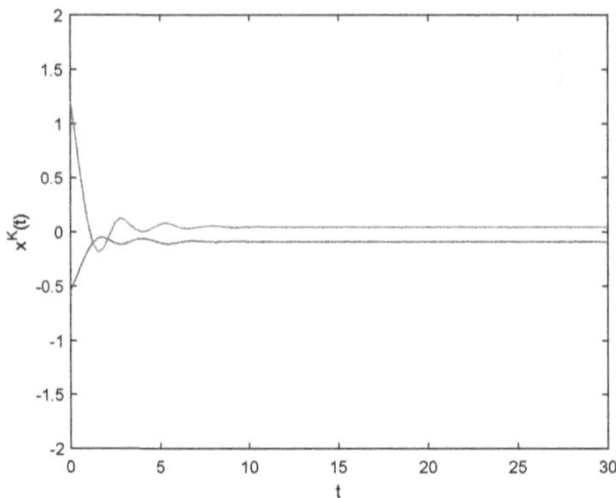

Fig. 4. The trajectories of $x^K(t)$ in the QVNN (24)

5 Conclusion

This paper discuss the equilibrium's uniqueness and the global μ-stability of the QVNN (quaternion-valued neural network) with leaky constant delay, non-differentiable discrete time-varying delay, distributed constant delay. Firstly, considering the QVNN as entirety, the uniqueness of the QVNN's equilibrium is obtained, via Homeomorphism mapping theorem and quaternion-valued linear matrix inequality. Next a novel Lyapunov-Krasovskii functional is derived according to the delayed state, and the sufficient condition of the global μ-stability is presented. Finally, we give a quaternion-valued numerical example to illustrate these new results.

Acknowledgements. The work described in this paper was supported by the Sichuan Province science and technology department application foundation project (2016JY0238), Sichuan Province Education Department Key Projects (18ZA0235), Sichuan Province Education Department General Project (18ZB0268, 18ZB0266), Research Fund of Leshan Normal University (JG2018-1-04, LZD003).

References

1. Wang, X.: Introduction to neural networks. China Sci. J. (2017)
2. Gong, W., Liang, J., Cao, J.: Global stability of complex-valued delayed neural networks with leakage delay. Neurocomputing **168**(C), 135–144 (2015)
3. Chen, X., Li, Z., Song, Q., Hu, J., Tan, Y.: Robust stability analysis of quaternion-valued neural networks with time delays and parameter uncertainties. Neural Netw. Offi. J. Int. Neural Netw. Soc. S0893608017300862 (2017)
4. Hamilton, W.R.: Elements of Quaternions. New York Chelsea, America (1969)
5. Liu, Y., Pei, X., Jianquan, L., Liang, J.: Global stability of Clifford-valued recurrent neural networks with time delays. Nonlinear Dyn. **84**(2), 767–777 (2016). https://doi.org/10.1007/s11071-015-2526-y

6. Zhu, J., Sun, J.: Global Exponential Stability of Clifford-Valued Recurrent Neural Networks. Elsevier Science Publishers B.V, Amsterdam (2016)
7. Yang, L., Zhang, D., Jianquan, L.: Global exponential stability for quaternion-valued recurrent neural networks with time-varying delays. Nonlinear Dyn. **87**(1), 1–13 (2016). https://doi.org/10.1007/s11071-016-3060-2
8. Liu, X., Chen, T.: Global exponential stability for complex-valued recurrent neural networks with asynchronous time delays. IEEE Trans. Neural Netw. Learn. Syst. **27**(3), 593 (2015)
9. Liu, J., Jian, J.: Global dissipativity of a class of quaternion-valued bam neural networks with time delay. Neuro-computing **349**(JUL.15), 123–132 (2019)
10. Liu, Y., Wang, Z., Liu, X.: Asymptotic stability for neural networks with mixed time-delays: the discrete-time case. Neural Netw. **22**(1), 67–74 (2009)
11. Xuemei, Yu., Wang, X., Zhong, S., Shi, K.: Further results on delay-dependent stability for continuous system with two additive time-varying delay components. ISA Trans. **65**, 9–18 (2016)
12. Tang, Q., Jian, J.: Global exponential convergence for impulsive inertial complex-valued neural networks with time-varying delays. Math. Comput. Simul. **159**, 39–56 (2018)
13. Jun, X., Zhong, S., Zhang, C.: Stability of Hopfield type neural network with distributed time delay. J. Univ. Electron. Sci. Technol. China **2**(2), 200–203 (2004)
14. Chen, X., Song, Q., Liu, Y., Zhao, Z.: Global μ-stability of impulsive complex-valued neural networks with leakage delay and mixed delays. Abstr. Appl. Anal. **2014**, 1–14 (2014)
15. Chen, X., Song, Q., Li, Z., Zhao, Z., Liu, Y.: Stability analysis of continuous-time and discrete-time quaternion-valued neural networks with linear threshold neurons. IEEE Trans. Neural Netw. Learn. Syst. **29**(7), 2769–2781 (2018)
16. Liu, L., You, X., Gao, X.: Global synchronization control of quaternion-valued neural networks with mixed delays. Control Theory Appl. (8) (2019)
17. Chen, T., Wang, L.: Global μ-stability of delayed neural networks with unbounded time-varying delays. IEEE Trans. Neural Netw. **18**(6), 1836–1840 (2007)
18. Liu, Y., Zhang, D., Lou, J., Jianquan, L., Cao, J.: Stability analysis of quaternion-valued neural networks: decomposition and direct approaches. IEEE Trans. Neural Netw. Learn. Syst. **29**, 1–11 (2017)
19. Liu, Y., Zhang, D., Jianquan, L., Cao, J.: Global μ-stability criteria for quaternion-valued neural networks with unbounded time-varying delays. Inf. Sci. **360**, 273–288 (2016)
20. Boyd, S., Ghaoui, L., Feron, E., Balakrishnan, V.: Linear Matrix Inequalities in System and Control Theory. Society for Industrial and Applied Mathematics, Philadephia (1994)

Learn to Rectify Label Through Kernel Extreme Learning Machine

Qiang Cai[1,2,3], Fenghai Li[1,2,3(✉)], Haisheng Li[1,2,3], Jian Cao[1,2,3], and Shanshan Li[1,2,3]

[1] School of Computer, Beijing Technology and Business University, Beijing, China
[2] National Engineering Laboratory for Agri-product Quality Traceability, Beijing Technology and Business University, Beijing, China
[3] Beijing Key Laboratory of Big Data Technology for Food Safety, Beijing Technology and Business University, Beijing, China

Abstract. Recent studies attempt to construct complicated and redundant Convolutional Neural Networks (CNNs) to improve image classification performance. In this paper, instead of painstakingly designing a CNN's architecture, we consider promoting classification performance by revising CNN's classification results. We therefore propose a novel image classification approach that Learns to Rectify Label (LRL) through Kernel Extreme Learning Machine (KELM). It includes two phases: (1) Pre classification, we put images into a trained CNN to generate corresponding incomplete labels. (2) Label Rectification, the incomplete labels are rectified by the KELM's high-dimensional mapping, so final classification results are acquired. Extensive experiments conducted on public datasets demonstrate the effectiveness of our method. At the meantime, our method has well generalizability that can be integrated with many popular networks.

Keywords: Convolutional Neural Networks · Kernel extreme learning machine · Image classification

1 Introduction

Image classification is a fundamental task in computer vision, which aims to distinguish the image categories according to their semantic information. It is widely involved in many real-world application, including face recognition [5], traffic sign detection [29] and brain image analysis [2]. An early typical approach is using the handcraft feature (e.g.SIFT [24], HOG [4]) and feature description combined with classical classifiers (e.g.SVM [3]).

Convolutional neural networks (CNNs) have exhibited strong learning capability on image classification [19]. Subsequent works [25,27] build deeper CNNs by designing small convolutional kernels. In [8,11], shortcut connections build relation of different convolutional layers and alleviate vanishing gradient problem

X. Wu et al. (Eds.): QShine 2020, LNICST 381, pp. 260–269, 2021.
https://doi.org/10.1007/978-3-030-77569-8_19

in deep networks. SKNet [21] propose a dynamic selection mechanism in CNNs that allows each neuron to adaptively adjust its receptive field size based on multiple scales of input information. Res2Net [6] represents multi-scale features at a granular level and increases the range of receptive fields for each network layer. All of above-mentioned methods try to improve CNN's design to boost classification performance, but complicated architectures and enormous parameters often lead to redundant complicated loads and poorly trained models.

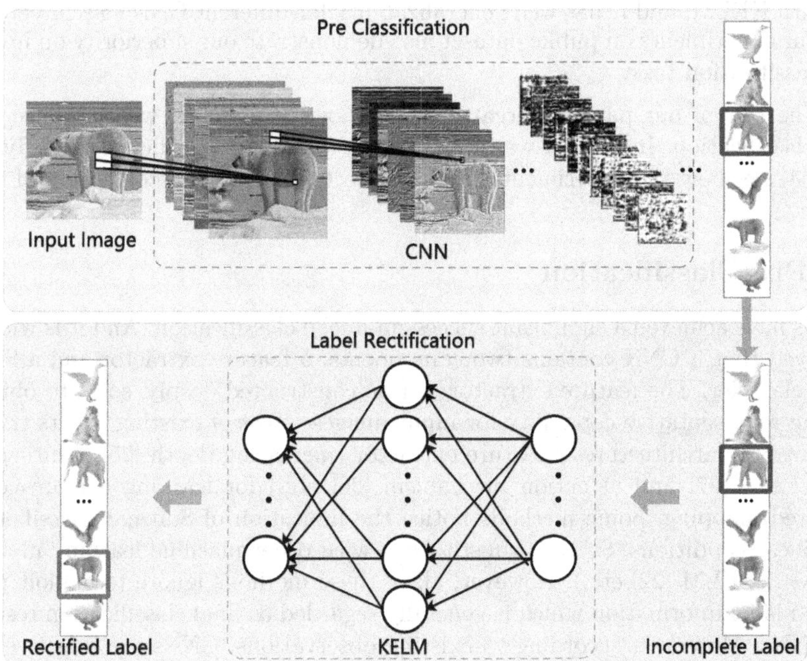

Fig. 1. Framework of the proposed approach.

On the other hand, some studies view CNNs as two main components, i.e. a feature extractor and a Softmax classifier. They consider improving discriminative ability of features by replacing Softmax with other machine learning algorithms [1,22]. e.g. Support Vector Machine (SVM) or Extreme Learning Machine (ELM). However, extracted high-dimensional features in matrix manipulation is time-consuming, which still remains a challenge.

In this paper, we consider a different view that tries to rectify labels output by CNN to a more correct distribution. We propose to learn to rectify label (LRL) through kernel extreme learning machine (KELM). Figure 1 illustrates our framework schematically. It involves two stages, i.e. pre classification and label rectification. In pre classification, we put images into a trained CNN and get corresponding classification results called incomplete labels. These labels may have large deviations with their ground truth. In label rectification, we aim to exploit label-wise relation, so the incomplete labels are fed into a KELM. By random kernel mapping and linear combination, we can get final classification results. Notably, compared with KSVM, KELM is more appropriate for mutil-class classification and it is higher efficiency [13].

In summary, the main contributions of this paper can be concluded as follows:

- To the best of our knowledge, it is the first time that a label rectification method is proposed for image classification.
- We present a novel image classification framework (LRL) that combines CNN with KELM, and it has well generalizability for different CNN's architecture.
- Our experiments on public dataset also demonstrate our superiority on image classification task.

The rest of our paper is organized as follows: In Sect. 2, we introduce the pre classification. In Sect. 3, we introduce the way label rectification briefly. In Sect. 4, we show the experiments result. In Sect. 5, we draw a conclusion of this paper.

2 Pre Classification

CNNs have achieved a significant success in image classification. And it is widely believed that a CNN contains two components: a feature extractor and a Softmax classifier. The feature extractor can be constructed deeply, so as to obtain strong representative capability for input images. Most of existing efforts try to improve the architecture of feature extractor (increasing depth [25], multi-scale kernel size [27] and attention mechanism [26] etc.) for learning a more complicated mapping. Some methods notice the limitation of Softmax classifier in nonlinear conditions. So they substitute it with other machine learning models (SVM [1], ELM [22] etc.). However, all of these methods ignore to exploit predicted label information which is generally regarded as final classification results of CNNs. Moreover, according to existing observations, CNNs are sensitive for hyper-parameters, so well training a network becomes hard, and it lacks adaptiveness in real-world application.

Above-mentioned illustrations motivate us to capture label information output by CNN to boost classification performance. This paper denotes the labels as incomplete labels because we conjecture these labels still have potential to be improved. In our method, we consequently extract labels of CNN instead of features. It can be formulated as Formula (1):

$$L = F_\theta(I) \tag{1}$$

where F is a well-trained CNN, I is the input image, L is extracted labels, θ is the parameters of CNN. Compared with features in fully connected layers [22], extracted labels contain predicted scores of each image category, and they are relatively low dimension (It depends on the number of all categories), so it is more efficient in subsequent operations. In addition, for a pre-trained CNN, the proposed method can drastically improve their performance according to our experiments. Notably, we do not use Softmax function to normalize the extracted labels in our method.

3 Label Rectification

Given incomplete labels L, we aim to revise them by a model f. For efficiently training f, we utilize ELM to rectify incomplete labels. ELM was proposed by Huang [15], it's a single-hidden layer feedforward neural networks (SLFNs) which randomly chooses hidden nodes and analytically determines the output weight of SLFNs. ELM tends to provide good generalization performance at extremely fast learning speed and the hidden layer need not be tuned. ELM consists of three layers: input layer, hidden layer and output layer. The structure of it is shown in Fig. 2.

Consider the incomplete labels L is passed through an ELM network, the hidden layer can map it to large dimensionality that increase the the universal infinite approximation ability of the ELM. The output function of ELM for generalized SLFNs can be described as Formula (2):

$$f_l(L) = \sum_{i=1}^{l} \beta_i h_i(L) = h(L)\beta \tag{2}$$

where $\beta = [\beta_1, ..., \beta_l]^T$ is output weights vector of hidden layer $h(L) = [h_1(L), ..., h_l(L)]$, and $h_i(L)$ is the output of the ith hidden node output, $h(L)$ maps the d-dimensional label L to the l-dimensional hidden layer feature. And the output functions of hidden nodes may not be unique. Different output functions may be used in different hidden neurons. In real applications, $h_i(L)$ can be defined as:

$$h_i(L) = G(a_i, b_i, L), a_i \in R^d, b_i \in R \tag{3}$$

where $G(a, b, L)$ is a nonlinear piecewise continuous function satisfying ELM universal approximation capability theorems and $(a_i, b_i)_{i=1}^{L}$ are randomly generated according to any continuous probability distribution. In our algorithm, we use the Sigmoid (4) which is used in feed forward networks and Gaussian kernel (5) function [14] which is used in RBF networks. In this paper, the sigmoid function is applied in ELM, and the Gaussian kernel is applied in KELM.

$$G(a, b, L) = \frac{1}{1 + exp(-(\alpha L + b))} \tag{4}$$

$$G(a, b, L) = exp(-b\|L - a\|^2) \tag{5}$$

ELM is to minimize the training error as well as the norm of the output weights. The ELM learning function use the minimal norm least square method, it can be formulated as:

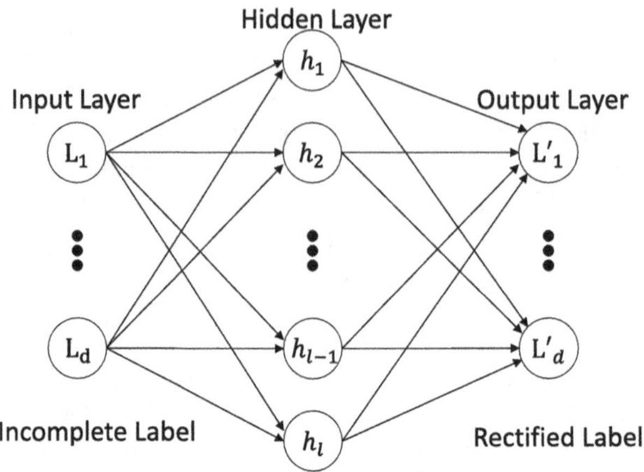

Fig. 2. Architecture of the ELM.

$$\arg\min_{\beta} \quad \|H\beta - Y\|. \tag{6}$$

where H is the hidden layer output matrix, Y is training data target matrix.

The minimal norm least square method was used in ELM, it can provide the solution of β:

$$\hat{\beta} = H^{\dagger}Y \tag{7}$$

where $H^{\dagger}$ is the Moore–Penrose generalized inverse of matrix H. The orthogonal projection method can be used to calculate $H\dagger$: when $H^{T}H$ is nonsingular, $H^{\dagger}$ can be defined as:

$$H^{\dagger} = (H^{T}H)^{-1}H^{T} \tag{8}$$

or when HH^{T} is nonsingular, $H^{\dagger}$ can be defined as:

$$H^{\dagger} = H^{T}(HH^{T})^{-1} \tag{9}$$

As far as we get the value of $\hat{\beta}$, we can pass through the incomplete labels L_{test} to well-trained ELM or KELM, and the rectified label feature matrix L' can be defined as:

$$L' = h(L_{test})\hat{\beta} \tag{10}$$

4 Experiments

4.1 Label Rectified Evaluation

To evaluate the effectiveness of the proposed, we conduct experiments on CIFAR-10 and CIFAR-100 [18]. Those two datasets are widely used in image classification benchmark and consist of colored natural scene images. The size of all images is 32×32 pixel. The training and test sets contain 50k and 10k images respectively. For adequately evaluating our method, we utilize two classical CNNs, i.e. VGG19 [25] and ResNet50 [8]. We set two baselines [12,23] for comparisons. Table 1 shows our experimental results.

LRL can obviously promote CNN's classification performance. On CIFAR-10, ResNet50 achieves 90.79% accuracy, while ResNet50-LRL (KELM) achieves 92.28% accuracy. On CIFAR-100, VGG19-LRL (KELM) outperforms VGG19 by 1.39% accuracy. From these comparisons, we can conclude the proposed method can effectively rectify incomplete labels output by CNNs. Compared with LRL (ELM), LRL (KELM) has more potential for image classification task, and it demonstrates RBF kernel is more appropriate to exploit label relations.

Table 1. Accuracy (%) on the CIFAR-10 and CIFAR-100 datasets.

Model	CIFAR-10	CIFAR-100
Gao et al. [12]	88.34	62.20
ResNet [9]	89.44	–
FractalNet [20]	89.92	64.66
Network in network [23]	89.59	64.32
VGG19	90.49	63.83
VGG19-LRL (ELM)	90.64	65.01
VGG19-LRL (KELM)	90.86	65.22
ResNet50	90.79	68.52
ResNet50-LRL (ELM)	92.11	71.44
ResNet50-LRL (KELM)	**92.28**	**72.27**

4.2 Compare with the State-of-the-arts

In this section, we compare our method with sate-of-the-art methods. we run experiments on the Caltech-256 dataset [7]. It contains more classes and less samples than CIFAR, with 256 classes and total of 30607 images. Each category has a minimum of 80 images. Following [22], we utilize 60 randomly selected images from per class as the training dataset, the rest as the testing dataset. We resize all images to 256×256 pixel. VGG19 [25], ResNet18 [8], Zhu et al. [28], SqueezeNet [16], VGG19-BN [17], Inception-V3 [10] are introduced for comparison, and we finally adopt ResNet50 for pre classification. Table 2 shows our experimental results.

Notably, ResNet50-LRL (KELM) also works well on Caltech-256 and outperforms all other state-of-the-art approaches. ResNet50 achieves to 80.40% Top-1 and 92.95% Top-5 accuracy. Compared with Zhu.et al. [28], our method outperforms it by 11.96% Top-1 and 2.4% Top-5 accuracy, ResNet50-LRL (KELM) outperforms base model by 1.46% Top-1 and 0.32% Top-5 accuracy. Those experimental results demonstrate the superiority of our approach on image classification.

Table 2. Accuracy (%) on the Caltech-256 dataset compared with other state-of-the-art methods.

Model	Top-1	Top-5
VGG19 [25]	70.54	87.81
ResNet18 [8]	73.22	86.96
SqueezeNet [16]	66.00	85.00
Zhu et al. [28]	70.00	89.00
Inception-V3 (with WCD) [10]	80.61	–
VGG19-BN [17]	74.83	89.85
ResNet50 [8]	80.40	92.95
ResNet50-LRL (KELM)	**81.96**	**93.27**

4.3 Classifier Comparisons

In our method, we adopt KELM to rectify incomplete labels. In order to evaluate it, we compare it with other different classifiers. Linear SVM, Kernel SVM (KSVM), Neural Network (NN) and Sigmoid function ELM (ELM). Besides, we introduce a method of Li et al. [22], which utilizes the KELM replace the Softmax classifier, that may cause two problem. First, features may contain more redundant information than labels. Second, the dimension of features is larger than labels, which consumes more times for training. In our method, we extract the labels whose dimension equal the images classes. We train the ResNet50 as pre classification. Table 3 exhibits our experimental results.

Apparently, Li et al. and LRL (KELM) are efficient in improving the performance of image classification. But other classifiers provide negative effect on improving accuracy. Li et al. can improve the accuracy of base model about 1.12%, and LRL (KELM) can improve about 1.07%. But, in contrast, experiment shows that the LRL (KELM) has advantage of high training speed. In conclusion, the LRL (KELM) achieve better balance between accuracy and training time.

4.4 Training Observation

In our training phase, we apply a pre-trained network (on ImageNet) into specific datasets (e.g. Caltech-256), so we can obtain a well-trained CNN for pre classification. But badly trained CNNs should be considered. So we apply LRL method into each ResNet50's training epoch. We plot the Top-1 accuracy of ResNet50, ResNet50-LRL (ELM) and ResNet50-LRL (KELM). The visual comparisons are provided in Fig. 3.

Obviously, KELM based LRL (green line) can substantially promote ResNet50 (blue line) performance in very early epochs. It manifests our method is effective for unwell-trained CNNs. Although the disparity of them becomes small, KELM based LRL still outperforms ResNet50. Besides, we notice the

Table 3. Accuracy (%) and training times on the Caltech-256 with different methods.

Classifier	Promotion (%)	Time (s)
Li et al. [22]	+1.12	18.02
Linear SVM	−5.06	16.26
KSVM	−3.03	47.07
NN	−5.70	120.31
ELM	−2.06	2.81
KELM	+1.07	4.58

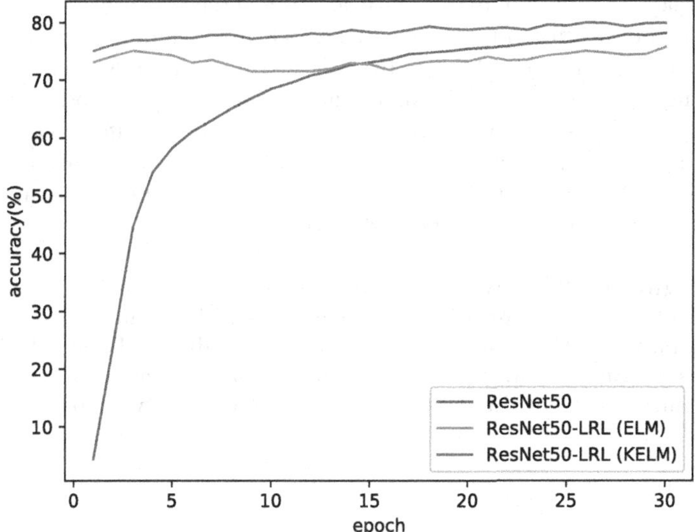

Fig. 3. Training observation of ResNet50 on Caltech-256 dataset.

ELM based LRL (orange line) cannot improve ResNet50 performance after the 13-th epoch.

4.5 Implementation Details

For Caltech-256 dataset, we modify the last layer of VGG19 and ResNet50 to 257 outputs. These CNNs are trained using a batch size of 32 for 50 epochs and the learning rate is set to 10^{-4}. For the CIFAR-10 and CIFAR-100 datasets, we replaced the fully-connected layer of ResNet50 to 11 and 101 outputs. ResNet50 are trained using a batch size of 128 for 300 epochs and the initial learning is set to 0.001 and is divided by 10 at 30 % and 75 % of the total number of training epochs. All CNNs are trained using stochastic gradient descent. And we use a weight decay of 0.01 and Nesterov moumentum of 0.9. In all experiments, the images are randomly flipped and cropped before passing into the networks. In all our simulations on ELM with Sigmoid additive hidden node and RBF hidden node, $l = 1000$. All the hidden node parameters are randomly generated based on uniform distribution. Experiments are performed on a NVIDIA Titan Xp GPU.

5 Conclusion

In this paper, instead of designing complicated and redundant CNNs, we explore the label-wise relation for label rectification, then propose a method (LRL) to learn label rectify through kernel extreme learning machine to improve accuracy of image classification. Compared with features, labels contain less redundant information and dimension is smaller. LRL can achieve similar accuracy but save more times for training. In addition, by training observation, we found that LRL shows strong advantages in rectifying the labels of pre-trained models. Extensive experiments conducted on public datasets demonstrate the effectiveness of LRL (KELM). To our best knowledge, this is the first labels rectification approach for image classification. In future, we will exploit a method to learn label rectify from source dataset to target dataset directly.

Acknowledgments. This work was supported by the National Natural Science Foundation of China (No. 61877002), Beijing Municipal Commission of Education PXM2019_014213_000007, Beijing Natural Science Foundation, Fengtai Rail Transit Frontier Research Joint Fund 19L00005, and Postgraduate Research Capacity Improvement Program from Beijing Technology and Business University in 2020.

References

1. Agarap, A.F.: An architecture combining convolutional neural network (CNN) and support vector machine (SVM) for image classification. arXiv preprint arXiv:1712.03541 (2017)
2. Bernal, J., et al.: Deep convolutional neural networks for brain image analysis on magnetic resonance imaging: a review. Artif. Intell. Med. **95**, 64–81 (2019)
3. Cortes, C., Vapnik, V.: Support-vector networks. Mach. Learn. **20**(3), 273–297 (1995)
4. Dalal, N., Triggs, B.: Histograms of oriented gradients for human detection (2005)
5. Deng, J., Guo, J., Xue, N., Zafeiriou, S.: Arcface: additive angular margin loss for deep face recognition. In: Proceedings of the IEEE Conference on Computer Vision and Pattern Recognition, pp. 4690–4699 (2019)
6. Gao, S.H., Cheng, M.M., Zhao, K., Zhang, X.Y., Yang, M.H., Torr, P.: Res2net: a new multi-scale backbone architecture. arXiv preprint arXiv:1904.01169 (2019)
7. Griffin, G., Holub, A., Perona, P.: Caltech-256 object category dataset (2007)
8. He, K., Zhang, X., Ren, S., Sun, J.: Deep residual learning for image recognition. In: Proceedings of the IEEE Conference on Computer Vision and Pattern Recognition, pp. 770–778 (2016)
9. He, K., Zhang, X., Ren, S., Sun, J.: Identity mappings in deep residual networks. In: Leibe, B., Matas, J., Sebe, N., Welling, M. (eds.) ECCV 2016. LNCS, vol. 9908, pp. 630–645. Springer, Cham (2016). https://doi.org/10.1007/978-3-319-46493-0_38

10. Hou, S., Wang, Z.: Weighted channel dropout for regularization of deep convolutional neural network. In: AAAI (2019)
11. Huang, G., Liu, Z., Van Der Maaten, L., Weinberger, K.Q.: Densely connected convolutional networks. In: Proceedings of the IEEE Conference on Computer Vision and Pattern Recognition, pp. 4700–4708 (2017)
12. Huang, G., Sun, Yu., Liu, Z., Sedra, D., Weinberger, K.Q.: Deep networks with stochastic depth. In: Leibe, B., Matas, J., Sebe, N., Welling, M. (eds.) ECCV 2016. LNCS, vol. 9908, pp. 646–661. Springer, Cham (2016). https://doi.org/10.1007/978-3-319-46493-0_39
13. Huang, G.B.: An insight into extreme learning machines: random neurons, random features and kernels. Cogn. Comput. **6**(3), 376–390 (2014)
14. Huang, G.B., Zhou, H., Ding, X., Zhang, R.: Extreme learning machine for regression and multiclass classification. IEEE Trans. Syst. Man Cybern. Part B (Cybern.) **42**(2), 513–529 (2011)
15. Huang, G.B., Zhu, Q.Y., Siew, C.K.: Extreme learning machine: theory and applications. Neurocomputing **70**(1–3), 489–501 (2006)
16. Iandola, F.N., Han, S., Moskewicz, M.W., Ashraf, K., Dally, W.J., Keutzer, K.: Squeezenet: Alexnet-level accuracy with 50x fewer parameters and <0.5 mb model size. arXiv preprint arXiv:1602.07360 (2016)
17. Ioffe, S., Szegedy, C.: Batch normalization: accelerating deep network training by reducing internal covariate shift. arXiv preprint arXiv:1502.03167 (2015)
18. Krizhevsky, A., Hinton, G., et al.: Learning multiple layers of features from tiny images. Technical Report, Citeseer (2009)
19. Krizhevsky, A., Sutskever, I., Hinton, G.E.: Imagenet classification with deep convolutional neural networks. In: Advances in Neural Information Processing Systems, pp. 1097–1105 (2012)
20. Larsson, G., Maire, M., Shakhnarovich, G.: Fractalnet: ultra-deep neural networks without residuals. arXiv preprint arXiv:1605.07648 (2016)
21. Li, X., Wang, W., Hu, X., Yang, J.: Selective kernel networks (2019)
22. Li, Z., Zhu, X., Wang, L., Guo, P.: Image classification using convolutional neural networks and kernel extreme learning machines. In: 2018 25th IEEE International Conference on Image Processing (ICIP), pp. 3009–3013. IEEE (2018)
23. Lin, M., Chen, Q., Yan, S.: Network in network. arXiv preprint arXiv:1312.4400 (2013)
24. Lowe, D.G., et al.: Object recognition from local scale-invariant features. In: ICCV, vol. 99, pp. 1150–1157 (1999)
25. Simonyan, K., Zisserman, A.: Very deep convolutional networks for large-scale image recognition. arXiv preprint arXiv:1409.1556 (2014)
26. Sutskever, I., Vinyals, O., Le, Q.: Sequence to sequence learning with neural networks. In: Advances in NIPS (2014)
27. Szegedy, C., et al.: Going deeper with convolutions. In: Proceedings of the IEEE Conference on Computer Vision and Pattern Recognition, pp. 1–9 (2015)
28. Zhu, C., Han, S., Mao, H., Dally, W.J.: Trained ternary quantization. arXiv preprint arXiv:1612.01064 (2016)
29. Zhu, Y., Zhang, C., Zhou, D., Wang, X., Bai, X., Liu, W.: Traffic sign detection and recognition using fully convolutional network guided proposals. Neurocomputing **214**, 758–766 (2016)

Author Index

Printed in the United States
by Baker & Taylor Publisher Services